MW00775165

A WIDENING SPHERE

A WIDENING SPHERE

Evolving Cultures at MIT

Philip N. Alexander

The MIT Press Cambridge, Massachusetts London, England

For information about special quantity discounts, please email special_sales@mitpress.mit.edu

This book was set in Adobe Garamond by the MIT Press. Printed and bound in the United States of America.

Library of Congress Cataloging-in-Publication Data

Alexander, Philip N., 1952–
A widening sphere : evolving cultures at MIT / Philip N. Alexander.
p. cm.
Includes index.
ISBN 978-0-262-01563-9 (hardcover : alk. paper) 1. Massachusetts Institute of Technology—Biography. 2. Cambridge (Mass.)—Intellectual life. I. Title.
T171.M49A85 2011
6071'17444—dc22

10 9 8 7 6 5 4 3 2 1

Contents

Throughout the coming ages, hers is a wid'ning sphere,
In never-ending progress, yet newer truths unfold:
While now we see her greatness, it does not all appear,
The future will develop what now is still untold.

From "Ode to Technology"
lyrics by Harry Hunt (class of 1897))

Acknowledgments

I am indebted to Paul Gray, MIT Corporation chair emeritus, president emeritus, and professor of electrical engineering emeritus, for encouraging me to undertake this independent, unofficial history of the Institute and for convincing president Susan Hockfield to help support it. Many thanks to Susan Hockfield and the President's Office for financial assistance.

During my thirty-five years at MIT, a number of colleagues and friends have taught me much about the Institute's unique history, culture, and community. I am particularly grateful to the following, some of whom are now deceased: Thomas Allen, Joseph Applegate, Ali Argon, Vera Ballard, Manson Benedict, Leo Beranek, Herman Branson, E. Cary Brown, Louis L. Bucciarelli Jr., Stephen Charles, Eulis Clarke, Norman Dahl, Richard Duffy and Adina Adler, Elzbieta Ettinger, Herman Feshbach, Michael Folsom, Ann Friedlaender, Thomas Goreau, Morris Halle, Harold and Ruth Hanham, Allan Henry, Kenneth Hoffman, Jarmila and Charles Hrbek, Marshall Hughes, Harold Isaacs, Jean Jackson, Howard Johnson, Frank Jones, Irving Kaplan, Marilyn Katz, Carl Kaysen, Phyllis Klein, Roy and Peggy Lamson, Joan Laws and Albert Gregory, Heather Lechtman, Reta Lee, Jerome Lettvin, Francis Low, Edward Lurie, Kenneth Manning, Loretta Mannix, William Ted Martin, Debbie Meinbresse, Elting Morison, Robert Morison, Philip and Phylis Morrison, Margaret Otto, James Paradis, Lynn Roberson, Janet Romaine, Walter and Judy Rosenblith, Vincent Rossano, Ann Serini, Constantine

Simonides, Louis Smullin, Arthur Steinberg, Julius and Catherine Stratton, Dirk Struik, Claudia von Canon, Phyllis Wallace, Charles Weiner and JoAnn Hughes, Clarence and Mildred Williams, James H. Williams Jr., Cynthia Wolff, John Wulff, and Muriel and Everett Zimmerman. It has been an extraordinary privilege and pleasure, over the years, to work on research and writing projects with Leo Beranek, Harold Hanham, Kenneth Manning, James Paradis, Walter Rosenblith, Charles Weiner, and Clarence Williams.

Staff members, past and present, of the Institute Archives and Special Collections helped identify, locate, and ease access to essential resources. Thanks to Elizabeth Andrews, Ewa Basinska, Lois Beattie, Bridget Carr, Myles Crowley, Margaret dePopolo, Paul Heffernan, Mikki Macdonald, Silvia Mejía, Jeffrey Mifflin, Nora Murphy, Thomas Rosko, Helen Samuels, Stephen Skuce, Sara Smith, Megan Sniffin-Marinoff, and Craig Thomas for their resourceful efforts to scout out often hard-to-find material. Thanks to Frank Conahan of the MIT Museum for his assistance with photographs. Library staffers Moses Carr, Dan Holland, Harolyn Hylton, Alan Rostoff, and Theresa Tobin accommodated many tricky, burdensome requests. I am grateful for the support of several current and former staffers of the Program in Writing and Humanistic Studies, especially Nicholas Altenbernd, Rebecca Chamberlain, Charles Fuller, Maya Jhangiani, Susanne Martin, and Magdaléna Rieb. Marguerite Avery, Gita Manaktala, and Ellen Faran of the MIT Press smoothed the publication process. Editor Michael Sims and designer Yasuyo Iguchi brought their usual flair and good humor; it is always a delight to work with them.

Thanks to Bert Singer, Phylis Morrison's son, for permission to reproduce on the front cover one of Phil and Phyl Morrison's creative illustrations. The back cover shows instruments used by my grandfather, Philip Hardie, as a mechanical engineering student at MIT in the early 1920s. Phil treasured and made good use of them until he and my grandmother, Emily Hare Hardie, retired in the mid-1960s. Images in the photo insert are reproduced courtesy of the MIT Museum.

Several friends—Karen Alphonse, Mager Benavidez, Katherine Ellins, the Ferrari Duff family (Diane, John, Juliana, and Nicole), William Haas, Carmen Lewis, Gloria Lyn, David Manning, Curtis McMillan and Mahalet Zewde, Nat Daniel Myles, Leonard Rieser,

Glenn Rigoff, Elizabeth Sandager, Ginny Schneider, Mary Van Winkle, and Myndilee Wong—supplied fellowship and strategic breathers along the way. Thanks, too, to the Hamilton-Wenham breakfast club—Stephen Boudreault, Nik Gregory, and Bruce Howell—and to friends at the Zesiger Sports and Fitness Center and the MIT Alumni Pool, especially Larry Anderson, Desirée Barrett, John Benedick, Brian Callahan, Wyatt Callander, Jimmy Carlson, Andy Carvalho, Renée Caso, Andrea Cohen, Munther Dahleh, Tracey Daley, Nancy Dalrymple, Dawn Dill, Philip Fisher, John Freitas, Richard Garner, Linda Greene, Jacob Gubbay, Mohammed Hannan, Evan Harlan, Robin Hill, Conan Hom, Flora Keumurian, Kathy and Rick Kosinski, Tunney Lee, Steve Lyons, Harold McCormack, Ken McRitchie, Ben McElhiney, Felix Mercado, Carrie and Tim Moore, Willis Negron, Bill Paine, Humberto Pereira, Robert Pinckney, Gerry and Nina Power, Paul Robertson, Bill Schoolcraft, Susanna Sung, Duncan Stewart, George Stiny, James Tangorra, Bill Ward, and Careaine Withock. Michael Kane, Christine McElroy, and Marcia Yousik helped keep my marbles intact.

Members of my family—Neville and Susan Alexander and Kristy, Paul, Rebecca, and Wendy Kahn—tolerated my too-frequent absences over the last several years. Neville, sadly, died not long after I started this work; Susan read early drafts and offered the wise perspective of an informed general reader.

My deepest gratitude goes to Kenneth Manning, without whose advice, encouragement, historical insight, and editorial skill this work could never have been written.

"A future full of promise"

William Barton Rogers, 1804–1882

The Massachusetts State House, Beacon Hill, Boston—hub of the solar system, declared Oliver Wendell Holmes. When Governor John Andrew showed up on April 10, 1861, to tackle his daily pile of papers, he signed one without giving it much thought. "An Act to incorporate the Massachusetts Institute of Technology. . . . William B. Rogers, James M. Beebe, E. S. Tobey, S. H. Gookin, E. B. Bigelow, M. D. Ross, J. D. Philbrick, F. H. Storer, J. D. Runkle, C. H. Dalton, J. B. Francis, J. C. Hoadley, M. P. Wilder, C. L. Flint, Thomas Rice, John Chase, J. P. Robinson, F. W. Lincoln Junr, Thomas Aspinwall, J. A. Dupee, E. C. Cabot, their associates and successors, are hereby made a body corporate, by the name of the Massachusetts Institute of Technology, for the purpose of instituting and maintaining a society of arts, a museum of arts, and a school of industrial science, and aiding generally, by suitable means, the advancement, development and practical application of science in connection with arts, agriculture, manufactures and commerce . . ." The governor had played a part in the legislative back-and-forth. House approval, April 8, 1861; Senate approval, April 9, 1861; governor's signature, April 10, 1861—a deceptively smooth, clean process, suggesting nothing of the weeks, months, even years of thought and reflection that had moved it this far along.

"I have very little of interest to tell," wrote William Barton Rogers to his brother Henry a week earlier. He and Henry had worked toward this goal, off and on, for nearly three decades; success now, finally in reach, although with much work ahead to get the enterprise off the ground, established, viable. Yet William sounded dispirited. Southern secessionists were gaining momentum and civil war looked all but inevitable. Confederate soldiers attacked Fort Sumter on April 12, just two days after the governor put his pen to paper. As the nation unraveled, the proposed Institute of Technology no longer seemed a high priority.

Rogers came from a family accustomed to facing down adversity. He knew what a hard fight was, and he never shied away from one that mattered. Born in Philadelphia, December 7, 1804, he was second-generation Irish. His father, Patrick Kerr Rogers, a native of County Tyrone, near Londonderry, was a hot-tempered, militant nationalist with ties to the United Irishmen, a follower of Irish patriot and martyr Robert Emmet and an admirer of both the American and French revolutions. After fleeing Ireland to escape arrest in 1798, he settled in Philadelphia, a thriving, cosmopolitan port receptive to immigrants, tolerant of religious differences, a magnet for artisans, tradesmen, Scotch-Irish Presbyterians. His interests lay in science, but as science could not guarantee a living wage, he studied medicine at the University of Pennsylvania and opened a medical practice after graduating in 1802. He and his wife Hannah (Blythe), also from a refugee family of United Irish partisans, produced a substantial brood—seven children—over the course of the next decade and a half. Three died in infancy, the youngest son, Alexander, and both daughters, Amelia and Matilda; four sons survived and followed in their father's footsteps as committed scientist-intellectuals.

Bored with medicine, Patrick Rogers sought academic jobs. When nothing turned up—Penn rejected his application for a professorship in chemistry—he went out on the independent lecture circuit, and in 1811 opened a lyceum, hoping to attract knowledge-seekers regardless of gender and from all stations in life: the genteel, the professional, the working classes. The family moved to Baltimore in 1812. In 1819, Patrick accepted the chair in chemistry and natural philosophy at the College of William and Mary in Williamsburg, Virginia. He may have had some help with this from Thomas Jefferson, whom he had approached about a post at the University of Virginia, then in its planning stages.

Patrick home-schooled his children. A confirmed agnostic in spite of his Presbyterian roots, he kept them away from sectarian influences. His values, distinctively age-of-enlightenment vintage, revolved around independent thinking, respect for human rights, scientific knowledge, and academic excellence. He took William as a teenager, perhaps the other brothers as well, to meet the sage of Monticello. Jefferson, statesman of science and former president of the American Philosophical Society, had done original research in paleontology and was quick to discern scientific talent. He once quizzed William, then turned to Patrick and said, "Your son has a future full of promise before him."

All four of the Rogers brothers attended William and Mary (without graduating, apparently), and followed careers that were remarkably similar. The eldest and youngest—James (born 1802), Robert (born 1813)—were most like Patrick, with science as their first love and medicine, teaching, or educational administration to earn their keep. James was professor of chemistry at Washington Medical College in Baltimore, then successively at the medical department of Cincinnati College, Franklin Institute (Philadelphia), and University of Pennsylvania; Robert was professor of chemistry and materia medica at the University of Virginia before succeeding to James's chair at the University of Pennsylvania in 1852, afterward serving as dean of medicine there and later on the faculty of Jefferson Medical College, also in Philadelphia. The middle two brothers—William, then Henry (born 1808)—looked to science, however, more as profession than as hobby. It was a time of change and self-definition for American scientists. The birth of the Association of American Geologists and Naturalists (later the American Association for the Advancement of Science) and the National Academy of Sciences ushered in an era of scientific professionalism, and the brothers were charter members of this movement.

They began as teachers and remained in academe. William and Henry opened a school in Windsor, Maryland, just outside Baltimore, in the fall of 1826. By January William was also lecturing part-time at the Maryland Institute in Baltimore, a lyceum that taught science only. His lectures proved so popular that in May 1828 the Institute allowed him to open a high school, along the lines of a college preparatory department. Henry joined him in this venture, whose purpose was "to impart such knowledge, and to induce such habits of mind as may be most beneficial to youth engaging in mechanical and mercantile pursuits."

In October 1828 William succeeded Patrick, who had died two months before, as professor of chemistry and natural philosophy at William and Mary, while Henry took a similar position at Dickinson College in Carlisle, Pennsylvania. William pushed the curriculum at William and Mary toward real-life applications and the commercial and industrial uses of knowledge. Henry, meanwhile, quickly grew restless at Dickinson. Drawn to radical social and political causes—a consequence, he said, of "the lofty spirit of my father in me"—he resigned and sailed for England in May 1832 to join up with Robert Owen, Frances Wright, and other utopian socialists. In London he was pulled, too, into scientific circles, cementing relationships with geologists Henry de la Beche and Charles Lyell, among others. In 1835, he became professor of geology and mineralogy at the University of Pennsylvania.

That same year William went to the University of Virginia, Charlottesville, as professor of natural philosophy. He was second choice; the position came to him when Joseph Henry, widely viewed as America's most brilliant scientist since Benjamin Franklin, decided not to leave Princeton. Henry put in a strong word for William: "I am confident he will do much towards elevating the scientific character of our country." At Charlottesville, in addition to teaching physics, William joined forces with the professor of mathematics to offer civil engineering. Influenced, however, by Henry's experience in England, his interests shifted rapidly toward geology.

Henry became New Jersey's first official state geologist in 1835, and, early in 1836, took on the parallel post for Pennsylvania as well. William was appointed geologist of Virginia, also in 1836, after drawing attention in a series of articles, published in *Farmer's Register* in 1834 and 1835, to greensand marls and gypsum for fertilizer in the eastern part of the commonwealth. William and Henry led teams of researchers into the wilds, then cataloged, discussed, analyzed, shared results at conferences and in published papers. The other two brothers joined in from time to time. Such collaborative efforts generated a remarkable cross-fertilization of ideas, mostly in geology, but often on that increasingly murky boundary between geology, physics, chemistry, and metallurgy. William and Robert coauthored no less than 19 articles; William and Henry no less than 20, all of these in the period before 1854. As individual authors, William and Henry put out dozens of articles in the major scientific journals: *Silliman's*, *American Journal of Science* (*Silliman's* successor), *Edinburgh New Philosophical Journal*, and in the published proceedings of various scientific societies.

While their geological work brought the brothers into close contact with emergent technologies—agriculture, mining, railway and canal transport, surveying, road construction, steam power—their deepest passion was reserved for what they called "true science and broad philosophical views." But the technological side fascinated them, too. As Francis Walker, MIT's third president, wrote about William in 1887: "He, of all men the least prosaic, gifted with a fervent imagination such as is rarely coupled with the disposition and capacity for patient and protracted research, valued science not more for the sake of truth than for the sake of the virtue which is to be found in it for the amelioration of the human condition." Science as basic knowledge, science in the service of humankind. In 1837, amid all their geologizing, William and Henry drafted a proposal to establish, under the auspices of the Franklin Institute, a science-based polytechnic school that emphasized "applications . . . by detailed practical lessons & especially by actual discipline in the workshop." They continued to refine a vision for science, technology, and education in the modern age.

The brothers quickly gained traction as professional scientists. The elite American Philosophical Society elected them to membership in 1835. In 1840, Henry helped organize the Association of American Geologists and Naturalists and presided over the second annual meeting in Philadelphia, April 1841. William was elected to the National Institution for the Promotion of Science in August 1840. Both became honorary members of the Boston Society of Natural History in 1842 and foreign members of the Geological Society of London in 1844; also in 1844, William got word of his election to the Royal Society of Northern Antiquaries, headquartered in Copenhagen. In 1845, they were elected fellows of the American Academy of Arts and Sciences, founded in 1780 in Cambridge, Massachusetts, "to cultivate every art and science which may tend to advance the interest, honor, dignity, and happiness of a free, independent, and virtuous people." They gave papers—sometimes jointly, other times separately but on the same program—at professional meetings in Philadelphia (American Philosophical Society, December 1841), Boston (Geologists and Naturalists, April 1842), Washington, D.C. (Geologists and Naturalists, May 1844), and elsewhere. When Amos Binney, president of the Geologists and Naturalists, died unexpectedly, William presided over the group right before it reconstituted itself, in 1848, as the American Association for the Advancement of Science (AAAS). The brothers worked hard to curb what they perceived as creeping elitism in organized science, the trend toward hierarchies separating professionals from nonprofessionals, the credentialized from the noncredentialized. They were as well regarded overseas as in America. On William's first trip to Europe in 1849, he was greeted warmly by a number of scientists, Charles Darwin among them; at the time Darwin was working through his theory of evolution, with *Origin of the Species* still a decade off.

The brothers felt restless, however, in their respective academic spheres. For William, life in the South became a source of increasing irritation. "Matters here as usual," he wrote to his brother James in March 1841, "too dull." The climate and recurrent epidemics of cholera and malaria kept him on edge and factored into the illnesses that would put him out of commission for long stretches later in life. But Southern culture bothered him more. He grew to think of the South as uncivilized, decadent, anti-intellectual. There was much violence and worse, *toleration* of violence; students at the University of Virginia toted firearms, drank to excess, rioted, and assaulted faculty members with near impunity. John A. G. Davis, chairman of the faculty and a close friend of William's, was shot and killed by a rampaging student in November 1840.

William, however, was far from a hardnosed, emotional radical; generally he projected balance, circumspection, a voice of calm reason. Elected chairman of the faculty in 1844, he garnered respect on all sides. His stance on slavery was moderate. While Henry had joined forces with abolitionists up north—Charles Sumner, the Howes (Samuel Gridley and Julia Ward), and others—William remained noncommittal until after the Civil War was well under way, by which time he was safely ensconced in Boston. One friend and colleague from Virginia days assumed, rightly or wrongly, that William was a slaveholder himself. Levi, this friend later recalled, was a "negro serving-man [who] drove . . . behind [William] on horseback, accompanying him on his geological rambles . . . learned to think as his master thought." Levi was said to have taken charge once, introducing Appalachian geology to a visiting English scientist, Charles Daubeny.

But William continued to think of Charlottesville as too provincial. He found little there, he told Henry, other than "stupid dullness and unvaried monotony. . . . I feel that I am but half-alive here, and am more than ever resolved, when able, to quit the scene for one more congenial to my tastes and more likely to promote my happiness." As early as 1833 William feared that the nation could well be on the brink of "fratricidal war," a theme he kept revisiting in the years leading up to the Civil War. The appointment of a Jew and a Catholic to the faculty in 1841 stirred up a frenzy of religious intolerance that troubled him deeply. He felt frustrated by the state legislature's ignorance of science, and by its lack of educability. On the plus side, as state geologist he gained negotiating experience that would prove useful later on. And, even though he had little positive to say about either colleagues or students at the university, he was among the most popular, inspiring teachers there. His lectures, especially those on astronomy, found the room packed up to an hour ahead of time, the aisles filled, "even the windows crowded from the outside with eager listeners." But he longed to join Henry up north.

No longer the provincial outpost it once was, Boston had emerged after 1800 as a mercantile and industrial power, and a seat of intellectual ferment. The city was as drawn to the Rogers brothers as they to it. A paper they gave on Appalachian geology became the talk of the Geologists and Naturalists' convention in Boston in April 1842. "A grander geological theme could hardly be imagined," one listener recalled. "The genius of the brothers Rogers . . . like the Egyptologist with the papyrus roll, unfolded the inverted and contorted strata, spread and smoothed them out, as it were, in an open book and showed them to the eye of science. . . . [in a] fluent and graceful oral statement of this hitherto mysterious mountain chain." The event marked their official entrée. John Amory Lowell, trustee of the Lowell Institute—Boston's vibrant experiment in public adult education, a lyceum-style effort not unlike the one begun by Patrick Rogers earlier in the century—invited Henry back in 1844 to give a popular course in geology. William came up from Virginia to hear Henry lecture, and afterward the brothers spent several weeks carrying out geological research in New Hampshire's White Mountains.

There they got to know the Savage family, on vacation at their country estate near Nashua. The Savages belonged to Boston's creamiest social elite. James Savage, the family patriarch, traced his roots back to an English settler, Thomas Savage, who arrived in 1635, a decade and a half after the Mayflower. A lawyer, James Savage was at different times a state legislator (House and Senate) and executive councilor, delegate to the Massachusetts constitutional convention, and Boston alderman and school committee member. As cofounder of the Provident Institution for Savings, he had the means and leisure time to pursue sidelines in politics and antiquarian scholarship. He helped found the Boston Athenaeum and was one-time president of the Massachusetts Historical Society. A four-volume set that he produced on early New England settlers kept him busy for nearly two decades. *Rich Men of Massachusetts* (1851) placed his net worth at $150,000, nowhere near the wealthiest but enough to make the grade as well-to-do.

William and Henry cultivated ties to Boston, to local scientists, educators, businessmen, and to the Savages. William courted one of James Savage's daughters, Emma, from a distance; Henry moved to Boston in 1846, intent on finding an academic position. Henry had his eyes set on the coveted Rumford professorship in applied science at Harvard. This did not materialize for him—his outspoken, radical views on science and politics offended Harvard's more conservative elements—but he took an active interest in discussions under way to create a school of science there, established in 1847 as the Lawrence Scientific School. On a related mission, he piqued John Amory Lowell's curiosity about possibly setting up a technical school under the

Lowell Institute, a venture similar to that which the brothers had proposed for the Franklin Institute in 1837.

Henry wrote William excitedly about this in March 1846. William, the more cautious of the two, was also enthusiastic; he drafted and sent Henry "A plan for a polytechnic school in Boston" for submission to Lowell. "The true and only practicable object of a Polytechnic School," he wrote to Henry, "is . . . the teaching not of manipulations and minute details of the arts, which can be done only in the workshop, but the inculcation of all the scientific principles which form the basis and explanation of them, and along with this a full and methodical review of all their leading processes and operations in connection with physical laws." Lowell, intrigued by the proposal, rejected it after discovering that his trust's fine print explicitly forbade use of funds "for bricks and mortar."

When William visited Henry in Boston, he often looked in on the Savages as well as on Harvard colleagues such as Joseph Lovering, Benjamin Peirce, Eben Horsford, and Josiah Cooke. In March 1848 he resigned his post at the University of Virginia, intending to move to Boston. When friends persuaded him that this was foolhardy without a firm job offer, he withdrew the resignation. But a year later, his marriage to Emma Savage on June 20, 1849 guaranteed closer ties to New England, relative freedom from financial worry, and a social, political, and financial network that promised to give his educational plans a lift. The Savages also saw advantages. This serious, mature scientist-educator, twenty years older than Emma, promised a solid, stable partnership for her and a close relationship with her father—and with her brother, James Jr.—based on shared values and intellectual interests. There was little doubt that the couple would settle in Boston; the only question was when. In 1853 they moved back for good, joining Emma's father and brother in their spacious, elegant town house at 1 Temple Place, across from the Boston Common (Emma's mother Elizabeth and sister Lucy had died in 1850). As if to cement the Rogers–Savage ties, in 1854 Henry married Emma's half-sister, Eliza Lincoln, daughter of Elizabeth Savage by her first marriage.

Temple Place sat near the heart of Boston's most elite residential enclave. Here, Rogers and the Savages lived first as a foursome, then as a threesome following James Jr.'s death from Civil War battle wounds in 1862. Summers were spent at Sunny Hill, the Savages' country estate in Lunenburg, about thirty miles outside Boston. As of 1869, the family spent more time in Newport, Rhode Island, first in a rented house, Castle Hill, corner of Bellevue Avenue and Bath Road; then, from 1872, in their own cottage—Morningside—on Gibbs Avenue. By 1871, Temple Place had become so overrun by commerce, noise, and traffic that the family abandoned it for the relative peace of the Back Bay: recently made land on the other side of the Boston Common. They took rooms, first, at the Hotel Berkeley, on the corner of Berkeley and Boylston Streets. Following James Savage's death in 1873, William and Emma spent winters at 117 Marlborough Street, first as renters then as owners.

Emma's life revolved around her three men. She was their constant companion and looked after their every need with the help of an entourage of retainers. Someone called her "a type of the best New England womanhood," combining common sense, charm, and tact. She used her winning ways to balance the clashing temperaments of her men-folk. Rogers—the one with Irish roots—was the quiet, unflappable one, while James Savage, the quintessential Anglo, was excitable, "given to rather extreme opinions and violent expression of them."

Once settled in Boston, William Rogers set about renewing old contacts and forging new ones. Science, he knew from experience, could go but so far without widespread public support. His struggle with the Virginia legislature over the geological survey had been distasteful yet necessary. He had stuck with it calmly, patiently, his skill as a consensus-builder serving him well as he negotiated with stubborn, uninformed public officials—unlike his brother Henry, who kept venting frustration over his parallel encounters with Pennsylvania officials. But in Boston, William found like-minded, like-tempered individuals and organizations eager to join forces in the interests of science. The state legislature was less convulsive, too, than Virginia's—or so it seemed to William at the outset—and open to progressive, forward-looking ideas.

He arrived at just the right time. Commercial interests in Massachusetts had long yearned for a school of applied science, one that would train men for regional, state, and local needs. Except for Rensselaer Polytechnic, founded by Stephen Van Rensselaer in 1824 "for the purpose of instructing persons . . . in the application of science to the common purposes of life," most efforts along these lines had proven disappointing. A wave of support had gone to Harvard, Yale, and other traditional colleges within the previous two decades, much of it from merchants and industrialists who wanted their donations used to train personnel for *technical* professions. Both Harvard and Yale established scientific schools in 1847. Harvard's, the Lawrence Scientific School, was funded by textile magnate Abbott Lawrence; Yale's, the Sheffield Scientific School, by railroad executive Joseph Sheffield. In 1851 Dartmouth College created its Chandler School of Science and the Arts with a bequest from commission merchant Abiel Chandler. But Lawrence's faculty—Louis Agassiz, a key figure there—was composed for the most part of men who lacked interest in the *applications* of science; their primary focus was knowledge for knowledge's sake, reinforced by institutional cultures where the useful, the practical, the vocational, the "merely" professional, were looked down on. Study science, yes, but its practical side belonged in the trade schools. Advocates of change at the regular colleges and universities often met with resistance. When Brown University's president, Francis Wayland, urged more emphasis on technical education, he drew little support from a faculty and governing board suspicious of his reformist vision, unwilling to rush toward all things (or anything) practical.

The field was wide open. Rogers needed to do little to raise consciousness—the understanding, the *moral* support were already in place—so he could concentrate on becoming better known outside the small group of scientists who knew him well and beyond the Savages' social circle. He made it a point never, if possible, to miss a meeting of the American Academy of Arts and Sciences, a prime venue to circulate, where he could talk science with Peirce, Agassiz, Horsford, and other academic bright lights—many from Harvard, but some from the worlds of politics, commerce, and public affairs as well. His father-in-law helped, too, by introducing him around.

By 1855, William was in demand as a public lecturer. He gave a series for the Lowell Institute at Tremont Temple, in 1856–57—riveting accounts of his and Henry's excursions into America's unexplored mountain regions, what they found there, what their findings revealed about earth's origins, untapped mineral wealth, prospects for new knowledge, discovery, and economic prosperity. His was an upbeat message that melded science with adventure, exploration, spirited travelogue. Meanwhile, Henry moved to Scotland in 1855 and, in 1858, accepted an appointment as Regius professor of natural history at the University of Glasgow.

The Lowell series reinforced William's growing local reputation. In February 1859, he joined a number of Bostonians—scholars, scientists, academic leaders, and businessmen—as a so-called Committee of Associated Institutions, which petitioned the legislature to set aside four squares of Boston's Back Bay "for the use of such public institutions as may associate together for the public good." The group included two or three dozen manufacturers, teachers, physicians, bankers, farmers, dry goods merchants, railroad men, shipbuilders, insurance executives, and import-exporters. Among the target interests were agriculture, horticulture, natural history, mechanics, manufacturing, commerce, fine arts, and public education. This rich but diffuse array made it difficult to identify or build a unified perspective. The group found common ground, however, in its proposal for a so-called Massachusetts Conservatory of Art and Science. Each square of land would group together related institutions. One, for example, would combine under a single roof (or several smaller roofs) collections of tools, models, and other items useful in agriculture and horticulture; another would connect natural history, geology, chemistry; yet another, mechanics, manufactures, commerce; and finally, history, ethnology, and fine arts. Legislators liked the educational side, especially its hands-on flavor.

In March, however, the plan was turned down—too large, too complex, too utopian. The legislature was put off, too, by a conjoint then competing proposal from William Emerson Baker, well-to-do manufacturer of sewing machines and partner in the firm of Grover and Baker (later to merge with Singer), who wanted space to construct a comprehensive museum, a so-called Conservatory of Arts, Science, and Historic Relics. While Rogers's role in this was minor—he was one of a number of petitioners, each representing a different constituency (his was the Boston Society of Natural History)—he felt drawn to Baker's ideas on technical education. "We need . . . a Polytechnic

Institute," Baker had written, "where the advancement of the useful arts may be noticed and practically described. Where may be properly organized a school of design to increase our supremacy as a manufacturing State. Wherein could be opened a Conversazione which would tend to disseminate useful knowledge upon subjects of every day life, upon domestic and political economy, etc." But Rogers understood that it would be best to let memories of the first effort fade before going back to the legislature. He worked quietly to keep the original group intact, minus Baker, and by January 1860 he was ready with a new proposal.

Meanwhile, a lucky set of circumstances added to his celebrity. The controversy over Charles Darwin's *On the Origin of Species*, which had raged ever since its appearance in November 1859, peaked early in Boston with a series of debates hosted by the Natural History Society in February, March, and April 1860. The public flocked to listen. The two main protagonists were Rogers, in support of Darwin, and Harvard's Louis Agassiz, opposed. Agassiz dismissed Darwin's theory of evolution as fanciful, citing scientific evidence but mostly falling back on nonscientific creationist dogma. Rogers, in contrast, played the role of the measured, thorough, open-minded scientist. While not prepared (yet) to concede the truth of everything proposed in *Origin*, he defended Darwin against Agassiz's ad-hominem attack. Stick to the facts, he prodded Agassiz; throw out emotion—"I *denounced* no doctrine which aims at honesty and truth, whatever might be its character, and . . . I thought no man of science would for a moment think of denouncing any scientific opinion whatever, much less the calm and candid arguments of so fair minded a philosopher as Darwin."

Rogers bested Agassiz not only on what constituted proper scientific temperament, but also when it came to evidence that both men, drawing on immense reservoirs of knowledge and experience, marshaled forth in detail. Even Agassiz's own students—Nathaniel Shaler, for one—gave the victory to Rogers. "Agassiz's . . . capacity for debate was small," Shaler recalled; "Rogers, on the other hand, was not only an able and learned geologist, but very skillful in argument, with a keen sense of the logic which should control statements." The debates were enjoyed as much for their entertainment value—this chance to witness a clash of wills between two titans of the scientific world—as for their content. Many in Boston, a hotbed of abolitionism, were predisposed to take Rogers's side because of Agassiz's popularity in the slaveholding south. Agassiz's reactionary racial views—"The more pity I felt at the sight of this degraded and degenerate race," he once wrote about blacks, "the more . . . impossible it becomes for me to repress the feeling that they are not of the same blood as we are"—were often cited in defense of slavery and became anathema to Rogers and other rights-of-man progressives in the North. When Agassiz died in 1873, he was still a vehement opponent of evolutionary theory, while Rogers turned into one of Darwin's most balanced, clear-headed advocates.

The *Origin* debates helped build support for Rogers's new proposal to the legislature, if for no other reason than that it increased his visibility, enhancing his reputation as unafraid to tangle with a scientist of Agassiz's stature and able to defeat him on his own turf. Rogers drafted all the necessary documents, canvassed representatives and senators, organized public meetings, gave newspaper interviews, and responded to the concerns of nay-sayers—those who worried that the proposal would adversely impact Back Bay land values and taxable income supporting the city's public schools. His efforts drew endorsements from powerful local groups like the American Academy of Arts and Sciences, Boston Board of Trade, Massachusetts Charitable Mechanics Association, and New England Society for Promotion of Manufactures and Mechanic Arts. The bill passed the House but foundered in the Senate and was voted down on March 30.

A self-critical Rogers saw that the proposal still suffered from imprecision, inadequate focus, over-ambition, trying to do too much at once. In May he started paring it to a manageable size. A new proposal emerged, *Objects and Plan of an Institute of Technology*, and this was approved, on October 5, at a meeting of interested parties. *Objects and Plan* laid out a bold yet coherent framework. First, a Society of Arts, which Rogers conceived as something of a cross between a lyceum, a salon, and a professional guild. Elected, dues-paying members would come together on a regular (perhaps monthly) basis to hear fellow members and invited guests speak on topics of scientific or technical interest. The second component, a Museum of Industrial Art and Science, or a Conservatory of Arts, would include departments of minerals, organic materials, manufacturing arts, textiles, implements and machinery, architecture, shipbuilding, and inland transport. Both the society and the museum promised, in Rogers's view, to advance practical knowledge and the industrial arts. But these, he said, would be incomplete without the third component, a School of Industrial Science and Art, to give shape and substance to the Institute's educational goals.

This time events moved rapidly. On January 11, 1861, several dozen enthusiasts met to adopt articles of association: "We the subscribers, feeling a deep interest in promoting the Industrial Arts and Sciences as well as practical education . . . hereby associate ourselves for the purpose of endeavoring to organize and establish in the city of Boston such an Institution under the title of the Mass: Institute of Technology." The statement attracted 54 signatories—incorporators, in effect—a number of whom would become long-term supporters and serve on the Institute's board of trustees (called the Government in its early years; then, as of March 1869, the Corporation). Before dispersing, the incorporators chose a committee of twenty to frame a constitution and bylaws, to press the legislature for an act of incorporation, and to secure a grant of land for this venture in the so-called practical sciences.

Rogers, oddly, was not among the original twenty; he was added as the committee's twenty-first member and elected chairman, while John Runkle, a prized former student of Benjamin Peirce's at Harvard, became secretary. Rogers, as usual, took responsibility for moving the petition through the legislature. Governor Andrew told him point-blank in March that he—and he alone—should present the case: "Between ourselves I know *you* would have a powerful effect, left to yourself, and I fear some one else might come in and weaken it." Legislative snafus materialized anyway, some members repeating earlier concerns about land values and public school funds. But in April, the local newspapers positively glowed; support came, too, from Peirce, Horsford, and others at Harvard, and an act of incorporation was passed and signed by the governor. The Institute would share with the Natural History Society one square of land on Boylston Street. Two conditions were tacked on: first, that the Institute raise $100,000 within the year; second, that it reimburse the state if land values declined (the latter, which Rogers described as ungracious, was later repealed).

The effort then promptly fizzled. One factor was Rogers's chronic poor health, worsened by nerves, overwork, and exhaustion. He did not have the stamina to go on, and others were too busy or lacked the requisite leadership skills. Rogers also grew increasingly preoccupied with the Civil War, filled with moral outrage toward the secessionists as well as admiration for President Lincoln's "patriotic and firm" leadership. The supporters he relied on were equally distracted, their distress centering as much on economic issues—disruption to trade, loss of assets down South—as on moral concerns, or on unionist idealism. A whole year went by with no progress on raising that critical $100,000. A promise of half the amount came from Ralph Huntington, wealthy landowner, industrialist, and president of the Boston and Roxbury Company (a major Boston thoroughfare, Huntington Avenue, is named for him), but no funds were actually in hand by April 1862. Just before the April deadline, the committee of twenty rushed to petition for a year's extension, which the legislature granted. Rogers, meanwhile, kept the institute idea afloat—money or no—by taking steps to formalize its existence. On April 8, 1862, he called the committee together to accept the charter, adopt bylaws, appoint a board of trustees, and schedule the first annual meeting for May 6, 1862. The membership elected him president on May 6, along with four vice presidents (John Amory Lowell, Jacob Bigelow, Marshall Wilder, and John Chase), a treasurer (Charles Dalton), and five to seven members for each of four committees (instruction, publication, museum, and finance). The day before, the Institute recorded its first donation: $3,000 from the estate of Mary P. Townsend.

Rogers projected optimism, even when he did not feel it. "The times are not favorable," he wrote to a friend in August 1862, "but we are not disheartened. The patriotism that is now so generously devoting itself to the safety of the nation and the promotion of liberty must erelong be released from its most urgent public duties, and be ready with deeper earnestness than ever to build up the peaceful structures of education and the arts." By the end of the year, still with little money in hand and only a few pledges to go on, the Institute went before the public for the first time. On December 17, 1862, the Society of Arts met at the Mercantile Library Association, in space leased by the Institute pending construction of a home of its own.

To start, Rogers laid out the society's goal: it would come together twice a month, keeping in view "as its leading object, the promotion of practical Arts and Sciences through the medium of written and oral Reports and Communications, and the exhibition of Models, Materials, Products, and other Objects relating to them . . . aim to secure a free communication and interchange of valuable thoughts on all matters relating to the Industrial Sciences and Arts." The museum and school would follow in due course, but these, unlike the society, he said, required substantial infrastructure and could not be rushed into operation; they "can be carried into effect only in an imperfect and rudimental way without . . . extensive Buildings and arrangements." A half dozen local inventors and industrialists, some of them Institute trustees, then stepped forward to speak on a range of topics: from guns fired under water and the use of wood and iron in shipbuilding, to ships' compasses, safety heating lamps for laboratory use, and cotton manufacture.

The museum idea went on hold indefinitely; the Institute would have no centralized museum until more than a century later, in the early 1970s. Whatever energy Rogers had left he put into the school, which he considered the heart of the enterprise. A large gift from William J. Walker, and smaller amounts from Nathaniel Thayer, Thomas Lee, and Henry Bromfield Rogers (a local businessman, no relation to William), boosted the tally to $100,000 in April 1863. On April 10, two years to the day after the charter was signed, the legislature gave the Institute permission to take possession of its part of the Back Bay square on Boylston Street, between Berkeley and Clarendon and backing onto Newbury Street. The Natural History Society, relatively plush with funds, had already taken its part.

Also granted was one-third of the annual income from federal land-grant legislation (Morrill Act of 1863), the other two-thirds going to the Massachusetts College of Agriculture. Governor Andrew, who had favored Harvard's bid for the land-grant funds, suggested that several local institutions—Harvard's Bussey farm, its Lawrence Scientific School, the Observatory, the Institute of Technology, and the Agricultural College—benefit under a single, unified, Harvard-dominated umbrella. But Rogers managed to have this proposal scrapped before it reached very far. Governor Andrew's proposal met, wrote Rogers to his largest benefactor, William Walker, with "the instant reply from myself and others that the Institute had from the beginning determined to stand alone, that its independence was essential to its success, and that it would accept no grant from the State, or from any other quarter, which should in the slightest interfere with this independence."

As momentum shifted in Rogers's favor, a building committee was appointed on May 6, 1863, architects J. & W. G. Preston were hired in August, and construction got under way by year's end. Hoping to start a few classes that fall, Rogers persuaded a generous friend—anonymous, but probably Henry Bromfield Rogers—to foot the bill for rent on a couple of rooms at the Mercantile, pending completion of the Institute's building, which progressed very slowly. A plan of instruction had yet to be mapped out, with an eye "more especially to the unbooked knowledge," so Rogers spent much of the winter, 1863–64, working up a curriculum. His loyal assistant, Runkle, joined with William Watson, a graduate of the Ecole des Ponts et Chausées in Paris, to frame a program in applied mathematics, one that would reach "from the very elements up to the fullest demands of the scientific engineer." In April 1864, Rogers put the final touches on Tech's magna carta, *Scope and Plan of the School of Industrial Science of the Massachusetts Institute of Technology*, which was adopted by the Corporation on May 30.

Scope and Plan differed from Rogers's earlier proposal, *Objects and Plan*, in laying out a precise program of *professional* education. *Objects and Plan* had focused on the general, popular angle: lectures and demonstrations, in the style of the lyceum, to satisfy a thirst for knowledge among the general public. *Scope and Plan* retained this feature—its so-called First Department: General or Popular Course—but added another. This Second Department: Special and Professional Instruction would rigorously prepare students for careers in the working world. Five courses were to be offered:

mechanical construction & engineering
civil & topographical engineering
building & architecture
practical & technical chemistry
practical geology & mining

The term course referred to a comprehensive program of instruction; units within a course were called subjects, not courses, a practice that continues at MIT to this day. The general department was lecture- or classroom-oriented, while the professional was practice- or laboratory-based. Fulltime professional students would follow a common core curriculum for the first two years, then specialize; part-timers—those who wanted a specific subject, or several subjects, rather than a full course—were welcome, too. Professional students, not general students, would be eligible for a diploma on completion of a full course.

While Rogers did his best to organize the departments in parallel, neither taking precedence over the other, the professional quickly emerged as the school's central feature. But to ensure that his broader goals would not get lost, he added a reminder: "In pursuing this object, it is intended to give to the teachings such scope and method, that while imparting a due measure of knowledge, and cultivating the habits of observation and exact thought,—so conducive to the progress of invention, and the development of an enlightened industry,—they may help to extend more widely the elevating influences of a generous scientific culture." While professional training took priority, then, he did not envision the Institute evolving as a narrow technical or vocational school. Through the end of 1864, both departments—he referred to the first, colloquially, as *popular* and to the second as *systematic*—stood on near-equal terms. A director of the Conservatoire des Arts et Métiers in Paris, and someone whom Rogers identified only as the ablest mathematical engineer in Britain, both endorsed the plan. Runkle captured the excitement close to home: "I knew how exceedingly able [the plan] was, yet am I more than ever delighted with it—I have analyzed it with the greatest care, carrying in imagination students through each of the courses from year to year, & I find it to my mind, perfect in all its parts."

Rogers went to Europe that summer to survey programs of technical education and to buy equipment, models, and apparatus for shipment back to Boston. He toured England, Scotland, France, and Germany, and came away most impressed by the polytechnic school at Karlsruhe, Germany—"nearer," he said, "what it is intended the Massachusetts Institute of Technology shall be than any other foreign institution. . . . Every part of the establishment is designed for use, and not for show." The trip both tired and energized him.

His next step: hire a good, solid faculty. There was not much time, so Rogers cobbled together the best group possible on short notice. Boston was full of capable educators, up-and-coming scientists, heads of firms eager to fill a growing number of slots in the civil, mechanical, and mining fields. John Runkle, Francis Storer, John Henck, and William Watson were natural choices because they had been Institute supporters early on. With Runkle (mathematics), Storer (chemistry), Watson (mechanical engineering), Henck (civil engineering), Ferdinand Bôcher (modern languages), W. T. Carlton (freehand drawing), and Rogers himself (physics and geology) all on board, the school was off to a fine start. Rogers brought in others that fall, to begin the first full academic year (1865–66). Among them were Charles Eliot, a young, up-and-coming chemist just back from Europe, who had given up on Harvard's Lawrence Scientific School, frustrated by its resistance to new ideas; James Hague, for mining engineering; William Atkinson, for English language and literature; and William Ware, for architecture. Ten in all, officially; but nine in effect, as Hague failed to show up for duty. The following two years saw the arrival of George Osborne (navigation and nautical astronomy, later mathematics), Edward Pickering (physics), Alfred Rockwell (mining engineering, to replace the perennially absent James Hague), and Cyrus Warren (organic chemistry).

The new year, 1865, dawned with a burst of optimism, the first in a long time. The Confederate army was sliding into confusion, then full retreat; Southern cities toppled domino-like to Union troops as the nation's four-year nightmare drew to a close. Boston's mayor glowed that the nation's sacrifices had begun to pay off, at least in his city, with a reenergized spirit, new businesses—signs, he said, of an America poised for fast, sustained recovery. Rogers predicted a time of "great and growing demand for Scientific Explorers, Mining Engineers, and Directors of Metallurgical Works. The vast field of industry, which is opening with the mineral resources of this Country . . . enlarged as it must soon be by the entrance of Northern enterprise and free labor on the richly endowed regions of the Southern States, calls for the services of men thoroughly instructed in the scientific principles and practical methods appertaining to mineral exploitations, and the working of Mines of Coal and Metals; and makes the present a most opportune period . . . to promote one of the leading industrial interests of this community."

Traffic was heavier than usual on February 20, a Monday. Hordes of merchants had descended on Boston for a sellers' convention, bringing wares from as far away as Illinois and California. On the second floor of the Mercantile Building, Rogers waited with the half-dozen men he had pulled together on short notice. The Mercantile, a long, two-story structure on Summer Street, sat between Hawley and Arch Streets, in the heart of Boston's fast-growing retail district. The Mercantile Library Association shared the second floor with the Mercantile Academy and the Musical Education Association. It took the grand share, however, slightly more than half, including the choice front-corner area overlooking Hawley and Summer Streets. Among its spare rooms were the ones let to Rogers for his educational experiment.

Eli Forbes, a freshly scrubbed 16-year-old, knocked at the door just after 9 a.m. His father Franklin Forbes, state legislator and spirited advocate of this innovative plan to groom young men for the practical professions, had introduced him to Rogers a few days before. What struck Eli most was Rogers's resemblance to the iconic Ralph Waldo Emerson—tall, gaunt, and pale, with longish hair and a rugged face. Other youngsters poked their heads in as the morning wore on. Abraham Bailey, Samuel Eastwood, Eben Stevens, Joseph Stone, Bryant Tilden. Some, like Eli, had heard about the school from friends or family; a few had seen ads in the local papers announcing a four-month course in "Mathematics, with practice in Geometrical Drawing, and Shading in India Ink,—Lessons in Descriptive Geometry, illustrated by a suite of models in relief. Physics, including elementary doctrine of Forces, and Mechanical properties of Solids and Fluids, accompanied by Manipulations. Chemistry of the Inorganic Elements, with Manipulations. Practice in the use of the Plane Table, Level, and Geodesic Circle. Free Hand Sketching. The French Language." These were practical fellows in search of a curriculum more useful, more marketable, than what Harvard had to offer.

One young man arrived a bit late. Robert Richards, older than the rest, was almost twenty-one and at loose ends. He had gone through a series of fine grammar schools, ending up at Phillips Exeter Academy, but the dull, conventional curriculum in Latin, Greek, and mathematics left him cold. Some of his teachers considered science, and certainly technology, as a lower order of learning than the classics, or literature, or history. But Richards took after his maternal grandfather, Benjamin Hallowell Gardiner, who had opened a lyceum in Gardiner, Maine, charged five dollars for a course in chemistry, and kept all "dead languages" out. The Richards family kept a residence on Beacon Hill and knew Rogers personally; they were distant cousins of the Savages, Rogers's in-laws. While young Richards found it hard to imagine what a scientific school offered, he was certain that it could only be better than the purgatory he had suffered through at Phillips Exeter.

Eventually, the group filed out and headed for the main hall. A few dozen folks gathered—friends and family, board members, local merchants, curiosity-seekers. The ceremony was over almost as soon as it began. Rogers, the only speaker, kept his remarks brief. He outlined subjects, teaching methods, and a few rules, closing with a comment about the value and dignity of the practical professions. That was it—efficient and businesslike, no pomp or circumstance, no benedictions. The afternoon off, then next morning all noses to the grindstone. A no-nonsense approach. Classes in mathematics and civil construction promptly at 9 a.m., physics at noon. Rogers kept his emotions in check until he got home, then jotted in his diary: "*Organized the School! Fifteen students entered! May not this prove a memorable day!*"

Rogers knew how to get things done, systematically and with a minimum of fuss—a talent that inspired confidence and eased the burden on Corporation and faculty members, who could focus on what they did best: fundraise and teach. By day he radiated courtesy, warmth, and sympathy; at night he would return to his aging father-in-law's residence where he, Emma, and James Savage would entertain members of their small social set, sometimes students and faculty, or pass quiet time reading, perhaps puttering over Mr. Savage's massive genealogical projects. No. 1 Temple Place was every bit the Tech president's house, even if not officially designated as such.

The school quickly became known as Tech or Boston Tech, local shorthand for the labored School of Industrial Science, Massachusetts Institute of Technology; "MIT" would come later, a twentieth-century moniker. The Institute's reputation for working students to death or desperation in the best—some would say worst—utilitarian spirit, consumed by a work ethic whose ends and means sometimes seemed indistinguishable, began to gel early. The epigram "Tech is Hell!" had yet to be coined, but the sentiment was already widespread. While most students wore the badge with pride, some needed a friendly ear, a comforting shoulder, a voice that soothed, a calming influence. Rogers offered all of these. "One day," Robert Richards recalled, "a student came into the lecture on physics, and finding his favorite seat already occupied, began to sputter. . . . Rogers, understanding the situation, set his gyroscope to spinning, carried it down and handed it to the sinner who was making the disturbance. The malcontent was so puzzled and diverted by the instrument that he forgot his grievance, his whole attention being focused on trying to keep up with the antics of the ungainly thing." Beyond that, his teaching skills were said to have been unparalleled. According to one observer, a class under Rogers was "a triumph of oratorical art. . . . whether treating of rocks, physical forces, or rigid principles of mathematics, he was always able to kindle the enthusiasm of the students, and make the most vivid and lasting impressions upon their minds." Charles Cross ('69) appreciated Rogers's unusual teaching style, one that mingled gravitas and scholarly intensity with playful pluck, the spirit of the proverbial kid in a candy shop. "He always," said Cross, "showed that freshness of appreciation which too often dies as experience grows deeper, and combined the wisdom of the sage with the enthusiastic appreciation of the child."

Other faculty favored the sink-or-swim approach, the dry, functional, impersonal style that students thrust into the working world would be faced with and must learn to handle. Charles Dickens's brutal satire on utilitarianism, *Hard Times*, was just a decade old; laissez-faire values prevailed, however, and Dickens's Josiah Bounderby—a man devoid of sentiment—was, to some readers, if not exactly a hero, not a villain either, more a lopsided exemplar of traits essential for commercial success. Some Tech faculty were Bounderby-like, as were many employers that students must face in the real world. Charles

Eliot came across as standoffish and authoritarian, a style brought over from Harvard, where he had a reputation for being "cold as an icicle." One student, convinced that Eliot hated nothing so much as human touch, would reach out to shake his hand whenever their paths crossed just to watch him recoil. Tech students found it difficult to reconcile Eliot's aloofness with the progressive values that he brought to the classroom, his insistence on the importance of challenging dogma, on direct observation, collection, and interpretation of data. While Francis Storer was better liked, his personality was none too pleasant either—irascible, impatient, overbearing, quick to label people dishonest, lazy, or stupid—but Tech students went for its relative warmth (heat, sometimes) over Eliot's coldness; and unlike Eliot, he made himself available at all hours and was never, apparently, too tired or busy to help a student or colleague in need.

But Rogers—and, from various reports, Runkle and Atkinson as well—embodied a different quality: compassion (Dickens would have approved). He listened patiently to each student's concerns, appealed to his sense of duty, and sent him on his way with renewed self-confidence. He kept balance, perspective, focus; rebelliousness melted away in his presence, not out of fear but in response to his "sweetness and sympathy." As James Tolman ('68), a member of the first graduating class, recalled: "Professor Rogers was always the student's friend. The lack of means and the immense amount of work assumed by the instructors sometimes caused implied promises to go unfulfilled, and inspired the restless pupils—a good proportion of whom were grown men, taking time for study from the practice of their professions—to feel rebellious against the direction of the school. I remember that some of these occasions resulted in visits to President Rogers, and that such was the invariable courtesy with which these complainants were treated, that we always came away feeling that, in so far as the means would allow, every need of the classes should be filled, and with our sense of manliness so appealed to that we were ready to recognize our duty as co-workers with the professors for the good of the School."

That first term, the half-dozen faculty and two dozen students who came together each day made do with just two cramped rooms—one for lectures, the other for lab exercises. The library's large hall could be hired on a separate, pay-as-you-go basis, but Tech used it only for public lectures and demonstrations of the Society of Arts, its announcements appearing regularly in the columns of the *Evening Transcript*. Tech's makeshift lab, as primitive as labs come, looked like space a poor inventor might have carved out for himself—very little equipment, glaring gaps, a single sample of each tool or instrument. One retort, one beaker, one condenser. Students wrangled over who would carry out the next experiment. Just one at a time, the others gawking from a few feet away. In a corner sat the prized tiny muffle furnace, where some lucky fellow might get to test or purify a sample of metal or a mineral deposit. What mattered most was the spirit inside those rooms.

The atmosphere outside inspired, too; not the street bustle so much, though that added a vitality of its own, more the feel of what went on elsewhere within the Mercantile. Tech's space was tucked off to one side. But activity swirled all around—merchants, industrialists, entrepreneurs of various stripes came, went, always on the move. Faculty and students brushed elbows with the captains of Boston commerce and industry. Lowells, Lawrences, Appletons, Cushings, Forbeses, and their assistants, agents, clients, legal counsel, advisers, rushed about talking, reviewing contracts, reading newspapers.

The Mercantile Library was Boston's central clearinghouse for commercial information, news, insights. Everyone who was anyone in business congregated there. Starting early each morning, and continuing through the day, a constant, energetic stream. The library was a quasi-public facility; members had special privileges, but the building and resources were open to all. A special side entrance that led directly into the hall, from Hawley Street rather than Summer Street, was suggested for ladies who preferred not to interrupt men conducting business in the reading and periodical rooms. But young male whippersnappers, including those starting out as students, could mingle unnoticed if dressed appropriately. Rogers encouraged this, confident that the exposure would help aspiring professionals adapt to business life, manners, and mores. Tech may have been dwarfed by its landlord, but the school benefited from just being where it was, in the midst of this frenetic activity.

Out went the old teaching styles—lecture, listen, learn by heart, drill, repeat, teachers ramming facts down students' throats; reward for repetition, penalty for lines not toed. "I have imagined," Charles Eliot later said, "that his [Rogers's] knowledge of the fact that I held the same opinion about teaching all sciences by the laboratory method, and not through lectures or books, probably encouraged him to offer the professorship of chemistry and metallurgy to a man only thirty-one years of age." Students touched, lifted, handled, worked directly with apparatus, got a feel for it, studied its innards. One endangered species was the teacher who preferred to hold forth from his desk with students looking on from a safe distance. Tech's instructors taught "through actual handling of the apparatus and by working on problems, shoulder to shoulder with the boys." Lab accidents turned into valuable lessons—a piece of glassware broken, a hydrogen generator blown up, a vial of acid spilt, disasters preventable and solvable in a spirit of calm, collected ingenuity.

All this a product of Rogers's lively imagination, backed by years of reflection on what worked well and what did not in a scientific curriculum. How to gather, record, and collate information; how to draw accurate, creative conclusions based on hard evidence; how to weigh and resolve conflicting conclusions; how to approach, tackle, and solve problems. Rogers brought an infectious enthusiasm to the lab and classroom. His "blackboards," according to James Tolman, were "filled with copious notes. . . . His absorption in his subject made him almost impatient of the restraint imposed by models and apparatus, and at times interfered with the smoothness of the experiments which he had always carefully prepared." Nature's laws, the only rules that counted. The classroom, a place not for laying out ideas-certain but for putting ideas-indefinite to the test. The lab, a place to identify, mix, and analyze chemicals with hydrogen explosions and nitrous oxide (laughing gas) as comic relief—fits of fun—between hours of serious learning. The perpetual-motion pendulum, a pure marvel. Geology with rock samples, three-dimensional models, and sketches from life. Geometry and mechanical drawing as tools of communication—common, universal languages. German and French not for show, or as marks of cultivation, but as tools to grasp concepts in the literature of foreign lands.

At Tech, good common sense always trumped book learning. It was an inversion experience, as much as a conversion; students worked because they wanted to, not because they felt they ought to, or because they were forced to. Wide young eyes opened wider each day; Robert Richards described the overall sensation as "a wonderful labyrinth" whose pathways twisted off in countless directions—some more predictable than others—and with truth as the end game. One problem was how to drag oneself away from texts, journals, drawing board, and laboratory long enough to grab a decent meal or a good night's sleep. A type of education, in other words, that "ceased to be a plague spot and became a delight."

Classes stretched from 9 to 5 six days a week, with two hours off at midday. This routine was later compressed, *not* relaxed—9 to 4:15 on weekdays, midday break from 1 to 2:15, half-day on Saturday—to allow for extra late-afternoon spare time. Students would congregate in small groups for lunch. Some ran tabs at local taverns or purchased board by the week at local rooming houses; others, commuters especially, would haul out tin lunch-pails brought from home, then scour out an empty lecture room or other secluded spot to consume their meal. It was a regimen designed to simulate a regular workday, and thus to ease transition to life in the working world, something that traditional colleges, relatively slow-paced, often failed to do. The difficulty, of course, was that school did not end at 5 or 4:15. Evenings spent studying, and preparing for the next day, left little time for rest or recreation—and this, for Tech, would become a sore point in years ahead, with frequent, highly publicized complaints about students driven to illness, drink, or worse. But many accepted the challenge willingly, with a sense that it would toughen them for any adversity.

The permanent building was nowhere near ready in time for the first full academic year, 1865–66. But with the size of the student body more than double that of the preliminary session (approaching 70, with more likely to appear), the two rooms in the Mercantile, already crowded, were now bursting at the seams. As a stop-gap, Rogers rented space in the home of a late justice of the Massachusetts supreme judicial court, Judge Charles Jackson, on Chauncy Street, a half block away. The crunch grew more acute when, in October 1865, John Lowell proposed that the Lowell Institute sponsor free evening courses for working adults, under Tech's supervision. Rogers gladly seized on the offer, not so much to generate extra income as to reinvigorate the public education component that he had laid out in *Scope and Plan*, then all but ditched. The Lowell courses were advertised in local newspapers a month later, and a month after that—on December 19, 1865—they met for the first time. The Tech faculty pitched in. Runkle lectured on math, Bôcher on French, Atkinson on English literature, Watson on practical science and mechanic arts. Chemistry under Eliot and Storer—among the most popular of the Lowell courses—had to be postponed until the Boylston Street building opened in 1866. The other courses, meanwhile, none of them requiring lab space, roamed between the Mercantile and Judge Jackson's house. Each course, typically, consisted of 18 lectures and met either once or twice a week; always after-hours, at nighttime.

Faculty and students spent several months that first full academic year stumbling along, quite literally, between three venues: Summer Street, Chauncy Street, and Boylston Street. Boylston was furthest off, a mile away from the other two, so classes met there, if possible, in the afternoons, following morning classes at Summer and Chauncy. "It was a long tramp," Ernest Bowditch ('69) recalled, "and a cold one as well, as there were no structures on Boylston Street below Arlington Street Church except the Natural History building, and a plank sidewalk only part of the way, west of the Public Garden. The few street cars available were drawn by horses, and ran from the Paddock Elms on Tremont Street along the Common to Boylston, through the latter to Clarendon, thence to Marlborough, where they terminated." The interval between cars was fairly regular, every fifteen minutes or so, but some students raced over on foot—even in bitter weather—rather than wait.

As Boston pushed west, road and sidewalk conditions remained primitive. Students who commuted from suburban towns—Brookline and elsewhere—came in by train and got off at the railroad terminus on Huntington Avenue, then tramped across vacant lots that were either dusty, muddy, icebound, or ankle-deep in sludge and water depending on the weather, maneuvering between river eddies where the reclamation project had not reached far enough. On the trip home, they would go a different way, down Columbus Avenue from Berkeley Street, a more convenient route for outbound departures. Much to be pitied were those who came from exurbs, like Fitchburg and Foxboro, as their day started *so* early and ended *so* late; those from Maine, or far western Massachusetts, had no choice but to scrounge for cheap digs in town. Bowditch remembered a couple of "down-easters" who boarded with a Mrs. Page on Berkeley Street, convenient to Tech but cramped: shared space "in what was intended to be the back entry of the house and was mostly doors and windows—a sort of general passageway during daylight hours."

The Boylston Street building—unnamed until 1883, when it became known as the Rogers Building—went up after its neighbor, the Natural History Building, was already in place. Both presented solid, elegant, neoclassical lines on a dignified, human scale, setting a tone for the neighborhood, still largely a wasteland in the process of development. Tech sat at the corner of Boylston and Clarendon Streets, Natural History at the corner of Boylston and Berkeley (the latter still survives, remodeled on the inside to house Bonwit Teller; Tech was torn down in 1938 to make way for the New England Life building).

An impressive, rectangular (90 × 156 ft.) design, Tech rose nearly 100 ft. high and consisted of four stories and a basement. "It stands," one contemporary account observed (ca. 1869), "upon about 1,500 spruce piles, twenty-four feet in length, driven to a firm bearing upon the solid clay. . . . Rusticated free-stone piers support a terrastyle portico, on a level with the second floor, which supports a richly-wrought entablature, crowned by a pediment, designed to contain an allegorical bas-relief representing the Genius of Art bestowing her favors upon inventors and mechanics, who are in the act of presenting the results of their skill for her consideration. The pediment is surmounted by a stone pedestal, intended for the support of a colossal statue of Minerva, as patroness of art, and typical of the purposes of the Institute." Rogers, however, wanted ornamentation kept to a minimum, which put him somewhat at odds with the architects, whose tastes ranged from baroque to neoclassical and styles in between.

As far as Rogers was concerned, the building was about functionality, first and foremost, and while efforts were made in ensuing years to decorate—a set of allegorical friezes, by Paul Nefflen, was later added as crown molding for the main hall—Minerva never showed up. Facilities were for work, not for show.

A pair of giant urns guarded each side of the main entrance, accessed by steep stone steps from street level. Inside, two administrative offices—president on the left, secretary on the right—sat off the main foyer. The rest of the first floor held lecture rooms and labs for physics and geology, plus space to store artifacts. The second floor, oriented around an enormous lecture hall that seated 900 (Huntington Hall, named for benefactor Ralph Huntington and completed in 1870), had rooms for mathematics, engineering, languages, English, and mechanics. The third floor housed architectural and engineering models. The basement held a chemistry lecture room, carpenter's shop, chemical storeroom, engine and boiler rooms; also chemical, mining, mineralogical, and metallurgical labs. Professors occupied small studies on the top floor. It all *seemed* like a lot of space.

Students quickly found that Tech's rigorous work schedule—and here work meant good old-fashioned manual labor, not just book learning—left little time for fun. The new Boylston Street building had nothing to offer besides classrooms, labs, and offices, so students looked elsewhere for recreation. A number joined the Tremont Gymnasium on Eliot Street, a few blocks away. This was a state-of-the-art facility built in 1859, run by John Doldt, a professional athlete, and frequented by local businessmen, or gentlemen-of-the-ledger as they were sometimes called. Here students sought a change of pace, a break from the grind, chance encounters with captains of commerce and industry. Whitney Conant, Eli Forbes, and Robert Richards (all '68) often arrived together and took turns at the dumbbells, swings, and parallel bars. There was no organized football team; fellows kicked the ball wherever they found space—a quiet back street, a stretch of grass on the Boston Common—and slugged it out without benefit of formal rules or referees. Injuries, some of them serious, became badges of honor, testimony to one's toughness.

With the Charles River just a few hundred yards away, Tech students also went in for rowing. Some joined the Union Boat Club.

Union sometimes competed against Harvard, which had a formal program of its own. "The Harvard men have been very helpful to us in regard to boats and in other ways," Robert Richards declared, "and as a result a very pleasant relationship has grown up between the two institutions." The Harvard group tried to coax one or two of Tech's best rowers to defect, promising various perks, not the least of which was a chance to compete against the celebrated Oxford and Cambridge teams, on the legendary River Thames. Thanks but no thanks, came the response. If Tech men had wanted a Harvard-style education, they would have gone there to begin with.

Students brought high jinks to class, if they felt they could get away with them. One faculty member, William Watson, was picked on mercilessly. Watson found himself at a disadvantage because of his age—he was not much older than the boys he taught—but, in addition, he carried his slight, elf-like frame with a pompous, dilettantish air picked up in Paris. He would arrive each day coiffed, smelling like a rose, decked out in suede or felt jacket and a brightly colored cravat; then, gingerly, he would remove his silk top hat and lay it next to his gold-tipped cane, *très élégant*. The idea that such an effete fellow would teach civil construction or mechanical engineering, his specialties, struck students as ludicrous. Would he venture into a muddy alleyway, much less a construction site? The boys were surprised, however, to find that he could tough out a field trip with the hardiest of them. Whether in the depth of winter or at the height of spring mud season, Watson would show them around gas and steel works as far away as Nashua, New Hampshire—all "in aid," he said, "of the practical studies of the School." But they mocked him anyway. He was tagged with the nickname Squirty, for the chemical wash-bottle he carried around to clean off desks or blackboard surfaces. One day students grabbed his prized hat, turned it top side down on the floor, and stood dripping umbrellas inside. Such pranks—precursors of Tech's famous *hacks*, its beloved ritual of student-driven mischief—helped relieve the rigors of the curriculum, providing an outlet for youthful energy at a school that did not (could not, at this point) offer organized extracurricular activities, or, in the middle of a crowded urban center, create a coherent campus feel.

Rogers brought sons of factory workers, laborers, blacksmiths, janitors, and hackney drivers together with upwardly mobile offspring of bank clerks and factory managers, together with heirs of vast family fortunes, privileged Brahmins whose families had built Boston from scratch. It was a true urban melting pot, the American fantasy played out on a small scale. The mix did not come easily, however, as these youngsters had grown up mutually hostile, separated by gulfs of class, ethnicity, and religion. The 1834 burning of the Ursuline Convent at Charlestown by a militant Protestant mob, inflamed by anti-Irish and anti-Catholic emotion, had been a watershed moment in the city's history, and in the 1860s was recent enough in memory to dredge up dark emotions. The elms and chestnuts of Boston Common provided cover for many a drawn-out skirmish, sons of Brahmins versus sons of "micks" (Irish immigrants, no Italian or Eastern European kids yet to muddy the fray further), each ranged on opposite sides hurling snowballs, some filled with rocks. Yet even with such tension in the air, Rogers drew youngsters into a setting where shared professional objectives overrode social, ethnic, and religious prejudices. Harvard had none of this flavor, nor did its feeder prep schools, Exeter and Andover. Among the Brahmins drawn to Rogers's leveling ideas were Cabots, Appletons, Lowells, Conants, Bowditches, and Forbeses, many of whom hailed from generation upon generation of Harvard loyalists. If they had not necessarily written off Harvard's curriculum as hopelessly old-fashioned or stagnant, they had grown concerned about its relevance in an age of rapid industrialization.

Tech, however, was far from broadly inclusive. At the start its reach stretched little if at all to nonwhites, and the number of admits from outside New England was relatively small; women were a special category. Rogers had voiced impassioned support for Lincoln's emancipation proclamation—"the slaves," he wrote to Henry in September 1862, would be "*forever free*" come the new year—and for expanding educational opportunities. He admired the work carried on by various commissions, forerunners of the Freedmen's Bureau; one of these, he said, "has brought within the folds of a free civilization for instruction and paid industry tens of thousands of fugitives, and of those deserted by their masters, and . . . give the fullest evidence of the capacity of these people for knowledge and training . . . to mitigate, if not remove, the prevailing belief in the hopeless degradation of our American negroes." Yet this passion failed to draw black students to Tech. There were few if any in Rogers's time; the first documented black student, Robert Taylor, was not admitted until 1888.

Tech evidently had no policy of inclusion or exclusion with respect to race, but gender was a different matter. Women faced official, explicit barriers for more than a decade. The Lowell Institute's night classes were open to "persons of either sex" from the start, in December 1865, but Tech's regular courses were closed to women. As the number (and generosity) of women donors grew, so did the pressure to accommodate women as students in the regular program. "We would add, as a matter of just pride," secretary Thomas Webb recorded in the Corporation minutes as early as May 30, 1864, "that on our roll will be found the names of several Ladies, who, by the liberal contributions they have made, evince the interest they take in our present efforts to increase and impart knowledge, and that they fully appreciate, and are ready generously to encourage by substantial gifts, a movement which . . . must materially advance and improve every industrial class in the community, thereby adding to the sum and usefulness and happiness for the benefit and enjoyment of all." But when women in the Lowell classes, particularly chemistry under Eliot and Storer, pressed for additional coursework in 1866 and 1867, their requests were denied. Not consistent, Rogers said in May 1867, "with the present condition of the school and organization of the classes." While he believed that integrating women was a valid goal, long term, it was impractical for the present: no way to comply without "seriously embarrassing"—disrupting, that is—the school's operations.

The only women in Tech's community at the time, aside from Emma Rogers and other faculty wives, were Margaret Stinson and Charlotte Thayer. Stinson, hired in February 1865 to take charge of the chemical supply room, was advised by Rogers that she would survive quite well in this male-centric universe—"They didn't want any woman around in those days," she later recalled—as long as she exercised "a little diplomacy." In response she tacked mothering onto her job description, soothing, comforting, tending to students' scrapes, scratches, and cuts until her retirement in 1911. Thayer became "lady assistant librarian" in 1866, and was succeeded by Augusta Curtis in

December 1870; their job was "to take charge of the library and study-room and keep order therein." Curtis was one of the Lowell Institute women who had applied for admission to the regular program, and this job may have been a consolation prize; she was also assigned to help the professor of English, William Atkinson, mark students' written exercises. Tech's first woman admit, Ellen Swallow, arrived in 1871, but only as a special student "in the nature of an experiment." In 1881, by which time the policy *had* changed and two young women—Marie Glover and Evelyn Walton—graduated in chemistry after four years alongside their male classmates, Rogers talked about how wonderful it was to see the Tech credential viewed not as a male prerogative, but as "belonging to any sex." A year later, with a new building for Tech on the horizon, he threw his support behind those who wanted "special accommodations for the use of women" included in the plans.

While the student population grew rapidly under Rogers in the mid- to late 1860s, more than fourfold in three years, the pool for the regular professional courses remained relatively confined: young white men, most of them born and bred in Boston and vicinity. A few came from other New England states, a smaller number from northeastern (non–New England) states such as New York and Pennsylvania. Smaller still was the group from the American heartland: Ohio, California, Illinois, Missouri, Kentucky, Colorado, Wisconsin, Maryland, Arkansas. The number from such far-flung places grew slowly but steadily; just 3 non–New England states were represented in the first year (1865–66), but 12 in 1866–67, 10 in 1867–68, 12 in 1868–69, and 11 in 1869–70. The first foreign students came from Nova Scotia, New Brunswick, and Upper and Lower Canada (as Ontario and Quebec were called, respectively, before Canada's confederation in 1868). One entrant from Cuba arrived as early as 1867–68. But the foreign numbers were minuscule. As state aid helped to underwrite the enterprise, and with the Corporation dominated by local businessmen, Rogers thought it a good idea if the benefits—the earliest ones, anyway—stayed close to home.

Mid-morning, Friday, October 24, 1868; near the start of Tech's fourth full academic year. Rogers gaveled open the faculty's regular weekly meeting. In the midst of business, around noon, he felt faint, struggled to speak, and lost control over one side of his body. A stroke, classic textbook symptoms. Runkle, Atkinson, and Storer rushed him home, where he was confined for the next two months. A speedy return to work looked doubtful and, by December, it was out of the question. Emma moved him to Philadelphia so that his brother Robert, dean of medicine at the University of Pennsylvania, could monitor his care. Physicians there, half-facetiously, diagnosed his condition as "Institute on the brain," which Emma took quite literally—"The Institute seems to be the one subject," she told Runkle, "most dangerous to his equanimity & it must therefore be some time before he can talk or think of the proposed plans in connection with it."

Tech had faced emergencies before, but this was its first crisis of leadership. While Rogers delegated authority from time to time, everyone—from the wealthiest Corporation member to the brightest-eyed first-year student—ultimately reported to him. There had been no time to groom a successor, nor did Rogers think in such terms. The Tech community took for granted that he would be around in full charge for the foreseeable future, a president for life, as it were, by general acclamation. Even if beyond his prime, he was not very old either, a few weeks short of 64. John Runkle, his closest friend on the faculty, took over as president pro tem for what everyone expected would be a few months. But the leave stretched to a year, then two years; and Rogers resigned, finally, in May 1870. He retained his seat on the Corporation so as to monitor developments from a not-too-distant vantage point, provide guidance as needed, and smooth the transition for Runkle, his handpicked successor.

Emma Rogers spent much of the next decade insulating him from stress, even at times from friends and allies—"his nerves," she kept telling Runkle, "are not yet strong enough to talk & think about the Inst." But Rogers also took his illness as an opportunity, a chance to recapture some of that broad-sweep drive that had kept him moving—and growing—in earlier years. While he had not felt confined at Tech, exactly, his role there had cut back on his time for other things:

research, teaching and lecturing outside Boston, playing a vigorous role in science, science education, and science policy on the national level.

He had done some of this even during the jam-packed early and mid-1860s, when Tech consumed much of his energy. In 1861 he helped out with the Illinois Geological Survey, probably as a remote adviser. In June that year Governor Andrew appointed him state inspector of gas meters, but the fumes brought on bronchial distress—this was no mere desk job—so he resigned in February 1864. In January 1862, he taught a Lowell Institute course on application of science to the arts, which dovetailed nicely, he thought, with his campaign to spread word about the new Institute of Technology. In March 1863 he and his brother Robert were invited to join the select group of 50 corporators (founding members) of the National Academy of Sciences. Neither took an active role in Academy affairs in these early years, not for lack of interest but because the Academy was dominated by a clique—the so-called Lazzaroni, led by Alexander Bache, great-grandson of Benjamin Franklin; Louis Agassiz, Benjamin Peirce, Joseph Henry, and a half-dozen or so other distinguished scientists made up the rest of the group—whom the brothers considered elitist, exclusionary, and antidemocratic. William Rogers's other outside activities during this period included a term as vice president of the Union Club, founded that year to bolster support for the federal cause; trusteeship of the Blind Institution, begun by Henry Rogers's abolitionist friend, Samuel Gridley Howe; corresponding member of the Essex Institute; corresponding secretary of the American Academy of Arts and Sciences; and member of the visiting committee to Harvard's Lawrence Scientific School. In May 1865, he traveled to New York to attend a convention of freedmen's aid societies. At its organizational meeting in Boston, in October 1865, the American Association for the Promotion of Social Science elected him its first president. Alexander Bullock, John Andrew's successor as governor of Massachusetts, persuaded him to lead the official state delegation to the Paris Exposition in the summer of 1867. Just about the only thing he did not do was get back to his research.

This schedule would have taxed the stamina of a younger, stronger, healthier man. Yet after his illness in 1868, Rogers added more commitments. He had time now, without the burden of the Tech presidency, to pursue other interests. Occasional winters were spent at 117 Marlborough Street or the Hotel Berkeley, but he avoided Boston and stayed in Lunenburg or Newport much of the time; by this point his sister-in-law, Eliza, Henry's widow (he died near Glasgow, Scotland, in 1866), and their daughter Mary had joined the household. Rogers traveled often, too. Depending on the season, he could be found in Philadelphia at his brother Robert's residence, 1004 Walnut Street, or in New York, Baltimore, Washington, D.C., even back at his old stomping grounds in Charlottesville, Virginia. His social life picked up after a hiatus. He grew active with several private clubs—the Thursday Evening Club (he was president at one point), the Saturday Club, and the Town and Country Club of Newport, formed in 1874 by Julia Ward Howe with the help of William, Emma, and other friends. In 1872, he sought a stronger role in the National Academy, now that Joseph Henry—a reasonable man, in Rogers's view, the least offensive of the Lazzaroni—had succeeded Bache as president.

While he published little original research after 1864, in the 1870s—his decade of illness—Rogers also began to think about getting back into science. He had wound down enough, in his mind, to donate the bulk of his scientific apparatus to Tech in May 1872, retaining—just in case—his microscope and one or two other critical pieces. Sure enough, a year-plus later, he was ready. "I am well enough to enjoy much," he wrote to a friend in December 1873, "and even to do a *little* scientific work." He resumed giving papers at professional meetings. He spoke on the geology of Newport at the Boston Natural History Society in May 1875; also that year, same venue, on gravel and cobblestone deposits in Virginia and the middle Atlantic states. His final scientific paper, on the geology of Virginia and West Virginia, appeared in 1880 in *Virginias: A Mining, Industrial, and Scientific Journal.*

This period of relative separation from Tech business also saw Rogers occupied as never before with the organizational side of science, on a national scale. Two events took center stage: his election first as AAAS president in 1875, then as president of the National Academy of Sciences in 1879. Rogers's term as AAAS president overlapped with the nation's centennial year, a distinction eclipsed only by his election as National Academy president, which Emma Rogers called "perhaps the crowning honor of his life." The Academy had just begun to establish itself as a force to be reckoned with when its second president, Joseph Henry, died in May 1878. That June, acting Academy president O. C. Marsh accepted Congress's request that the Academy help plan surveys of still-unmapped territories out west. Rogers's background for this work was unmatched, and Marsh appointed him—"the Nestor of American geology," in Marsh's words—along with five other scientists (geologists John Newberry and James Dana, engineer William Trowbridge, mining expert Alexander Agassiz, Louis Agassiz's son) to carry out a study. The committee's report, submitted in the fall, gave Congress a basis on which to create the U.S. Geological Survey, under the Department of the Interior. Rogers's central contribution to this effort played a part in his election on April 16, 1879, to a six-year term as the Academy's third president. To everyone's surprise, he left his sickbed in Boston to rush to Washington. "[He] had been informed by telegraph," recalled George Brush, head of Yale's Sheffield Scientific School, "and although in feeble health, he responded at once by taking the night train . . . arriving early on the morning of the last day of the session, almost exhausted by lack of sleep and the fatigue of the journey. We hardly expected that he would be able to attend."

Rogers spent the next three years commuting regularly to Washington. He guided Academy business with the help of Marsh, who sometimes filled in when Rogers was ill. Rogers positioned the Academy as quasi-official adviser to Congress on yellow fever epidemics and other national health emergencies. A smaller, but still important, project, undertaken at the request of President Rutherford Hayes's interior secretary, Carl Schurz, in May 1880, involved a study of "the question of restoration of the faded writing of the original manuscript of the Declaration of Independence." Rogers was often called on to intervene with high public officials. "I know how much influence a few words from you will have," one scientist wrote in May 1881, urging Rogers to take up a particular cause with President James Garfield. The month before, Garfield had hosted a reception for Academy members at the White House and promised Rogers, schedule permitting, to put in an appearance at scientific meetings then in session. Rogers helped persuade Garfield to appoint Julius Hilgard to succeed Benjamin Peirce as superintendent of the U.S. Coast and Geodetic Survey. Garfield's successor, President Chester Arthur, entertained Academy members at the White House in April 1882, also at Rogers's behest.

Rogers involved himself in Institute affairs only when necessary, to help fend off calamities. He often excused himself from Corporation meetings, citing not illness but business elsewhere. As a mark of appreciation, but also to encourage him not to drift too far, the Corporation named the physics laboratory the Rogers Laboratory of Physics in 1872; in 1876 a detachment of Tech students attending the nation's centennial expo in Philadelphia named their camp for him.

One crisis called for his intervention in the winter of 1869–70, while he was recuperating in Philadelphia. Charles Eliot, by this time president of Harvard (he had resigned as Tech's professor of chemistry in July 1869), floated a plan to establish formal ties between Tech and Harvard. Rogers took prompt steps to thwart it. "I am convinced that such a connection would be a decided disadvantage to the Inst," he wrote Eliot, "which owes its success in great measure to the fact that it has stood entirely unconnected with other institutions." But when Eliot refused to take no for an answer—the proposed change, he wrote back, would not make the Institute "any less independent in reality than it is now"—Rogers persuaded Emma, adamant about keeping him away from Tech business, that he must see Eliot to set him straight. Eliot went to Philadelphia in February 1870, expecting to gain Rogers's approval for some sort of amalgamation (Rogers called it an annexation). They met for a little over an hour. "I could not," Rogers told him, "see any advantage to the Institute from the proposed change but the gain of some funds—but that the Institute would be a *great loser* by relinquishing its present independence." What Rogers found particularly distasteful was Eliot's suggestion that the Institute, as a Harvard school or department, be named in honor of the Rogers family. "I expressed my repugnance to all such names," he noted, taken aback that Eliot would imagine that a bribe of any kind—much less one that appealed to personal vanity—could sway him on what course he would recommend for the Institute.

But Rogers, by and large, was content to let Tech move along on its own steam. He had done his bit; it was time for others to step in. "The Institute of Technology," he told a friend in December 1873, "is now very prosperous, and both as to extent and thoroughness of teaching and number of students is at the head of the scientific schools of the country." He wrote this just a few months after the so-called Panic of 1873, precipitated by a huge bank failure, and with the nation on the verge of a deep, decade-long depression. But while Rogers acknowledged the impact of such economic forces, he never reconciled himself to the unraveling that took place at Tech during this period. His handpicked successor, Runkle, proved unequal to the job and drew little help from the Corporation, whose members either turned on him or retreated to attend to their own precarious finances. A disappointed Rogers watched from the sidelines as student enrollments dropped, key faculty left (for greener pastures, they hoped), and donations fell off.

By the fall of 1877, with Tech in growing jeopardy, he returned to meetings on a more regular basis—not full attendance, but enough to signal readiness for duty, to find ways to stabilize a faculty and Corporation whose mood ranged from dispirited, to angry, to downright unruly. Everyone turned to him for a magic bullet. Runkle pleaded for advice on which programs to consolidate, trim, or do away with; some had to go—the deficit was just too large—but the dilemma was how to reorganize without imperiling Tech's overall mission. The Corporation pressed him with suggestions that were at times constructive, at times defeatist: recruit more members, appoint an executive committee to help with policy and management, close temporarily, shut down altogether. Institute secretary Samuel Kneeland captured the despair felt by many, and a sense of Rogers as Tech's last, best hope: "It appears to me," he wrote in October 1877, "that you alone can save the Institute from impending danger, and save it from decline, perhaps from death. . . . I think you are the only Palinarus who can steer the bark of the Institute between this Scylla and Charybdis—having so successfully launched this bark, do not let her perish on the rocks of parsimony and business red tape."

When Runkle resigned under duress in June 1878, and all eyes turned to Rogers again, he resisted the call. But he agreed to take over as interim president on three conditions: that the Corporation make efforts to raise $100,000 within three months, an amount he felt "satisfied . . . would place the School in a perfectly safe and satisfactory condition"; that a faculty chair be elected to relieve him of certain executive functions; and that a new president be hired as soon as possible.

The first condition, impossibly ambitious, saw some progress within the year: $75,000 pulled together from designated "friends of Industrial Education." Few if any of these were alumni—still struggling, early-stage professionals, for the most part—but in anticipation of future largesse, Corporation members urged that an alumnus be appointed to their ranks so as to draw on "the cordial good will of the graduates and more and more to secure the benefit of their experience in the conduct of its affairs." The election of John Ordway, professor of metallurgy and industrial chemistry, as faculty chair met Rogers's second condition. But the third condition—and, in Rogers's mind, the most essential—proved problematic. He accepted personal responsibility for this one, as the part that he had played, with near-disastrous consequences, in pushing Runkle into the presidency weighed heavily on his conscience.

Rogers served as president pro tem from June 1878 to December 1879; and then, with no successor on the horizon, as president for nearly two more years. He spent much of this period doing damage control, smoothing frayed relations between Runkle, Corporation members, and a demoralized faculty. In September 1878, William Ware slammed the Corporation for dereliction of duty. "It was mutually understood between us," he wrote, referring to the original faculty hired in 1865, "and to the projectors of the school that a first class establishment was in contemplation . . . that money would be needed to this end; the names of the Corporation were intended to be a sufficient guaranty that no pains would be spared on their part to procure the necessary endowments and benefactions, and that their efforts would be successful; they were not men who were in the habit of failing in anything they seriously undertook. . . . We have performed our part, but we do not feel that the other party have performed theirs. They obviously have not." Ware charged that the Corporation persistently asked the wrong question—"not what a good school would need, but what reductions the school we have got can endure and yet survive."

This critique triggered outrage among Corporation members—Matthias Ross called it a declaration of "war before he is ready to meet the Enemy"—some of whom weighed ways to take it out on Ware personally, or even on his own department (architecture). The ever-loyal Robert Richards jumped to the Corporation's defense, assuring them that Ware did not speak for the faculty. "I cannot think," he wrote, "that the majority . . . hold such views. . . . It seems to me especially unkind not to say impolitic if there is any coolness of feeling in any direction among members of the Corporation to stir it up by such bombshells as this." Richards told Rogers that Ware's position—a minority of one, he implied—was at odds with the rest of the faculty, who had "never returned from vacation with a warmer zest for work or more encouraged by the state of things than at the present time."

Through all this, Rogers filled the role of chief pacifier while keeping other duties to a minimum. Finances and janitorial oversight were handled by Tech's bursar, correspondence by the Corporation secretary, "matters of discipline & in whatever relates to the interests of the students" by the faculty chair. The president's sole official duty, as Rogers stipulated, was to preside at faculty meetings whenever he "sees fit to be present but shall not be charged with any of the business details heretofore entrusted to the President."

He found it difficult, however, to confine himself in this way. The Corporation pulled him into drafting and editing correspondence, particularly letters to donors. John Ordway asked him for help in resolving a ticklish problem with the Chauncy Hall School, whose headmaster complained that Runkle had charged outrageous fees for part-time use of Tech's gymnasium. Rogers paid more than usual attention to public relations issues—Tech's growing role, for instance, as a model for other institutions; how to spread the word through paid advertising in key city newspapers from Providence to Chicago to Baltimore, and places in between; how to neutralize bad publicity, as when a professor from Illinois Industrial University complained that he had been rudely treated on a visit to Tech. A commercial group in Richmond, Virginia, solicited Rogers's expert advice on how to go about establishing an industrial institute; the trustees of Columbian University (later George Washington University) asked for guidance in setting up a scientific school, "polytechnic in its character—looking to professional life, and not to abstract science, except as auxiliary to the former." Charles Venable, a former student and colleague of Rogers's at the University of Virginia, sought ideas on how to "divert into a scientific direction a greater part of that bright intellect of the South which wastes itself in law & politics." In December 1880 came pleas from Boston University president William Warren for help in lobbying Congress on behalf of higher education, before "the political bush-whacking & interminable debates of spring" began. The usual invitations had to be either accepted or deflected. One that arrived in February 1881, from Corporation member Charles Flint, chairman of Boston Latin and Boston English High Schools, was too important to pass up, even in the dead of winter: the dedication of the schools' new buildings. "Many of our graduates," observed Flint, "go to complete and 'round off' their education at the Institute. . . . I think we are certain to have a very large audience of the very best people of Boston & vicinity [and] your presence will be worth more to the Institute than several hundred dollars spent in advertising in the newspapers."

Much as Rogers would have liked to accept Augustus Lowell's invitation to prepare a new set of Lowell Institute lectures for the winter of 1878–79—two decades after his previous set—he realized the pressure would be more than he could handle. He had to conserve his energy to deal with crises: how to keep Harvard at bay again, for instance, when rumors circulated in March 1881 that the Lawrence Scientific School might consider relocating from Cambridge to a site near Tech in downtown Boston. One alumnus urged him to write a history of the Institute: "Not one among us is so preeminently well fitted to tell so much about his children and their home as their father"—a challenge that Rogers never accepted, partly because it would have required this most self-effacing of men to write extensively about himself.

It was a busy time, and some worried that the strain would overwhelm him. "Though seeming weak," a former student wrote to Mrs. Rogers, "he has the energy and vigor of a lion, but I often wish he would give up his labors of presiding and managing, and rest on his well earned laurels." Still, Rogers kept his eyes trained on finding a suitable successor. His position as National Academy president gave him unusual access to information on who might be qualified, available, and recruitable. By May 1880 he lit on Francis Amasa Walker, economist, statistician, and military war hero—a proven leader in academe, as professor of political economy at Yale's Sheffield Scientific School; and in government, as head of the federal census bureau. It took more than a year of patient maneuvering, but in May 1881 Walker was elected Tech's third president and assumed office on November 1 that year.

May 30, 1882. Huntington Hall began filling up around ten thirty. The faculty gathered on stage, their new leader—this, his inaugural commencement—on duty as master of ceremonies. Students and instructors occupied the front rows; behind them ranged family, friends, curiosity-seekers. William Rogers was present, too, a special guest, to give what he hoped, finally, would be his valedictory address. No flowers, no ornaments, no music; a simple, low-key setting, the kind Rogers preferred. He disliked rituals of this sort; but when students had begged for one a few years earlier (in 1879), he relented, insisting only that Tech's be different: spare, businesslike, nothing fancy. This graduating class, May 1882, was Tech's fifteenth—24 graduates in all, including 2 women—yet only its fourth commencement ceremony. In the old days, students had simply stopped by the secretary's office, picked up their diplomas, and gone about their business.

The ceremony opened with the usual introductions, followed by a file of eager (some shy) students stepping forward to read from their theses. Next up was Rogers, to present diplomas. Walker led in with a glowing tribute. "In a high sense," he said, "Professor Rogers will always remain President of the Institute of Technology. Present or absent, his spirit will preside over it. No man can succeed him in his fame; no man can hope to do more than successfully administer the school which he alone could have created. Founder and father is his title perpetual, by a patent indefeasible." The kind of florid, semi-idolatrous homage that Rogers loathed, but he accepted it this time, making allowances for Walker's notoriously enthusiastic style.

As he rose to speak, his frame looked stooped by age, illness, and fatigue. Yet his eyes gleamed. They had been fading to gray for a while, but the familiar blue twinkle shone through again—his spirits lifted, in part, because Walker's arrival promised new, solid, creative leadership. He aimed to capture a bit of Tech's history for an audience too young to remember. "It is true," he said, "that we commenced in a small way, with a few earnest students, in some rooms fitted up in Summer Street, while . . . the tides rose and fell twice daily where we now are. Our early labors with the legislature . . . were sometimes met not only with repulse but with ridicule, yet we were encouraged and sustained by the great interest manifested by many in the enterprise. Formerly a wide separation existed between theory and practice; now in every fabric that is made, in every structure that is reared, they are closely united into one interlocking system,—the practical is based upon the scientific, and the scientific is solidly built upon the practical." From there he moved on to outline how modern sources of energy had emerged, expanded, fed into new technologies. "Stephen Hales published a pamphlet on the subject of illuminating gas, in which he stated that his researches had demonstrated that 128 grams of bituminous coal—" And here, mid-sentence, he crumpled to the floor. Those nearby rushed to his side, but he was gone.

Among the pallbearers at the memorial, on June 2, were Runkle and William Atkinson (for the faculty), Henry Bromfield Rogers and John Forbes (for the Corporation), O. C. Marsh (for the National Academy), and Edward Pickering (formerly on Tech's faculty, now on Harvard's). Burial took place at Mount Auburn Cemetery, Cambridge, in the Savage family lot (no. 178) on Walnut Avenue, where Eagle, Magnolia, Mountain, and Spruce meet. Rogers shares a headstone with his father-in-law, James Savage; Savage's inscription faces outward, toward the street, Rogers's inward. At commencement a year later, Walker eulogized him in purplish prose—"his expositions of scientific truth radiant with a light which scarcely seemed to come from earth"—that the simple, direct, understated Rogers would probably have cringed at, but that the bereft found comforting. Emma Rogers lived to a grand old age, 87; she died on May 18, 1911, and her ashes were buried on October 11 beside her two favorite men. She spent her last years shaping her husband's personal legacy; her monumental *Life and Letters of William Barton Rogers*, issued in two volumes in 1896, was a mark of deep affection for both him and Tech. Much of her estate, the residue of her father's fortune, was willed to the Institute.

"Sailing seas not well charted"

John Daniel Runkle, 1822–1902

Following his collapse in October 1868, William Rogers chose John Runkle to fill in as president. Runkle had been his trusted right hand, a reliable, organized attendant to detail, while Rogers concentrated on the larger picture—relations with the state legislature, fundraising, faculty and student recruitment, the creative side of educational policy. They had bonded, too, on a personal level. Rogers depended on Runkle's readiness for duty day or night, but also on his loyalty as a friend. Runkle had been there from the start: secretary to the founders' group in 1861, aide in mapping out the school's inaugural curriculum, faculty member ever since the school opened its doors in February 1865. When Rogers fell ill, it was Runkle who reached his side first, led the convoy home, summoned medical help, monitored his progress in the ensuing months, and, when Rogers's wife Emma would permit, kept him posted on Institute affairs. "I need not tell you," Emma Rogers told Runkle in March 1869, "with how much affection Mr. Rogers regards *you* particularly among his colleagues."

Runkle became acting president on December 3, 1868, although for legal reasons the term *president pro tempore* was quickly substituted. A number of problems faced him at the outset. With finances precarious, Tech petitioned the legislature in February 1869 for a grant of $50,000. The appeal, denied in April, was the first of several setbacks for which Runkle would be blamed. As years went by, he became a favorite scapegoat. For one thing, his connection to the city's powerbrokers was limited. Unlike Rogers, he did not marry into local influence, and some of the Institute's supporters lost interest once Rogers faded from view. But Runkle's personality—nervous, highly strung—failed, also, to build confidence. He wrote breathless, pleading letters to Emma Rogers. "Tell Dr Rogers that I want to see him—I suppose it is out of the question for the President to talk over Institute matters; but I want to see some one to compare notes with & get advice." His deep dependency on Rogers overrode his cheeriest tone. "I am not discouraged by this failure—We shall find some way to meet the coming year; I am certain that a kind Providence will watch over us in the future, as in the past. . . . I am very tired and anxious for the vacation. . . . Give my best love to the President, & tell him that I am looking forward with the greatest anticipations to loafing on the beach with him this summer."

Everyone, Runkle included, waited impatiently for Rogers to take back the reins of leadership as Rogers's brother and attending physician Robert kept counseling him not to. Rogers took Robert's advice, eventually, convinced that Tech could do well under Runkle. Few shared his optimism. The only faculty members to voice support were William Atkinson and Edward Pickering. According to Atkinson, Runkle would keep Tech on a steady course. "It must be a great pleasure to you," Pickering told Rogers, "to leave the Institute in the care of so able a man as Prof. Runkle if (as I trust there is no doubt) he is to be elected President." But some Corporation members—George Bigelow, for one—went on record against, even before the election. Henry Bromfield Rogers wondered if William Rogers might retain the title of president, without duties, so that the Institute could at least "reap advantage of having his name at the head." Henry Fuller suggested keeping Runkle on as *pro tem* until a replacement could be found, perhaps from the outside. When the votes were cast, 18 of 21 went for Runkle, plus 3 abstentions. A majority, but not enough for election as the bylaws required a majority of *all* members of the Corporation (42 total), not just of members *present.* At a special meeting, called four days later, 28 members showed up. Rogers left his sick bed to vote and to coax resisters. This time Runkle squeaked by with 21 votes.

His presidency began on this edgy note and ended under a shadow, plagued by a chronic deficit of confidence. Yet over the course of a decade, 1868–78, he pushed forward on a number of fronts. The 1870s saw Tech's student numbers fluctuate, mostly downward, but graduation rates rise. Students came from further away, a small but growing contingent from outside New England, a few from overseas. Steps were taken to create more of a sense of community, to reduce Tech's commuter-school ambiance. The faculty almost doubled in size; instructors, those with nonfaculty status, grew by half. A formal summer school emerged in 1871. Laboratories were founded for mining, biology, industrial chemistry, physics, and mineralogy. New courses appeared—natural history (combining geology and biology) in 1871, followed by physics, philosophy, and metallurgy in 1873. Among new subjects offered were vocal culture and elocution, international and business law, money, banking, and currency. In 1872, a framework for postgraduate studies was laid. The Lowell School of Design joined the Institute, also in 1872, and a School of Mechanic Arts for training in shop and foundry work was organized in 1876. The Institute started admitting women to its regular courses. The technology museum, an idea dormant but central to Tech's original three-pronged structure, acquired new life, at least in the talking stages. The other prong, the Society of Arts, hosted events with growing popular appeal, including, in 1876, Alexander Graham Bell's first demonstration of his yet-to-be-perfected telephone. Two new buildings went up—one a gymnasium-cum-drill hall, with a cafeteria attached; the other, called the annex, housed mechanic arts, the laboratory for microscopic analytical chemistry (forerunner of the biological laboratory), the laboratory for industrial chemistry, and the women's laboratory.

These initiatives seemed bold to some, but struck others as unfocused and foolhardy in the context of a nationwide depression following the Panic of 1873. Even optimists saw that while Rogers had laid out a fine path for the Institute, there was no predicting the future—"it was a ship," someone observed, "sailing seas not well charted, with many chances of shipwreck even without a change of navigator." And, by most accounts, Runkle was not an ideal leader for such volatile times, which required rapid, forceful, sustainable action, management skills of no mean order. His talents lay elsewhere: as a careful, thorough educator and a dedicated, persuasive spokesman for Tech's value to the nation. It was that problematic middle ground—administration—that gave him trouble, and, compared to his predecessor, he always came up short. Rogers's long shadow proved difficult to get out from under.

Runkle came from a simple, modest farm background. Born on October 11, 1822, he grew up in Root, a settlement near the Mohawk River in rural Montgomery County, New York, west of Schenectady. His parents, Daniel and Sarah (Gordon) Runkle, owned and operated a farm adjacent to farmland run by other Runkles—Cornelius, James, and Earnest—probably Daniel's brothers, certainly relatives. Daniel was third-generation Dutch, his grandfather, Cornelius, having come over from Holland with an older brother, Hance or Johan, around 1750. Daniel and Sarah had five children, all boys. John, the eldest, was overseer, accountant, and field laborer rolled into one. The others, his assistants, were Henry, Daniel Jr., Cornelius, and Jacob Gebhard Runkle.

The family had little time or money for formal education. John attended district schools seasonally, between crop rotations, and at age 16 went to a private school run by a student saving to put himself through Union College in Schenectady. He continued to work the farm, meanwhile, and saved money to further his education by teaching at a district school and for three years (1844–47) in the science department of the Onondaga Academy. In 1847, he corresponded with Harvard's professor of mathematics, Benjamin Peirce, who urged him to come to the Lawrence Scientific School, just opened at Harvard.

John, now 25—older, then, than the typical undergraduate—took Peirce's advice. His name appears in Lawrence's 1848–49 catalog, its first, as "student in mathematics." Peirce was impressed by his patient, methodical approach, and in 1849 recruited him for the staff of the Nautical Almanac Office, which ran predictive calculations of planetary positions and orbits for the *American Ephemeris and Nautical Almanac*, a reference tool for astronomers, navigators, and surveyors. After graduating in 1851, John stayed on in Cambridge and continued working for Peirce. Among others who joined the group in the 1850s was a young Simon Newcomb. "I date my birth into the world of sweetness and light," Newcomb recalled, "on one frosty morning in January,

1857, when I took my seat between two well-known mathematicians, before a blazing fire in the office of the 'Nautical Almanac' at Cambridge, Mass. . . . The men beside me were Professor Joseph Winlock, the superintendent, and Mr. John D. Runkle, the senior assistant in the office." Besides Newcomb, several of Peirce's employees went on to distinguished careers in astronomy: Asaph Hall, George Hill, Truman Safford, Chauncey Wright, F. W. Bardwell, John Van Vleck. During this period, Runkle brought two of his brothers to Harvard. Cornelius and Jacob, both part-time district schoolteachers, between farm chores, graduated in 1855 and 1857, respectively, Cornelius from Harvard Law School and Jacob from Harvard College.

In 1852, a couple of Runkle's papers—orbital calculations of the asteroids Psyche and Thetis—were accepted for publication in *Astronomical Journal.* He was drawn to this type of work, nonabstract mathematics applied to real-time events, and stayed connected to the Almanac Office for more than thirty years, long after leaving Harvard for Tech. In 1872, Simon Newcomb involved him in expeditionary arrangements to observe the 1874 transit of Venus. Runkle was instrumental, too, in persuading his mentor, Peirce, to publish a textbook (*A System of Analytical Mechanics*, 1855) crafted out of a series of Harvard lectures. Then, at Peirce's suggestion, Runkle compiled *New Tables for Determining the Values of the Coefficients, in the Perturbative Function of the Planetary Motion*, which appeared in 1856 as part of the Smithsonian Institution's Contributions to Knowledge series.

His scholarly reputation on the rise, Runkle founded a journal, *Mathematical Monthly*, to encourage the sharing of ideas, theorems, and proofs. William Rogers was among those whose support he sought at the outset. The goals of the *Monthly*, launched in 1858, were modest compared to the analogous European literature—"Care should be taken," Runkle wrote, "not to graduate the magazine as a whole too high above the average attainments of mathematical students"—but it *was* a start. ("You have one astronomer, Professor Peirce, and no mathematicians," quipped a visiting German scholar in the 1850s, looking down his nose at American practitioners of the exact sciences.) Each number of the journal—not to be confused with *American Mathematical Monthly*, founded in 1894, later the official organ of the Mathematical Association of America—printed a series of problems for students to try their hand at solving. Answers appeared two numbers later and prizes were awarded for ingenious solutions, the first going to George Hill, just out of Rutgers College, who was invited to join Peirce's group.

Among the contributors to *Mathematical Monthly* were Runkle's almanac colleagues—Peirce, Newcomb, Hill, Wright, George Bartlett—along with William Chauvenet, professor of mathematics at the U.S. Naval Academy, and others delighted to find a homegrown channel for mathematical ideas. Feature articles leaned toward the applied, with Peirce submitting work on the comet of Donati, Newcomb several notes on probabilities, Wright a study of the economy and symmetry of honey-bees' cells. Peirce's influence was evident throughout, thematically and in the mathematical notation used. Newcomb and Bartlett, who served as volunteer assistant editors, helped the gentle, diffident Runkle deal with recalcitrant, sometimes dishonest authors. One math professor "from a distant state," according to Newcomb, submitted a paper plagiarized from William Walton's *Collection of Problems in Illustration of the Principles of Theoretical Mechanics* (1842), which the contributor assumed he could get by with since the work, an English one, was little known in America. Runkle wanted to "let him down easy," while Bartlett urged a more blunt response: "Just write to the fellow that we don't publish stolen articles, that's all you need to say." But Runkle, typically, chose a milder course proposed by Newcomb: "inflict all the necessary humiliation on him by letting him know in the gentlest manner possible that we saw the fraud."

Mathematical Monthly was considered the best American mathematics journal of its day. Its predecessors—*Mathematical Miscellany* and *Cambridge Miscellany of Mathematics, Physics, and Astronomy*, the latter edited by Peirce and his Harvard colleague, physicist Joseph Lovering—had been spotty, irregular, short-lived. The *Monthly* did not last long either, folding after just three volumes. Runkle had a hard time retaining subscribers. When about a third failed to renew after the first volume, he pleaded with them: "I have supposed that those who continue their subscription to the second volume would not be so likely to discontinue it to the third volume, and I have made my arrangements accordingly. If, however, any considerable number should discontinue now, it will be subject to a very serious loss. . . . I ask a favor for all to continue to Volume III., and notify me during the year if they intend to discontinue at its close. I shall then know whether to begin the fourth volume. I shall not realize a dollar."

The journal shut down in 1861—too few subscribers, as Runkle had feared, but chaos owing to the Civil War was also to blame. During its brief history, it sparked positive responses at home and abroad, particularly in Great Britain. "In my own country," William Walton wrote to Runkle from Trinity College, Cambridge, in October 1860, "there is no mathematical periodical of an educational character which can be compared with yours, and I feel sure that all true lovers of logical science in my own university will regard themselves as your debtors." George Boole, inventor of Boolean logic and professor of mathematics at Queen's College, Cork, recommended the journal to his students. After it went out of business, there would be no other such American periodical for more than a decade—not until *The Analyst: A Monthly Journal of Pure and Applied Mathematics*, edited and published by self-taught mathematician Joel Hendricks, appeared in 1874, followed by *American Journal of Mathematics*, founded in 1878 by James Joseph Sylvester, the brilliant Jewish mathematician who, persecuted for his faith, left his native Britain to serve as Johns Hopkins University's first professor of mathematics.

Around the time that Runkle founded his journal, he and Peirce joined with community activists excited about the filling of Boston's Back Bay, which looked to transform the city into a thriving cosmopolitan center. Peirce, Runkle, Eben Horsford, William Rogers, and other mostly Harvard-connected academics hoped to establish, as part of this ambitious plan, a progressive, practically oriented educational institution, a polytechnic outside the classical tradition. Peirce, in conversation with Rogers, talked up Runkle's prime qualities: reliability, consistency, diligence. Rogers, in turn, brought him on as a detail-manager behind the scenes. Runkle became the fledgling Institute of Technology's first secretary in January 1861, charged with scheduling meetings, laying out agendas, and organizing records. It fell to him to notify Rogers of his election as president, in April 1862, and, in May, to assist his own successor as secretary, Thomas Webb, in coordinating the election and appointment of officers. He was elected chairman of the Corporation's committee on publication, backed by his experience as founder and editor of *Mathematical Monthly*.

Runkle put excess copies of his by-now defunct journal to good use, circulating them in order to establish a claim to faculty status—more, that is, than a useful, small-time functionary. Corporation members were impressed, even though the journal's content went over their heads. "You have," George Emerson wrote to him in January 1864, "what we should find it difficult to meet united in any other individual, experience & success in a similar labor, and that thorough knowledge of mathematics which will make all the difficulties in the applications of science comparatively easy to you." Then, when Rogers asked Runkle to help lay out the Institute's first mathematics curriculum, there was little doubt that he would be invited aboard as its first professor of mathematics.

Runkle, meanwhile, continued to work for Peirce in the Almanac Office. During the Civil War, he volunteered as organizer and fund-raiser for a local chapter of the Freedmen's Aid Society, helping to mobilize a mass migration of teachers to educate the newly freed slaves. He went door-to-door, canvassing local businessmen and others for donations. His brother, Cornelius, by then a successful lawyer, kept him abreast of war-related activities in his own New York City neighborhood. When the dreaded Copperheads—Democrats opposed to President Lincoln's policies—stirred up antiwar riots there in July 1863, Cornelius mobilized for action. "An attack was made by the Riders," he told John, "and although we were here then we were powerless to prevent it, not one of us having a musket pistol or anything else to use." The Runkle brothers, both avid prounion partisans, approached the issues somewhat differently. John, the mild-mannered educator, stuck with his freedmen's aid work, while Cornelius, the feisty attorney, grew more bellicose, rhetorically anyway, as the war pressed on.

Runkle resided in Dedham, a far-flung suburb of Boston. In 1851, he had married Sarah Willard Hodges, of Dedham, and when she died in 1856, he remained with her mother, Mary Hodges. Mrs. Hodges's household consisted of three other daughters and a young ward or relative, Catherine Robbins Bird. Runkle married Catherine in 1862 and the couple moved to Brookline, on Boston's immediate outskirts, when his commitments at Tech made the trek from Dedham by rail or post chaise too difficult to handle on a daily basis.

The Runkles raised a family in Brookline. Catherine Bird (often called Kittie) was born in February 1863, John Cornelius in December 1866. Two other children—William Bird and Emma Rogers—died in infancy, while two more, Eleanor Winslow (born March 1881) and Gordon Taylor (born July 1882), arrived much later. Runkle's former sister-in-law, Emily Hodges, joined the household to help care for the children and another Hodges sister, Lucy, also moved in at some point. By 1870, the Runkles were renting part of a house on Harvard Street owned by George Tyler, a civil engineer, who occupied the other part with his wife and four children. The following year they moved a short distance away, to a rental on Kent Street. Runkle owned no real estate yet—his net worth was estimated at just $2,000—but in 1874, he bought a 10,000-foot lot, 82 High Street, and built a brick dwelling completed in 1875. He later bought an adjacent lot, 84 High Street, where he put up another house. The family would live on High Street for over twenty years.

The Runkles survived on modest means and often looked to Cornelius, John's more entrepreneurial brother, for assistance. Cornelius sent money—$150, for example, in 1867—and grew offended when John tried to repay him. The gifts kept coming, and in 1873 Cornelius helped finance a $12,500 mortgage on John's Brookline property. He relished this role of guardian angel, keeping his poorer, less financially astute, brother out of the clutches of creditors—the least he could do, after John had helped put him through law school.

John Runkle loved to teach. He was good at it, too, a persistent, caring educator in his role as Tech's first professor of mathematics. "A kinder gentleman treads not the earth," wrote one student. At the start of term, he would size up students' abilities, pay special attention to those who were slowest (without losing sight of the fastest), and carry the group forward gradually, if necessary, so that each step—like a building block—was understood by everyone before the class moved on. Edward Rollins ('71) remembered one student with whom Runkle took special pains. "[He] would demonstrate a problem, walk up and down the platform, then turn to this particular student with the question, 'See it, Mr. ***?' If *** said that he understood the problem, Mr. Runkle would quickly proceed to the next one, assuming that the whole class understood it. If Mr. *** said that he did not, President Runkle would go over the demonstration again, with the utmost care and patience, and continue this until the particular student was satisfied. It was all done in a good natured way; it never seemed to annoy the particular student . . . and always amused the class."

On some matters Runkle was loose, on others quite strict. While he disapproved of smoking in the hallways and lounging about on the Boylston Street steps, he did little to stop either, aware that students had few alternatives on a crowded urban campus. But when it came to academic dishonesty—plagiarism, cheating—and behavior that spilled over from mischief to violence, he could come down hard on culprits. "I remember very well," Rollins wrote, "his calling my class before him once, after we had been engaged for about two months in a deeply laid line of mischievous performances, and telling us very frankly and flatly that the whole thing must stop. . . . I think this talk impressed itself very fully on all the members and no further trouble occurred." Sometimes an unplanned appearance—right place, right time, nothing said—was the only remedy required. "The other morning," wrote one student in November 1873, "some of the men in '76 enjoyed quite a lively little game of foot-ball in the lower hall. A well-known form appeared. *Exeunt omnes.*"

In the classroom, Runkle struck an informal, relaxed pose. Often he would perch on the edge of his desk, legs dangling and blackboard pointer swinging lazily to and fro. One day Edwin Blashfield ('69), later a well-known muralist, put the mysteries of calculus aside and chalked a humorous little allegory onto his slate. The sketch showed Runkle with his legs hanging over a pond rather than a desk, fishing pole in hand and a calculus problem for bait at the end of his line. A couple of fish, with the remarkable likenesses of Blashfield's classmates Ernest Bowditch and William Tryon, sniffed at the bait; fish-Bowditch turned up his nose in disdain while fish-Tryon positioned him(it)self to swallow the bait whole. The slate made the rounds under students' desks until Runkle snatched it from one fellow, looked it over, paused, then broke into a broad smile.

Beneath his quiet, dignified air, some students detected "a man with a streak of fun," an "occasional twinkle" in his eye. Just one or two, out of many hundreds of students over the years, interpreted his gravitas as "repressed" or "distant." "Possessed an ability," wrote one student, "to get much work out of [us] with no friction." "He put us," said another, "through our mathematics in a thoroughly practical way." Charles Eliot, one of Tech's first two chemistry professors and later president of Harvard, called him the best mathematics teacher he ever knew. Runkle, said Eliot, had perfected "the concrete method of teaching." Eliot marveled at his ability to integrate mathematical concepts—from calculus, algebra, geometry, arithmetic—into a single lucid unit, without confusing students.

Few faculty members were as well liked. Only Rogers, perhaps; students found Eliot cold, standoffish, an indifferent lecturer, while a number of other faculty struck them as overbearing or authoritarian. But while Runkle's approach to mathematics was clear and straightforward, its scope was limited, confined to elementary concepts. Runkle rarely touched on higher, more abstract, complex branches, the kind that the finest German and French mathematicians were working in, partly because he lacked background but also because mathematics at Tech was a so-called service subject—essentials for engineers—and remained so into the early twentieth century. Runkle's title went from professor of mathematics to professor of higher mathematics in 1875, but no special significance was attached to this. The change was recognized as honorific, and the word "higher" was withdrawn in 1880. Yet even when it came to fundamentals, Runkle could come across as tentative, feeling his way alongside his students. He was known to conclude coursework in differential and integral calculus by lamenting how much he had missed, what he hoped to accomplish next time around. Students wondered if this was an apology for being disorganized or a confession about needing to brush up on his own. Perhaps it was a little of both.

Runkle wrote a couple of textbooks intended for beginning or intermediate college students, with a slant toward engineering applications. The first, *Elements of Plane and Solid Analytic Geometry* (1886), started as an in-house teaching tool, then gained limited notice outside Tech. *Science* magazine gave it a short write-up, the reviewer observing that its numerous examples would help clear the beginner's pathway through tricky concepts. Runkle once told a class that the book included some very basic explanations "not because they are necessary, but just to help you fellows." The last three chapters were modeled on two classic texts on conic sections, one by George Salmon of Trinity College, Dublin, and the other by Charles Smith, of Cambridge University. Runkle borrowed problems and solutions from Smith's work. Because it was so largely derivative, the book never achieved the status of either Salmon's or Smith's. Runkle promised a third book, on solid analytic geometry, but this never materialized.

It was tragic, in a way, when Runkle was foisted out of the classroom and into the presidency. Besides money woes, his first major challenge came from Charles Eliot, a man whom Runkle had helped to nurture as a young Tech faculty member, whose values—commitment to the laboratory method, a progressive outlook on modern education—matched the Institute's, and who had been building, alongside Francis Storer, a fine program in chemistry there. Eliot was the first of many high-profile losses to Harvard. Storer, his brother-in-law (married to Eliot's sister), followed him in 1870, upset over Runkle's elevation to the presidency. John Trowbridge, a talented young physicist, went too, coaxed by Eliot in what Runkle considered a rather underhanded fashion. "I learn that Eliot told Trowbridge," Runkle complained to Rogers in July 1870, "that, if he decided to make the change, not to speak to me about it till it was settled." Still others—Ferdinand Bôcher, Edward Pickering—followed in due course.

But Eliot's designs went beyond siphoning off Tech's best faculty. In the fall of 1869, he intimated that the larger interests of higher education would be served by some kind of merger or consolidation between the institutions, a formal attachment, something beyond loose cooperation. This was the first of three campaigns orchestrated by Eliot along these lines (the other two came in 1897 and 1904). Each appeared at a moment of stress for the Institute—a financial crisis, or uncertainty surrounding the death, illness, or departure of a president—and some suspected that they were planned this way, to strike at a target whose defenses were down.

One evening in January 1870, Eliot cornered Runkle at a meeting of the American Academy of Arts and Sciences in Boston. He pushed hard, pointing to a groundswell of support for the merger idea from the Cambridge side. Runkle went home convinced that Eliot, as he put it to Rogers, was "agitating the question." Rogers's response was unequivocal: "I can see nothing but injury to the Institute from the projected change. The Institute has already taken the first place among the Scientific Schools of the U.S. and if untrammeled will evidently continue to grow in reputation & numbers. . . . No kind of Co-operation can be admitted by the Institute which trenches in the least degree upon its independence. What alone is desirable is a friendly working of the two Insts. in their respective spheres." Runkle agreed.

Eliot refused to give in, however, even after insinuating himself into Rogers's sickroom in Philadelphia a month later and being met with the same firm no. He made various promises: Harvard would give up mining and engineering, transfer all related funds to the Institute, make its Bussey outfit (agricultural emphasis) a Tech department, and "if its friends prefer to retain the name 'Institute of Technology'—why Amen." In March, he tried to ease Runkle into Harvard's orbit by appointing him to a visiting committee. In June, he urged Runkle to jumpstart talks. Meanwhile, Tech Corporation member Edward Atkinson (brother of William Atkinson) and a few others on the Institute side carried out a survey that found not a single technical school, anywhere, benefiting from a union of this kind. These findings essentially doomed Eliot's proposal. But a committee comprised of Atkinson, four other Corporation members, and Runkle went ahead with negotiations anyway, later that summer and early fall. Proposals and counterproposals made the rounds, the major sticking point always degree of autonomy—control of funds, faculty appointments—with Harvard insisting on small but from Tech's perspective significant riders that undermined the Institute's identity. The new institution, for example, was to be called "the Technological School" with a status more independent or self-governing, perhaps, than Harvard's Medical School or School of Law but still under Harvard, beholden to it, subordinate.

From the sidelines, Rogers's rhetoric grew more intense—the Institute, he said, "cannot without a kind of suicide merge itself into any other institution."

A final nail was driven in Eliot's plan by Alexis Caswell, the shrewd, experienced president of Brown University. "The rule of letting what is well alone is a very safe one," he advised Runkle. "Unless the College has some tangible & decided benefits to bestow, I should say that your school would find its 'fullest & most complete development as an independent school.'" A merger, in Caswell's view, would benefit Harvard more than Tech. Negotiations ground to a halt after Eliot fired off a snippy note to Runkle on March 22, 1871: "The Committee on the part of the President and Fellows of Harvard College do not desire further conference with the Committee on the part of the Institute." By then, the Tech community regarded his maneuverings as brazen, threatening, noncollegial. Even William Atkinson, like Runkle a fairly mild fellow, lashed out at him—"I am not afraid of Eliot any longer," he confided to Rogers in August 1871; "he is a shallow fellow and is cutting his own throat, though he thinks he is running us down." Tech could stand on its own. The Lawrence School was a potential drag rather than a gain. "Our only real rival thus far is Sheffield [at Yale]," wrote Atkinson. "I am sure we have nothing to fear [from Cornell] . . . Worcester [Polytechnic] is quite insignificant, and the 'Stevens' at Hoboken has everything *but* men." At Tech, then, lines of resistance and self-confidence were drawn for when Eliot revived his plan nearly three decades later.

Even though Tech survived Eliot's maneuver unscathed, the episode tarnished Runkle's leadership image. As he pulled together information, he also fretted, avoided, waited for others, especially Rogers, to weigh in, wasted time on negotiations that went nowhere, and wilted under pressure. Runkle's placid, low-key style resembled Rogers's on the surface, but he lacked Rogers's confidence, his strength of character, and his winning way with dominant types like Eliot. He projected benign, shy awkwardness. Soft where firm was needed, he came across as indecisive, a bit of a bumbler. Students called him Kind Old Johnny, or Uncle Johnny.

Yet he brought tremendous vigor to certain aspects of the job, a physical energy that the frail, cerebral Rogers did not have. Runkle had grown up on a farm and was used to long hours in the field. In the summer of 1870, at the tail end of his term as president pro tem, he undertook a grueling cross-country tour. His aim, to fire up interest in Tech in out-of-the-way places. It was not too soon, he felt, for the Institute to look west—far west, beyond Chicago even—with its reputation on the east coast already strong and on the rise.

The transcontinental railway had just opened, the ceremonial last spike linking the Union Pacific and Central Pacific Railroads driven at Promontory, Utah, on May 10, 1869. Runkle rode hot, dusty rail cars out to Chicago, stopping along the way in New York, Pennsylvania, Ohio, Michigan, and Indiana. Then it was on to Omaha, eastern terminus of the transcontinental line, to begin the final leg along the Platte River, over the Rockies at South Pass, Wyoming, down through northern Utah and Nevada, across the Sierras, and into Sacramento. Along the way, he branched off to Denver and Salt Lake City, reachable only by small feeder rails. He stopped at mines, smelters, and manufacturing plants to spread word of this new institute out east, poised to propel the nation to unheard-of levels of innovation. Tech graduates, he bragged, were already showing their prowess in the heartland. Some worked the very route he traveled, or adjacent routes. From the class of '68, Ellery Appleton was assistant engineer on the Indiana and Illinois Central Railroad stretch, Frank Firth in Kansas, Charles Gilman in Iowa, William Hoyt in Illinois, Charles Smith in Missouri; and from '70, Samuel Mason worked the Indiana and Illinois lines. Among the rest from the first three classes, a number of those not out on the tracks as levelers, surveyors, back flagmen and chainmen, engineers, superintendents, or consultants of one kind or another, worked in related industries such as mining and steam-pump and steam-engine production. As he made his way westward, Runkle touched base with these former students and they, in turn, boosted his publicity campaign.

The trip laid the groundwork for a summer school. Runkle's plan was to expose students to the wilds, build stamina, give them a taste of hard manual labor—a bit like boot camp with mining, metallurgy, surveying, and rail-laying thrown in. But it was a way to encourage fellowship, too, to forge relationships with peers and outside communities, to network, and to get in some healthy fun and recreation on the side. The first expedition, a party of a dozen or more mining engineering students, set out in June 1871 accompanied by Runkle and several instructors: John Henck, John Ordway, Alfred Rockwell, Samuel Kneeland, and Robert Richards. They passed through some of the same places that Runkle had stopped at the year before. He twisted the arms of railroad executives to offer the group discounted rates, and on certain stretches to let them ride for free.

It was a hectic month-long journey, full of activity, adventure, and learning. First stop, Chicago. Everyone was in high spirits. "It has been a great pleasure," Runkle wrote to his wife Catherine on June 17, "to see all the students so happy. Even our friend J.B.H. [Henck] seems like a new man, & amuses us by his awful puns." Next up St. Louis, with a brief stopover at Washington University, whose cofounder and chancellor, W. G. Eliot, a relative of Charles Eliot and grandfather of poet T. S. Eliot, looked to Runkle for ideas on technical education following the death, in 1870, of professor of mathematics William Chauvenet. Runkle declined Eliot's offer of a job offer there—"I owe too much," he said, " to my friends, if not to the Institute, to be tempted." From St. Louis the party headed into Kansas, stopping at cheap hotels along the way, three or four to a room. At one spot, where the postwar mood still ranged from dejection to outrage, Tech boys roiled the locals with choruses of the popular song "Marching through Georgia," written to commemorate a key advance by Union soldiers under General Sherman. They meant no harm—it was the tune they liked, no lording over intended—but the hotel lobby emptied out, with patrons grumbling about insults from carpetbaggers.

In Colorado, the group moved between Central City, Caribou, Boulder, Georgetown, and Denver—their goal, to gather (literally) tons of ore for Tech's planned metallurgical laboratory. "We have made large collections of ores," Runkle wrote to Catherine, "& all of us have gained a fund of information which will be of great value to us in the future. . . . This is of the utmost importance to us, & another such opportunity is not likely to offer again for some time if ever." One excursion took them to a silver mill owned and run by Francis Crosby, assisted by his son William, just turned 21. The Crosbys had been testing their new furnace, the Ballard, for ore smelting. Runkle proposed that William come to Boston to build furnaces for Tech. He offered train fare east, free tuition if Crosby worked toward a degree, and a salaried post once he got the furnaces up and running. "Mr. Runkle seems to have taken quite a fancy to me," a surprised Crosby wrote in his journal. "Now, considering that he had known me but four days and had not had more than fifteen minutes conversation with me, that was certainly a remarkable proposition & I could not but consider it quite a compliment to myself. . . . he undoubtedly sees that with my practical experience in those things I could be of considerable use to them." Crosby accepted the offer, came to Tech, and became a mainstay of its geology program.

After Colorado, Runkle guided the group through the Sierra Madres and into Yosemite, where they paid a call on John Muir. Muir, not yet famous as the father of America's environmental movement—he was instrumental in founding the Sierra Club two decades later—had just begun to spread his gospel according to the natural world. One of his heroes, transcendentalist Ralph Waldo Emerson, had come to Yosemite a few weeks earlier, and it was likely Emerson's account of this visit that aroused Runkle's interest in Muir. Muir, however, had been put off by the fussy Boston Brahmins that Emerson brought along with him. They were "full of indoor philosophy," he remarked, and, when they refused to venture far into the wild and kept Emerson from going on guided hikes and camping out under giant sequoias, this was the last straw. No such problem, Muir was relieved to find, when Runkle—another Bostonian, thankfully not Brahmin—arrived with his group of eager students and faculty. Runkle was so impressed by Muir's knowledge of glaciers that he offered him Tech's chair in geology, now vacant owing to Rogers's retirement. Muir declined. "He soon said 'It is true,'" as Muir described their encounter, "receiving knowledge by absorption past all his flinty mathematics quite forgetful of his ghaunt [sic] rawboned Euclidal 'Wherefors.' He thinks that if the damp mosses and lichens were scraped off I might make a teacher—a professor faggot to burn beneath their Technological furnaces. All in kindness but I'd rather grow green in the sky."

The travelers returned to Boston by mid-August, tired but exhilarated. Next summer it was the civil engineers' turn. John Henck and his assistant, William Hoyt ('68), led a party of twelve students on a western swing, this time to study railway bridges over the Hudson, Ohio, Mississippi, and Missouri rivers. The summer after that (1873), the miners had another chance with John Ordway and Robert Richards in charge. The group scoured mining regions in northern New England and eastern New York state, particularly the Adirondacks, then wended their way back through western Massachusetts. "A little reflection," wrote Ordway afterward, "will convince every friend of sound education that there should be chances given for studying operations as they are carried on in real life. It is clear that no course of instruction can claim a reasonable degree of completeness which does not make provision for rendering the student's knowledge in some measure specific, and thus overcoming the not uncommon repulsion between theory and practice."

Summer school became an annual event, much anticipated, much looked forward to. In years to come, Richards and other faculty would lead excursions to Maine, Nova Scotia, Michigan, and elsewhere. Tech would eventually, in 1912, acquire property in East Machias, Maine, as a base for such operations. The year 1876, however, saw summer school with a difference. Runkle pulled off what amounted to a mass migration from Tech to Philadelphia for the Centennial Exposition. Some Corporation members objected, citing financial and tactical concerns, but Runkle stood his ground: Tech *had* to go, he argued, not just to celebrate the nation's hundredth, a decade after it had come so close to falling apart, but because the overall theme—industrial progress—mirrored Tech's primary mission and promised to draw widespread publicity for its contributions. With formal delegations from 26 states and 11 foreign countries, and many thousands of casual visitors expected to attend, the opportunity was too good to pass up.

Here was a case where Runkle's sense of Tech's best interests overrode his inclination to retreat, to let others decide for him. The state offered a small grant, which helped, and Runkle pieced together additional resources: gifts in kind (space, donated services) as well as cash. Tech's group was large, over 300 in all, including Corporation members and faculty, their wives, past graduates, current students, and friends and relatives. But Runkle kept the excursion from turning into a logistical nightmare. The University of Pennsylvania let the

Institute pitch tents on campus. Robert Rogers, Penn's medical school dean, turned over a large basement area for use as a makeshift dining room, along with a number of smaller, private spaces for Corporation members and faculty traveling with wives.

The two-week trip went off with hardly a hitch. Runkle assigned Lieutenant E. L. Zalinski, Tech's efficient man-in-charge of military instruction, as coordinator from start to finish: the trek down, accommodations, order, discipline, the journey back. A Brookline neighbor, Dr. Sabin, volunteered as physician, treating cases of heat stroke—Philadelphia was in the throes of a brutal heat wave at the time—and other ailments with remarkable efficiency. Runkle came away pleased, on the whole, although disappointed that the overwhelming size, range, and variety of exhibits had not allowed students to absorb well. "The excitements and fatigue of each day," he observed, "left little power to take full notes, and still less to write them out for the instruction or inspection of others at the end of the day. Notwithstanding this apparent, not real, failure . . . the exhibition proved a grand lesson to all of us, and I am happy in believing that the Institute is only just beginning to reap the fruits which in succeeding years will reach greater maturity and value." The expo ran through November 1876. Runkle brought Catherine and the children down to see it, just before the new academic year began.

Runkle's expansive vision, this urge of his to spread news of Tech, to gain public attention, paid off in several ways. The Institute's western adventures had been written up by the widely syndicated Associated Press, and in Philadelphia a reporter for the *New York Tribune*, Horace Greeley's popular newspaper, prepared a piece on Tech's exhibit at the Centennial Expo. The nation was getting to know the Institute—a trend that showed up, initially, in shifting student demographics. While the first students from outside the northeast had arrived on Rogers's watch, from Kentucky, Illinois, Arkansas, Colorado, Ohio, and California, this geographic range widened under Runkle. The earliest students from Maryland, Wisconsin, and Tennessee entered in the fall of 1868; District of Columbia, 1870; Iowa, 1871; Minnesota,

Indiana, and Kansas, 1872; Nebraska and Missouri, 1873; Michigan, 1874; Virginia, 1875; Texas and Georgia, 1876; and South Carolina, 1877. The numbers were not huge—overall enrollment was in decline, anyway, owing to the nationwide depression—but they signaled a shift outward, a conscious strategy on Runkle's part to draw folks from faraway places, to avoid settling into a provincial mindset.

Students came from overseas, too, Japan in particular. The first Japanese, Aechirau Hongma, arrived in 1870 from Fukuoka. He was joined a year later by Fuyouki Tominaga, of Tokyo (Edo, the old city), then by Haryoshi Mori, of Tokyo, in 1874, and Takuma Dan of Fukuoka, Joseph Batchelder of Yokohama (an American residing in Japan), and Takeo Miztuoka, all in 1875. Runkle corresponded with Japanese government officials. "I was so much occupied before my departure to my country," E. S. Hiraka wrote to him, after settling Hongma in at Tech, "that I could not see you to my great regret. Mr. Hongma will remain at your Institute at least four years, I presume." When Hongma was ordered home by the Japanese government in April 1874, shortly before graduation, Runkle worked with him and his sponsors to ensure that he left degree in hand. Runkle met with Japanese dignitaries at the Philadelphia Expo in 1876 and presented souvenir samples from the Lowell School of Design. The Japanese reciprocated three years later—"I have now the honor to send you a case containing specimens of our wall papers, silk-fabrics &c," wrote senior vice-minister of education Tanaka Fujimaro in April 1879—as ties with Japan grew closer. In 1878, freshly minted Tech graduate Cecil Peabody ('77) was recruited as professor of mathematics for the new Imperial Agricultural College in Sapporo. He served there for three years, returning in 1881 to teach at Illinois Industrial University and then, by 1883, back at Tech in mechanical engineering.

Besides Japan, students came from the Kingdom of Hawaii as early as 1870, the Azores, Canary Islands, and Panama in 1871, and China in 1877. Frank and Herbert Dabney (cousins, both '75) were from an American family that had staked out territory in the Azores in the early part of the century, made a fortune there in wine and fruit exports, and served as U.S. consular officers to the region. Another cousin, William Dabney (also '75), was a resident of Tenerife, in the Canaries. The first Chinese student, Mon Cham Cheong, came from Hong Kong. Efforts were made to overcome barriers of culture and language, as when one student from Panama, described in March 1882 as "earnest and faithful but suffering from poor preparation and lack of knowledge of the English language," needed special attention to survive the rigors of the curriculum. Through the 1870s, however, the student body consisted mostly of native New Englanders—the bulk of these from Massachusetts—so that in spite of Runkle's energetic forays, Tech continued, for the time being, to function as a regional school.

Runkle acted on impulse in recruiting faculty, a style that worked to Tech's advantage sometimes, sometimes not. Robert Richards, one of his hires—for mining engineering; a promotion, actually, as Richards was already on the instructing staff in chemistry—got his offer almost as spontaneously as Muir and Crosby got theirs. The challenge was not merely to attract skilled, forward-looking experts in their respective fields, but to retain them, to reverse a pattern of attrition from Tech's ranks. Of the faculty recruited by Rogers, twice as many left as stayed through Runkle's presidency. A good chunk of this outflow stemmed from Eliot's efforts to pull Tech's brightest and best over to Harvard. Storer was the first to follow Eliot. Others—Ferdinand Bôcher (modern languages), John Trowbridge (physics)—went in quick succession, Bôcher awkwardly in the middle of the academic year, February 1871, with Edward Pickering (physics) heading over a bit later, in 1877. But Eliot was only partly to blame. Faculty left for opportunities elsewhere, not Harvard-related. Cyrus Warren, organic chemistry, moved on to pursue commercial interests, while Alfred Rockwell, mining engineering, went off to chair Boston's board of fire commissioners after the devastating, citywide fire of 1872. Runkle benefited from the stick-with-it-ness of a small group—William Atkinson (English and history), John Henck (civil and topographical engineering), George Osborne (mathematics), and William Ware (architecture)—that stayed through his administration, although Henck and Ware left shortly afterward. Storer's description of Tech's faculty as a "strong and sometimes turbulent corps" was, deep self-irony and all, remarkably apt.

Geology sat atop Runkle's target list for hires, alongside allied disciplines such as metallurgy, mining, paleontology, hydrography, topography, and physical chemistry. The quest for gold, silver, and other precious metals that kept capturing the public's imagination, starting with the mid-century California Gold Rush, relied on experts to find the best veins or outcrops and to advise on ore extraction. Also, with westward migration showing little sign of abatement by the 1870s, geology had become increasingly important in civil and mechanical engineering—fields that attracted most Tech students at the time—as roads and railroads were laid across difficult, unfamiliar terrain, towns built, sewage systems installed. Runkle's focus was in some sense a nod to his mentor Rogers, himself a geologist, but the field had taken on a life of its own, promising adventure, discovery, even (for the lucky ones) profit. Runkle knew from working with Rogers, from employment opportunities that recent Tech graduates had found most attractive and secure, and from contacts made on his trips out west, that geology held enormous promise.

Over three years (1869–72), Runkle brought in Henry Mitchell for physical hydrography, Henry Whiting for topography, John Ordway for metallurgy and industrial chemistry, Alpheus Hyatt for paleontology, T. Sterry Hunt for geology, and William Niles for geology and geography to reinforce Tech's program in the field sciences. All but Ordway were specialists in one or another discipline. Ordway doubled up, partly in metallurgy but also as a hoped-for substitute for Eliot and Storer in building a solid, all-around program in industrial chemistry. Rounding out the new recruits for chemistry were James Crafts, who arrived in 1870, a year after Ordway, and Charles Wing in 1875. Among others hired under Runkle were S. Edward Warren for descriptive geometry, stereotomy, and drawing in 1872, George Howison to develop a new program in philosophy, and Charles Otis to replace Ferdinand Bôcher as professor of modern languages.

Some of these new faces had local ties. Hyatt and Niles had been students of Louis Agassiz at Harvard. Hyatt, while teaching at Tech, was also curator of the Boston Society of Natural History. Ordway was a Boston-based chemical consultant who had served on Tech's Corporation, 1866–69. But more than half were outsiders, a further sign of Runkle's expansive reach. Warren came from Rensselaer Polytechnic,

Otis from Yale, Howison from Washington University, St. Louis, Crafts and Wing from Cornell. Mitchell and Whiting were volunteers, without pay, from the U.S. Coast Survey, their regular work ranging up and down the New England coast. Runkle cajoled an agreement on them out of his old mentor Peirce, now survey director, who struck a hard bargain. No objection, said Peirce, "provided that this position involves no occupation of time which shall interfere with their discharge of their duties which claims *all their time*." Runkle's star outside recruit, however, was T. Sterry Hunt, a brilliant chemist and geologist whose reputation exceeded that of any other Tech faculty member thus far, except perhaps Rogers. A native of Connecticut, Hunt had spent much of his career outside the United States as chemist and mineralogist to Canada's geological survey, also at universities in Quebec City and Montreal (Laval and McGill, respectively). He had been elected fellow of Britain's exclusive Royal Society in 1859, and in 1870 was chosen to serve out the late William Chauvenet's term as president of the American Association for the Advancement of Science.

The faculty, however, was a volatile, unpredictable group whose loyalties remained fluid. Many recruits felt out Tech as much as Tech tested them for signs of dependability, for long-term compatibility. A kind of revolving-door syndrome emerged—faculty in, faculty out—exacerbated by the Institute's economic woes, its inability to guarantee a living wage much less steady employment. To offset this problem, Runkle increased the Institute's reliance on instructors and assistants. Almost all were recent Institute graduates. Of 53 such hires during Runkle's administration, 44 were Tech men just graduated. The 9 from outside offered expertise peripheral to the Institute's central curriculum, particularly in modern languages (Jules Lévy, Jules Luquiens, J. Nöroth), photography (John Whipple), practical design (Charles Kastner), vocal culture and elocution (Lewis Monroe), and military tactics (Hobart Moore). A few non-Tech graduates were hired in core areas. Ernest Schubert came aboard for freehand and machine drawing in 1868, Gaetano Lanza for mathematics and mechanics in 1871, Eugène Létang for architecture in 1871, and W. Bugbee Smith for mechanical engineering in 1877. Lanza and Létang later joined the faculty.

There were several advantages to this approach: more staff, smaller payroll, and, in the case of Tech graduates, built-in allegiance to alma mater and grasp of the Institute's special program, mission, and culture. From the graduates' standpoint, an assistantship or instructorship offered a chance for further study and a crack at supervisory experience that could prove useful in their transition to the working world. While such appointments were for one year (renewable), the average stay lasting one to three years, several staffers were promoted into the faculty ranks. These included William Ripley Nichols ('69) in chemistry, Charles Cross ('70) in physics, Thomas Pope ('71) in chemistry, Webster Wells ('73) in mathematics, William Crosby ('76) in geology, Silas Holman ('76) in physics, and Lewis Norton ('76) in chemistry. The only Tech graduate appointed to the faculty during this period without a prior instructorship or assistantship was Channing Whitaker ('69), recruited from the commercial sector on short notice, in 1873, to take charge of mechanical engineering. As mechanical, civil, and mining were Tech's bread-and-butter fields—through 1878, well over 70 percent of graduates took their degrees in one of those three disciplines—many breathed a sigh of relief when Whitaker agreed to fill the gap left by William Watson's sudden exit. "I am glad to know," Rogers told Runkle, "that Channing Whitaker would be willing to relinquish his present business for that of a Prof. on trial at the Ins^{t}. I have thought of him in this connection often . . . but have been led to believe that he would not give up his lucrative profession for a chair at the Institute & that he really had a disrelish for teaching. It gratifies me to find that his inclination & his tastes are in favor of a Professorship, for this harmonizes with the love of study & the noble ambition which seemed to characterize him while with us at the Institute. I do not believe that he can fail to make an efficient and acceptable teacher of Mech. Eng. in practice & in theory." Whitaker served through 1883, reinforcing a growing sense that Tech grads often made the best Tech faculty, especially insofar as staples like mechanical and civil engineering were concerned.

Under Runkle's administration, the curriculum broadened beyond the civil-mechanical-mining orbit that pulled most students to Tech during the 1870s. The first to graduate in chemistry was William Ripley Nichols ('69); in architecture, Henry Phillips ('73); in physics, Samuel Mixter ('75); and in natural history (geology) William Crosby ('76), Runkle's cherished recruit from the western frontier. But choice ranged more widely, even, than this. Rogers kept pushing the big-tent view over and beyond what the Institute already offered in William Atkinson's English and history classes. In 1871, Corporation member Joseph White, secretary of the state board of education, suggested inviting outside lecturers to speak on political economy, so that the Institute might further meet the "obligation . . . to give its students a liberal as well as a technical education." George Emerson proposed adding American and English history to the admission requirements. These sentiments were echoed in the public press, which referenced Tech's mandate to "secure to every student a liberal mental development and culture, as well as the more technical education which may be his chief object."

Yet George Howison, brought in to teach philosophy, met with stiff opposition from faculty and students concerned that this subject distracted from Tech's primary mission. If given the choice, one student wrote in December 1873, 95 percent of his classmates would drop it "not only because the subject is dry and abstruse, although we have no doubt that Prof. Howison does his best to make it interesting and attractive, but also because the students who come here want practical, useful knowledge, which will be of more actual service to them than all the Logic, Philology, or Mental Philosophy which they could obtain in tenfold the amount of time now spent on logic." Howison defended his territory with limited success. In December 1876, chemistry professor Charles Wing moved to scratch logic from the schedule for third-year students, a motion that was lost before the faculty, although not by much—Runkle voted against, in a tie-breaker. William Atkinson, who as professor of English and history would, under normal circumstances, have been a Howison ally, became his fiercest detractor. This was a classic turf struggle—Atkinson worried about Howison as a competitive threat—but it also mirrored mismatched views on how Tech students should be exposed to humanistic, nontechnical subjects. "His high *a priori* abstract methods," wrote Atkinson in February 1878, "have no place with us except in mathematics. . . . What our young men need is History and Literature in the *concrete.* They need above all things, to have a taste for reading created in their minds. They are drilled and disciplined enough and too much. . . . His lessons are dreaded and disliked except by the exceptionally mature among the students. . . . The simple truth is that we have got hold of a clever and rather conceited man who is out of place and whom it is difficult for us to utilize." Following his departure in 1878, Howison went on to a fine career as founding head of the philosophy program at the University of California, Berkeley, where he taught for twenty-five years. Among his students was the influential philosopher and intellectual historian Arthur O. Lovejoy.

This tension over the role of the humanities and social sciences would linger long after Howison's time. The consensus was that the Institute ought to avoid getting sidetracked by such tangential areas. "We are not a University," wrote Atkinson, "and cannot undertake the teaching of general subjects on a university scale. Our limited resources forbid the attempt to occupy a wider field than that which strictly belongs to us, namely, the sound training of young men destined for technical professions or for active business life." Yet others on the faculty—scholars like Crafts and Pickering, in fields of pure science—pulled in another direction, proposing university-style opportunities for research and graduate work. A doctoral degree program was placed on the books in 1872, but the first doctorates were not awarded until 1907. Tech's faculty stood united against what they saw as a cheapening of postgraduate credentials—170 Ph.D.'s were awarded by 79 American colleges between 1872 and 1879—with William Rogers serving on a joint committee of the American Association for the Advancement of Science and the American Philological Association in the early 1880s to find ways to halt this trend.

Runkle's sympathies lay primarily with those who wanted to keep Tech focused on professional course matter. But, in yet another move that would prove controversial, he pushed this preference toward the vocational, the paraprofessional, the manual training or trade school concept—and away, some worried, from the higher professions. One such scheme involved the Lowell School of Practical Design, formally attached to the Institute in September 1872 with the support of Tech Corporation members John Amory Lowell and James Little, agent for textile conglomerate Pacific Mills. The school offered instruction in "the art of making patterns for Prints, Delaines, Silks, Paper-Hangings, Carpets, Oil-Cloths, etc." A kind of fashion school, in other words, which Bostonians found intriguing as an "eminently useful addition to the *practical* branches of this popular institute—the *only* school of its kind in America." Charles Kastner, a designer well known in fashion circles, came over from Paris to head the program. Its students, men and women in roughly equal proportions—this Lowell school, like its sister, the Lowell Institute, made a point of recruiting women—exhibited samples of their work at Horticultural Hall and other venues around the city to the delight of designers, mill owners, and the general public. "Nothing more delicate and tasteful can well be imagined," a columnist for the *Boston Globe* raved in 1876. Demand for graduates outpaced supply. But in spite of its success, the school grew increasingly marginalized at Tech, viewed as a Runkle-inspired distraction from more pressing areas. Kastner, never a regular faculty member (his title began and remained Lowell instructor in practical design), eventually retired back to France and the school was merged into the Massachusetts School of Design, managed by the Museum of Fine Arts School.

Another of Runkle's pet projects in this style was the School of Mechanic Arts, modeled on a Russian exhibit that caught his eye at the Philadelphia Exposition in 1876. This exhibit showcased the work of St. Petersburg and Moscow technical schools that emphasized instruction rather than construction, the end result—production of machines and tools for sale, service, and workability—being less important than the process, how students moved through all stages from start to finish: concept, creation, production, use. Among the topics covered were woodworking, lathing, joinery, forging, and metal fitting. So keen was Runkle on what he saw that he cobbled together a version for the start of Tech's new academic year, 1876–77. The program opened as the School of Practical Mechanism (changed to Mechanic Arts in 1877). But like the design school, it was never integrated into the regular curriculum. Students were younger (they could enter at age 15), admissions requirements were lower, the program lasted two years rather than four. Runkle hoped it would become a feeder-source for regular Tech coursework—a kind of preparatory school—but this never worked out as many observers remained skeptical. Corporation member Stephen Ruggles, local manufacturer John Newell, and ex-professor of mechanical engineering William Watson ganged up on Runkle quite publicly, criticizing the program as cumbersome, cost-inefficient, obsolete, a foreign import with little if any relevance to American needs. Even supporters, like John Ordway, believed that a separate-and-unequal status was the only viable route: "We need to . . . keep its students apart from ours as far as possible. They must be treated more like mere schoolboys." Runkle fought a losing battle with Corporation members who were convinced that the mechanic arts school threatened to drag down the Institute's standards and reputation.

As doubts continued to grow, Runkle turned to local officials in Boston and Brookline to apply the concept in primary and secondary schools and to promote its strategic value for the state's industrial health. A report, *The Manual Element in Education*, that he prepared for the state board of education in 1877, served as a blueprint for further efforts in this area. In 1887 Runkle joined with Edward Atkinson and two other Corporation members, Percival Lowell and Matthias Ross, in "regretfully" accepting the Corporation's decision to shutter the school, and placing on record their belief that mechanic arts—as a function basic to technical instruction—would "not be among the least of the benefits" that Tech would be remembered for in years to come. Atkinson and Ross revisited the issue with a strongly worded protest some months later, pointing out that the school had been a model for successful experiments elsewhere, including the manual training school at Washington University, St. Louis, and the Chicago Training School. Tech's school closed its doors in 1889, but its workshops remained a vital adjunct to the regular curriculum, especially engineering subjects.

Runkle's successes were often viewed, unfairly, as failures. This happened not only with programs on the educational side, like design and mechanic arts, but also with the Institute's two other prongs: the Society of Arts and the ever-elusive Industrial Museum. The museum fell by the wayside early. Runkle, however, got the Society of Natural History, Tech's helpful next-door neighbor, to agree in November 1870 to care for the Institute's mineral collections and other artifacts. The Society of Arts, on the other hand, grew in stature as a lively, informative forum, Tech's main avenue to the public. The society benefited, as Tech did overall, by its location near the heart of a vibrant, accessible, rapidly expanding neighborhood, with the Back Bay's muddy swamplands fast receding into distant memory.

Bostonians were captivated by the society's proceedings: inventions of all sorts announced, described, demonstrated; quirky phenomena illustrated ("electricity in the female," for instance); practical technologies outlined (how to prevent boiler explosions, minimize ship collisions, avoid icebergs); inventions of dubious quality—"mechanical humbugs," so called—exposed to protect unsuspecting consumers. On May 24, 1876, Alexander Graham Bell gave his inaugural demo of telephony, using a clock-spring transmitter to communicate between Tech's building and a neighboring house. When someone brought in a typewriter on April 15, 1875, Edward Pickering called himself "confident that some such machine would very soon for most purposes entirely supersede the pen." Participants exchanged frank, thoughtful views, too, on larger issues of technology development, potential, risks, social and economic impact. At a meeting in February 1877, H. P. Langley worried about "the indisposition in this community, exhibited by parents to train their sons to mechanical pursuits; almost all the workmen are foreigners . . . . Young men rush into the store and the counting room, carried away by the idea that manual labor, if not dishonorable, is not genteel, and that the path to social position lies not through the workshop, but through the banks and the exchange."

Yet in spite of this vitality, a steady drumbeat of criticism—if not downright disapproval—emanated from Corporation members and some faculty. The society, they said, was not living up to its promise, exploited by some for crude advertising, trivial content, failure "through inadvertence . . . to keep up its organization." A poor reflection, as usual, on Runkle's leadership.

Students, however, consistently regarded him as their champion. In October 1871, when faculty denied them access to the main building after-hours, Runkle reversed the policy on condition that order be preserved at all times. He supported their campaign for suspension of classes on the Friday and Saturday after Thanksgiving, although the faculty kept revisiting the issue year after year. When parents began to fret, in 1872, about the impact of Tech's demanding regimen on the health of their sons, Runkle initiated a review that led to reforms. S. S. Greeley, a well-known surveyor and engineer in Chicago (father of Frederick Greeley '76, who left Tech after three years without taking a degree), argued that the program "required too much study for the health of his son, and for boys of average ability." The faculty, given their druthers, would have kept working Greeley and company half to death. What made the Institute unique, they held, was its regimen that toughened gentle hides for life's rigors. But Runkle countered that there were better ways to accomplish this, more humanely, and without sacrificing standards. A committee appointed by him to address the problem recommended a limit, for each student, of 5 hours a week class-time per subject plus 4 hours per day (Monday through Friday) and 2 hours on Saturday of outside preparation. The faculty signed off on this with the caveat "if possible." Runkle also appointed a committee, in April 1873, to consider ways to assist students struggling to pay tuition fees.

Nowhere was his reputation as a friend and ally of students more manifest than in his concern for social adjustment, lifestyle concerns outside the classroom. Students needed a better environment than that offered by Tech's outer stairwell, or by the barrooms, pool halls, theaters, music halls, betting parlors, boxing rings, and other shady venues springing up all over Boston's South End. Boylston Street started to look more prosperous by the mid-1870s—just a couple of gaps were left between the Public Garden and Tech's building, a new hotel (the Berkeley) stood across the street, Trinity Church was going up diagonally across—but none of this helped. In some ways the bleak, open fields and marshlands were preferable to new edifices as they, at least, had afforded space to roam about in, to throw a football, to run impromptu footraces. Now everything was more cramped. Students looked in vain for outlets and, when they found none, reverted to sheer silliness—writing on walls, breaking furniture, stealing signs, defacing blackboards, loitering, borrowing (pilfering, really) personal property. John Ripley Freeman ('76) cited a string of such pranks pulled by classmates, whom he described as "a fine lot of young fellows . . . a democratic mixture of a few sons of wealthy manufacturers, a few boys from the farm, a few sons of clergymen, a few sons of teachers, a few sons of ambitious mechanics." One presumptuous freshman, he recalled, was thrust into a potato sack and dumped, squirming, at his next class. At a friendly bout between Freeman and Frank Hodgdon, each pushing the other's head under an open sink faucet, twenty or so fellow students crowded around egging them on, placing bets. Professor Ware of architecture raced in to thwart a drowning, or at least to keep the drafting tables from getting wet. Nitrogen triiodide was sprinkled on the floor at the start of assembly, its loud zap! and purplish cloud scattering folks left and right. These and other infractions—tardiness, noise, chronic absence, hallway disturbances—were often referred to Runkle, who interviewed culprits and gently elicited promises to reform.

The military side of the curriculum, mandated under terms of the original land-grant act, kept students from sliding deeper into anarchy, and this spirit of regimentation was not confined to drill exercises. Classrooms, drawing rooms, drafting rooms, anyplace students congregated, all simulated a kind of martial discipline. Each class had a homeroom, so to speak, which Freeman referred to as "our station or abiding place." Here students stored books, studied, chatted in the off hours. Each student had a half-desk to himself, actually "a cumbrous combination of drafting table and desk"; the other half, opposite, belonged to a classmate. Students sat alphabetically, so they often got to know best those whose last names began with letters not far off their own. Freeman, thus, grew close to several F's, G's, and H's. On the other side of his half-desk sat Charles Fox, an architecture student. Others nearby included Martin Gay, at Freeman's left elbow, who arrived with work experience under his belt, as a rodman surveying railroads in Nebraska. Frank Hodgdon, a desk or two over, came from a family of leading physicians but broke tradition to study civil engineering. Sumner Hollingsworth, whose father was a paper manufacturer, joined the mechanical course. Alfred Hunt studied mining.

The system did not create order, exactly—these students were too rambunctious for that—so much as the appearance of orderliness. Tech had a love-hate relationship with the military. Students went for the pomp and ceremony—parades, flashy uniforms, gold brocade, muskets—but rebelled against the harsher rules, while the faculty, concerned about the vast amount of time drawn from the academic side, kept chipping away at the requirement. Runkle worked to strike a balance that met needs and satisfied obligations. He was conscious, too, of the military's role in filling a social void. He managed to persuade Lieutenant Zalinski—Tech's military commander known to all, not so affectionately, as the General, coordinator of the Institute's excursion to Philadelphia in 1876—to relax discipline enough so that students could have some fun. On the final night in Philadelphia, June 21, students were given leave to host a grand soiree. "We were favored," a loosened-up Zalinski wrote, "by the presence of a large number of ladies. The camp was illuminated and some tent floors having been placed together, out of doors, dancing took place on these and in the dining room." Freeman, however, recounted how his class organized a grand funeral parade down Boylston Street to "bury" military drill. Zalinski, beckoned to an upper window to watch, was impressed by the dignity of the ceremony until he learned the corpse's identity. The satire touched too close for comfort and Zalinski, never satisfied with Tech's attitude toward military instruction, resigned at the end of the 1875–76 academic year, replaced by Henry Hubbell Jr.

Aside from military exercises, students found few beneficial distractions from their regular coursework. The excitement aroused over a spelling bee, which matched Tech against Tufts College at the Boston Music Hall in April 1875, illustrates just how starved they were for entertainment. Competitors kept the audience in stitches of laughter by making fools of themselves, sheepishly tackling words such as "hemorrhoids," "prognosticated," and "demagogue" (spelled "demmygog" by one humiliated fellow). But students also began to ask about more serious activities. A petition to form a debating society in March 1873 drew tepid support; fine idea, the faculty said, but no room and little opportunity for supervision, given how much needed to be accomplished in the formal curriculum. In October the issue of fraternities, secret societies, and professional groups was broached, but without a specific proposal. No action resulted. A chapter of Chi Phi, chartered a few months earlier, failed to gain traction.

The faculty, however, with Runkle's backing, approved a plan in 1873 for second-year students ('75) to start a newspaper, *The Spectrum.* This was a risky move, considering that a couple of years earlier Tech's reputation had been sullied, so the faculty felt, by a scurrilous article in the *Boston Herald* allegedly written by a student. The article had described certain end-of-term pranks: torpedoes, firecrackers, alarm bells rigged under chairs and in closets, "a large black tom-cat" sent chasing from room to room, professors presented with "gifts"—"a huge cabbage-head to the Assistant Professor of Physics, and a tombstone, suitably engraved to another." The suspected author, Frederick Fellows, was granted a hearing before Runkle and left the Institute quietly; the messenger, in this case, appears to have been punished more severely than the perpetrators. But as part of his support for *The Spectrum,* Runkle argued that pranks of this kind—not to mention sensationalist write-ups about them—would wane if students had better ways to let off steam. And he was right. While *The Spectrum* was by no means uncritical, editorially, it was also responsible, careful, literate, well produced.

The first issue appeared on February 22, 1873. Intended for biweekly publication, every other Saturday, the paper stuck to schedule during its first two semesters, spring 1873 and fall 1873, then grew erratic in the third semester (spring 1874) before a final number appeared on May 16, 1874. It focused on Tech-related issues: curriculum, admissions, exams, tuition costs, extracurricular needs, physical environment, faculty, space, plans for new buildings, student behavior, field trips, school spirit, the Institute's standing vis-à-vis other institutions. Why it folded so quickly remains a mystery. As final-year students in the fall of 1874, its editors may have found little time to spare and could come up with no one from the lower classes able or willing to take over (its successor *The Tech,* founded in 1881, was still several years off). But as a formal, organized, student-run pastime, the first of its kind at the Institute, it filled a void, and the editors hoped to see it inspire similar efforts. "One cannot," they wrote, "remain long a student at the Institute without noticing the lack of united class feeling. . . . It seems to us that there might be some means devised which would draw the students more into each other's society while here." A plea, in other words, for some atmospherics of the traditional college without actually turning into one.

Students were of two minds about how far to go with this. One step, the founding of an alumni association in January 1875, met with broad-based approval—a useful, noncontroversial way to move the Institute's prospects forward, to generate fellowship, to build school spirit. The notion of an annual commencement ceremony, however, elicited strong views pro and con. Supporters played up its feel-good side, opponents argued that it strayed too far from Tech's modern, spare, professionalized culture toward a kind of Gothic mumbo-jumbo best left to classical institutions. In response to an appeal from about half the student body in January 1874, Runkle opened a discussion among faculty and Corporation members. George Emerson and John Philbrick gave their go-ahead; William Rogers and Edward Atkinson said no, favoring—as a rather dry alternative—a simple list of graduates by department, with thesis abstracts, printed in the president's annual report. Runkle, as usual, went along with his mentor Rogers. But when a portion of the student body revisited the issue a year later, Runkle moved it a bit further along. He helped lay the groundwork for a graduation ceremony that would involve little expense, a benediction (clergyman to be selected by the class), an outside speaker (also selected by the class), degrees conferred by the president, but no music, flowers, or garish decorations. Even this minimalist, austere observance fell victim, however, to concerns about undermining Tech's inherent difference from the run-of-the-mill college. The faculty voiced lukewarm support on March 19, 1875, then quickly reversed itself a week later "in conformity with the wishes of the greater part of the class *not* to have such exercises." The first public commencement ceremony was held, finally, on May 29, 1879, minus the medieval garb, floral arrangements, and other trappings that typified such events at Harvard and other colleges.

There was one extracurricular issue that most students agreed on: the value of athletics. By 1874, Tech had a baseball team—an informal one, neither faculty sanctioned nor particularly competitive as college teams went. "Our . . . Club has been in the field once or twice this spring," *Spectrum* editors noted on May 16 that year, "but has played no regular matches; they were to have played the Harvard Freshmen a few Saturdays ago, but were prevented by the illness of some of the latter." Yet what students wanted, more than organized sports, was a way for everyone, athlete and nonathlete alike, to benefit from systematic physical exercise, aside from twice-a-week required military drill. This, they argued, would reduce idle time and motivate higher scholastic performance: "If there were a gymnasium close to the Institute, so that only a few minutes would have to be spent in going and coming, all would have a chance to exercise. We could go for half an hour or so before the noon meal, at all events, and any hour during the day, when we were not in recitation, could be spent there. . . . This is a matter which deserves the careful attention of every member of the M.I.T., and we sincerely hope to hear of some plan, and see it carried out before next winter." In response, Runkle put together a plan for a drill hall that doubled as a gymnasium, along with something students had not bargained for but certainly appreciated—a lunch room, "perhaps a kitchen, in which under management of responsible ladies, the students might obtain dinners, warm, wholesome, and at cost price" (the ladies turned into gents, under a Mr. Jones, but the basic idea was the same). The Corporation threw cold water on the plan, citing financial concerns, but Runkle vowed to collect funds by subscription and, if necessary, to foot the bill himself for whatever could not be raised.

Not all students favored the idea. "The Institute is supposed to be in rather straitened circumstances," one wrote in April 1874, "and is known to be in need of a building for the use of the Chemical Departments. Now if any money can be obtained for supplying more room, it were infinitely more profitable to the students to put it into a permanent and necessary structure than to use it for putting up a shed for a purpose, at least partially, unnecessary." The majority, however, stood with Runkle, and he with them. His subscription drive raised over $3,000 and, as promised, he went into hock personally for the balance ($5,000), although the Corporation—perhaps shamed into it—eventually agreed to assume the debt.

The new building, Tech's second, was a simple affair—one story, 50 × 160 ft.—at the corner of Boylston and Clarendon Streets: compact, not built to last, dwarfed in every way by the solid, majestic main building, but perfectly serviceable. Tech's third building, called the annex, went up in 1876—also on Runkle's initiative—at the corner of Clarendon and Newbury Streets, behind the drill hall. It was slightly smaller (40 × 150 ft.), T-shaped, single-storied, and more solid, with 12-foot brick walls, tar-and-gravel roof, ample light, heat, and ventilation. The annex freed up space in the crowded main building and, like the drill hall, was financed partly out of Runkle's own pocket. The School of Mechanic Arts moved in, along with other untried programs such as the women's laboratory, but existing labs in need of more room—organic chemistry, industrial chemistry, microscopy and spectroscopy—also found a provisional home there.

A number of positive developments at Tech emerged not from any conscious choice, or deliberate shifts in policy, but by sheer chance. One such involved the admission of women. The Institute's first woman student—Ellen Swallow, a Vassar College graduate—entered in January 1871. Tech had never been part of her plans. Right out of Vassar she looked for work in the chemical industry and was turned down by two firms, one of which (Merrick & Gray, chemists, of Boston) told her they might have hired her had she possessed better technical skills. Go to Tech, they said, and bone up on your lab work, then come back and we'll take another look. One of Swallow's professors at Vassar, Maria Mitchell, a well-known astronomer, pointed her to two personal contacts at the Institute. Mitchell's brother, Henry, was professor of physical hydrography there. Her other contact, Runkle, she knew through Benjamin Peirce and the ephemeris work.

When Runkle heard from Swallow, and probably from Mitchell as well, he wondered what to do. Tech had never let in a woman outside the Lowell Institute night classes, even though several had asked over the years. The Lowell classes, in fact, had helped Tech project an illusion of opportunity for women. Opportunity it offered, to be sure, except not at the level sought by some. This was the status quo when Swallow's application arrived in December 1870. Runkle felt torn. He was not daring enough to set a precedent on his own, nor could he bring himself to rebuff his old colleague Maria Mitchell. He took what seemed a noncontroversial middle ground, telling

Swallow that she could come and make use of Tech's facilities free of charge. His response was dictated in part by the faculty, who voted to admit Swallow with one important stipulation—"that the admission of women as special students is as yet in the nature of an experiment, that each application should be acted on upon its own merits, and that no general action or change of the former policy of the Institute is at present expedient." Swallow, however, read Runkle's invitation another way: that Tech had admitted her on scholarship. "I thought the President," she wrote, "realizing I was a poor girl, remitted my fees out of the goodness of his heart, but I later learned that it was because he could say I was not a student, if anyone should raise a fuss about my presence in the laboratories."

For a woman, the lone woman student, life at Tech was hard. From the moment she arrived, Swallow was nudged aside, kept out of view. "I was shut up," she recalled, "in the Professors' private laboratory very much as a dangerous animal might have been. Whenever classes came into the 1st year laboratory, the door was kept carefully shut and I was expected to stay in. I was not allowed to attend classes." If not excluded, then hidden away. That fall the faculty voted to omit her name from the list of students to appear in the annual catalog, an action reconsidered two months later just in time for her name to be added. A different woman might have rebelled, or given up, but Swallow played along. She steered clear of conflict, aware that any misstep might not only ruin her own chances but set back prospects for other women. The first opportunity she had to mix, to edge her way further in, came when a new student—William Crosby, Runkle's recruit from Colorado—arrived in the fall of 1872. Crosby was the only other student with a serious interest in mineralogy, and since part of Tech's philosophy involved students working together, hands-on, Swallow was invited to link up with him. The professor in charge, Robert Richards, left them alone for two hours every morning, three days a week, to work on rock samples—an almost unheard-of liberty, in light of prevailing views on gender-mixing. But, Swallow later joked, "if we two got a reputation of any kind, it was only for our subsequent work in our fields . . . which I think did much for the advancement of coeducation."

Swallow was hardly a women's-rights activist even as she began, in her quiet, deliberate way, to alter the face of the Institute. "I hope I am winning a way which others will keep open," she wrote to a friend in 1872. "Perhaps the fact that I am not a radical—and that I do not scorn womanly duties, but deem it a privilege to clean up and supervise a room and sew things, etc., is winning me stronger allies than anything else." She tended men's sore fingers, dusted their work areas, mended their suspenders; always close at hand lay her trusty stock of needles, pins, thread, and scissors. The strategy worked, for her. She graduated in 1873 (her application to sit the final exams was not granted until May 14, just a few days before they were held), married one of her professors (Robert Richards), became the trusted confidant of several other faculty members, produced pioneering work in public health and sanitation, helped build a new field (home economics), encouraged opportunities for women, and served on Tech's teaching and research staff—although never with faculty status—for nearly four decades, until her death in 1911. The women of Tech, she once remarked, "have proved as no other college could better do that the most severe training did not make them repulsive, nor unfit them for housewifely duties." Swallow knew her place, so to speak. She went about her business quietly, deferring to her male colleagues. Leadership, she always held, was a man's prerogative.

Runkle had to contend with a steady trickle of women seeking admission. A number were Massachusetts residents, but some hailed from other New England states—high-school science teachers, for the most part, eager to brush up on their knowledge, lab skills, the latest methods. "I have heard," wrote Miss A. R. Wadleigh from East Berkshire, Vermont, in July 1872, "that instruction in certain branches is given to women, in connection with the Institute of Technology. . . . Could a lady familiar with elementary chemistry obtain tuition and practice in chemical analysis?" In August 1872, prominent Boston architect Edward Cabot wrote on behalf of Louisa Sewall, a pupil and friend of his, who had applied for admission to Tech's architecture course. "It really seems hard," Cabot wrote to Runkle, "that any obstacle should be allowed to stand in the way of one, (merely on account of sex), to the attainment of a good education in a specialty." Runkle deflected such requests, pointing Wadleigh, Sewall, and others, as Rogers had done, to the Lowell Institute night classes. But these were geared primarily toward beginners and some women wanted more advanced work.

The matter gained public notoriety when abolitionist and feminist Julia Ward Howe weighed in. Howe was a national icon, a symbol of grace and strength whose Battle Hymn of the Republic had inspired a nation. Since the Civil War, she had turned her attention toward women's suffrage, access to higher education, and other human-rights issues. In February 1873, she organized a petition drive to persuade legislators not to grant Tech any further land rights until it stopped denying women admission. Runkle had applied for about 13,000 square feet, between Huntington Avenue and Boylston Street, to put up a new building for mining and chemistry. But Howe, along with thirty-plus female sympathizers (including Augusta Curtis, whose application had been denied by Rogers some years earlier), sought to stop or delay the transfer. "The undersigned," their petition read, "respectfully remonstrate against any further grant of land to the Massachusetts Institute of Technology except on condition that Women shall be admitted to its advantages on the same terms as Men. They feel it to be unjust that land which should be the common birthright of all the inhabitants of the state should be given to any institution which does not open its doors on equal terms to all citizens." Howe and her fellow petitioners underscored the hypocrisy of Tech's claim that it would admit women but for lack of space, while at the same time finding space for male foreigners (Tech's catalog for 1872–73 listed students from Japan, Hawaii, the Azores, Tenerife, and Canada).

Runkle was pulled into a scorching public debate, arguing that Tech had no choice but to exclude women as there were not enough places even for men. This, Howe caustically observed, was analogous to "a man bringing home half a loaf of bread, and refusing any to his wife, on the ground that there was not enough for himself." Runkle assured her that while he sympathized with her views on principle, he was faced with a practical reality: that to change Tech's policy would "disturb the harmony of the corporation," place a damper on fundraising, estrange the local community (unprepared, he said, for such a radical step), and throw the whole enterprise into disarray. The Institute, claimed Runkle, could not afford to martyr itself over this issue regardless of its merits. Howe voiced outrage that, for expediency's sake, Runkle would sacrifice core values of justice, fairness, and equality.

The legislature rejected Howe's appeal and granted Tech's request. Nothing was ever built on the land, however, and it was returned to the state in due course. Ironically, this was the only time that Runkle found a willing ear among state legislators—not on account of any persuasive powers of his own, but because Howe, some felt, overplayed her radical agenda. Women's suffrage was still half a century off, after all, while Swallow, Tech's only woman student, had resigned herself to gradual, timid inroads, a survival strategy dependent on male sufferance. Legislators denied Howe's appeal without comment, but Tech's students—all male, but one—went on the attack, slamming the position taken by these "extreme women" and their "rash and ungenerous way of attempting to attain the desired end." For them, the Institute's existing policy was quite fair enough.

Howe was no stranger to Tech. She and her husband, fellow abolitionist Samuel Gridley Howe, were good friends of William and Emma Rogers. They moved in the same social circles, and cofounded a club in Newport, Rhode Island, in 1874. The Howes had encouraged their son Henry, the third of their six children, to attend Tech for professional training after his graduation from Harvard in 1869. Henry Howe took his degree at Tech in metallurgy ('71) and went on to a career with various iron, steel, and copper concerns, then as an independent consultant based in Boston (part-time lecturer in metallurgy, too, at Tech), finally as professor of metallurgy at Columbia University. Mrs. Howe, in fact, was a strong Institute supporter and intended her petition, which acknowledged the Institute's great merit, not to undermine what went on there but to hold its feet to the fire on a fundamental principle.

Tech moved slowly, but deliberately, in response. A request from the Women's Education Association in 1876 led to the establishment of a women's laboratory under the direction of John Ordway, assisted by Ellen Swallow (Ellen Richards by now, having married her former professor, Robert Richards, in June 1875). Also in 1876, the Corporation authorized Runkle to open "such departments of the School as they saw fit to advanced special students of either sex, or to special classes, where it could be done without interfering with the regular work of the School." Further progress occurred in 1879, when the Corporation voted that "women who may have been or may hereafter be admitted to departments of instruction in the School may be, if they so desire, examined for a degree and if found qualified to pass under the same conditions that are applied in the examination of male students shall be entitled to graduation and to the usual diploma." Limits were eased, but not lifted right away. In 1879, Edward Atkinson remarked that integration made sense under certain conditions—academic standards, record keeping, financial accounting—but that it was "not expedient that the women should go into the regular laboratory; we have provided fitly for them [otherwise]."

Marie Glover and Evelyn Walton, the first women to earn degrees since Ellen Richards, graduated under these terms in 1881, followed by Clara Ames and Carrie Rice in 1882. Walton, the second woman (after Richards) hired onto the technical staff, became assistant in chemistry on the recommendation of John Ordway, whom she would marry. The Ordways moved to New Orleans in 1884, where he taught at Tulane University and, after Tulane's women's branch—Sophie Newcomb College—was founded in 1886, she joined its faculty as professor of chemistry and physics. The first woman to go through the entire four-year program alongside the rest of her class was Alice Brown, who graduated in 1884. Amy Stantial also graduated that year. All six, like Ellen Richards, earned degrees in chemistry. Two, besides Walton, followed in Richards's footsteps by marrying faculty members. Glover married Silas Holman (physics) and Brown married Harry Tyler (mathematics).

Ellen Richards's views on gender remained quite conservative. She opposed admitting women as first- or second-year students, citing "the dangers of intermittent coeducation." She also doubted women's adaptability to the rigors of Tech's curriculum. "The present state of public opinion among women themselves," she observed, "does not give reason to believe that, of one hundred girls of sixteen [years old] who might enter if the opportunity was offered, ten would carry the course through. It is demoralizing to have such results in the early stage of scientific education for women." In some respects, then, a few of her male counterparts stood ahead of her on this issue. And, ironically, a number of opportunities that came her way were unusual for anyone, male or female.

In May 1875, her proposal to become Tech's first candidate for a doctoral degree was accepted. When the plan failed to move forward, her husband, Robert Richards, suspected bias. "Some of the difficulties," he wrote, "may have arisen from the fact that the heads of the department did not wish a woman to receive the first D.S. in chemistry." But the evidence suggests otherwise, that the faculty was quite supportive. It approved Swallow's study plan; no obstacles were placed in her way, not, at least, by the faculty itself. Other factors may have worked to her disadvantage. Robert Richards occupied a large chunk of her time (after their marriage, she joined him and his students on summer field trips, serving as chief cook and bottle-washer); she grew busy with the work of the women's laboratory and as Tech's unofficial, unpaid dean of women; and, as her subsequent career showed, she was less interested in pursuing large, independent research projects in pure chemistry than in carrying out smaller bits of applied chemical analysis. Two male students—William Crosby ('73) and William Nickerson ('76)—followed her as Tech's second and third doctoral candidates in 1876 and 1877, respectively, and their programs also fell by the wayside.

The 1870s was a lean decade and the Institute could ill afford to alienate donors, many of whom happened to be women. Gifts came from wealthy spinsters, widows, legatees of deceased fathers, siblings, cousins, or husbands, all on the lookout for worthy causes. Elizabeth Thompson of New York hoped to spend part of her fortune helping institutions, like Tech, "do something to benefit the world with *ideas*, or inventions . . . to help themselves and the world to more intellectual and material wealth." Not all such donors worried over Tech's treatment of women, but some did. Marian Hovey, a well-to-do heiress—the Hoveys had built a department-store empire—asked in 1878 whether Tech would accept scholarships earmarked for women only. The answer was no or, more accurately, not yet. When the Institute liberalized its policy on women's admissions, Hovey's comfort level rose. In December 1878, a gift of $10,000 from the estate of George Hovey, her father, was announced, following assurances not just that the funds could be targeted by gender but also that Tech would open "a broad field . . . for women in which they have not yet had an opportunity to exercise their genuine capacity." Only industrial chemistry, mechanical drawing, and architectural design were mentioned—no reference to any engineering fields—but Hovey thought this a good start anyway. There was one precondition, that the money be used in a way "best suited to promote the most thorough education of students of both sexes." In the end, she agreed not to restrict the gift to women only. "Our donation," Hovey wrote, "was made from appreciation of the great practical, scientific work which the Institute is doing, as well as from admiration for the justice and common sense with which women students are treated." She offered a considerable sum, too, to the Harvard Medical School if it agreed to open its doors to women. Harvard declined, the Hovey money went elsewhere, and another half century would pass before the first woman would walk the wards at Harvard.

Throughout his presidency, Runkle suffered from nervous strain brought on by anxiety, fear, and self-doubt. In January 1875, he approached Rogers for a frank evaluation. "Can some one else," he asked, "do better for the Inst.? Is there the least feeling . . . that it would be better for its future success for me to withdraw? If so, I am more than anxious to know it. I have never intended to remain one day longer than it was for the interest of the Inst. that I should. I dread nothing but failure. I am sometimes despondent, & doubtful of my duty, to myself, as well as to the Inst. I know that now I have health and strength enough left for some years of usefulness in the right place; & I should at once seek this place if I am not in it now." Rogers's response does not survive.

This period marked the start of Runkle's unraveling, set off, ironically, by one of his more noteworthy successes—the recruitment of famed geologist T. Sterry Hunt. Early warning signs went undetected. Hunt had asked Runkle whether, if he accepted Tech's offer, he could take advantage of ways to supplement the paltry salary, find other professional avenues, consulting opportunities elsewhere—hardly a word about the Institute itself. Then he played Tech off against other institutions that were also keen to hire him. "I had much rather," he told Runkle in June 1871, referring to Rensselaer Polytechnic's more lucrative offer, "if other things permitted, take a subordinate place in your Institute than 'reign' at Troy." So when Hunt came to Tech and proved derelict in his duties—rarely around, often traveling, and at one point, behind everyone's back, giving away parts of Tech's valuable mineral collection to a Philadelphia museum—Runkle tried to fire him. But some Corporation members balked. Hunt, they said, was too valuable an asset to give up, and they blamed Runkle for mishandling the affair. "I understood," Edward Atkinson grumbled to Rogers, "that [Prof. Hunt] was not to be dismissed but admonished that he would be unless he gave more attention to the duties of his office. Fearing that Prest Runkle had not the tact to manage the case I tried to have the matter placed in your charge but . . . the Prest interposed and took it in charge. I next heard that Prof Hunt was dismissed—I therefore sent for him to come and see me. . . . We need the utmost *tact, care* and *comprehension* at the head of our Institute, else we shall fall behind." Hunt was reinstated, over Runkle's objections, and remained at the Institute for three more years. In 1878, he resigned and went back to Canada. A brief, tempestuous marriage to Anna Gale, daughter of a wealthy Montreal jurist, ended with Hunt making out well enough in the divorce settlement to establish himself as a scientist of independent means. To show that he had no hard feelings, and as a mark of respect for his friend and fellow geologist Rogers, Hunt left money in his will to support scholarships at Tech.

For Runkle, however, the damage was done. The episode reinforced doubts about his ability to function as chief executive. The Corporation, increasingly uneasy, lost a number of members. Charles Dalton, J. Ingersoll Bowditch, Henry Fuller, and Edward Tobey all resigned in 1875, followed by Joseph Fay (1876), Erastus Bigelow (1877), and Otis Norcross and J. Baxter Upham (1878). Rogers, called in to stem the tide—deeply worried, he said, about "the somewhat flagging zeal of our Corporation"—managed to persuade Dalton and Bowditch to stay, but only briefly, as Bowditch left for good in 1877, Dalton in 1879. Bowditch had resigned once before (in 1872), along with George Bigelow and Jacob Bigelow. Richard Greenleaf, George Tuxbury, and William Endicott followed in 1873, with Endicott eventually agreeing to stay. Lack of confidence in Runkle was just one factor. Corporation members felt weighed down, too, by depression-era worry over the state of their own finances.

Runkle muddled along, then, with the Corporation ranks dwindling and those who stayed ranged increasingly against him. Edward Atkinson played the role of lead critic. A financial hawk even in good times, Atkinson openly blamed Runkle for many of the Institute's troubles. Runkle was not, he said, "a man of quite sufficient breadth and culture." Matthias Ross thought so, too, while several others agreed without going on the record. Poor leadership, inaction, "want of . . . a firm will and a clear purpose"—these, Atkinson said, had brought Tech to the brink of ruin. The alerts he had sounded fell on deaf ears and with near-tragic consequences. His suggestion that Tech establish an executive committee to help out with management was not acted on. Next he turned to Rogers for help, without success, urging him to come back as president and leave Runkle—with the title of principal, perhaps—to supervise *only* the school.

In October 1877, Runkle fretted about having to cut salaries, eliminate positions, slash courses (metallurgy, physics, natural history, and philosophy), and double the work of the remaining faculty. "It is this or bankruptcy," he wrote to Rogers in despair, "if we do not get at once from some source a large increase of our invested funds." The Institute faced a massive drop-off in receipts from tuition, declining cash reserves, skyrocketing expenses, outflows that overran income by a wide margin. To cut salaries, Runkle urged those who could manage on less to volunteer for reduced compensation, on account of "the present financial embarrassment of the Institute." The faculty declined, however, citing sacrifices already made. Meanwhile, Runkle was raked over the coals for issuing a new catalog (1877–78) that promised courses likely to be discontinued for lack of resources. He was criticized, too, for soliciting funds for his pet project, the School of Mechanic Arts, from his few remaining allies on the Corporation when he should have been attending to the interests of the Institute as a whole.

The crisis for Tech, as Samuel Kneeland put it, was "full of peril, impending danger, perhaps death." In March 1878, Runkle helped frame a desperate appeal that the state legislature rejected in April by unanimous vote of its finance committee "on the ground of the present financial depression." This, at least, was the reason cited by the Corporation. There appears to have been more to it than that, however, as the committee carried out an inspection and site interviews on March 28 with a view to weighing Tech's petition against other demands on the state's shrinking resources.

Worried about just how far the Runkle presidency had damaged public trust, Atkinson launched into a campaign for his removal. Donors were needed more than ever now, to narrow or close the budget gap, and Atkinson told prospects not to worry, that Runkle would soon be gone. He apologized for whatever "unfavorable impression" Runkle had made and promised that, in selecting a new president, the Institute would make "no mistake . . . this time." Atkinson must have known that while Runkle was not a particularly gifted administrator, a different president—even a brilliant financier like himself, or a more dynamic leader—probably could not have done better in the dire economic climate of the 1870s. But if Runkle were out of the way, Atkinson imagined, Tech could at least begin to chart a new course with fresh blood at the top.

The state's snub in April 1878 all but sealed Runkle's fate. He hung on, beleaguered, for two more months. In mid-May he talked about fundraising prospects, with Marian Hovey in mind. On May 23 he attended an alumni association meeting at Young's Hotel—"a very successful & pleasant affair," he said—and on May 28 he went to work as usual, hoping that next fall's entering class would prove "a large one, for our encouragement." By the end of the week, however, he had made up his mind not to go on. As he confided to Rogers on May 31:

> Institute matters have for a long time given me great anxiety, & at last I have come to the decision to tender my resignation as President. This is not a hasty step on my part, but one which I have long contemplated. During the past year I have made all possible sacrifices, & now I must make the last and greatest. I hope to be allowed to retain my Professorship, at least till I can find something to do; for I am almost wholly dependent on my daily work. . . . I shall ask to have the resignation take effect at the close of the present school year. I shall not resign from the Corporation, but continue to do all in my power to aid the Corporation in carrying on the work— & as I said, I hope to be allowed to retain my professorship. If

> allowed, I shall continue to serve the Inst. with all my heart & strength in this way, & I hope my Dear Prof that you will aid me in this desire, if you think I have earned such consideration.
>
> I take this step because I am sure that this anxiety & burden will soon break me down, & make my remaining days not only useless but a burden. If we had means to go on comfortably, & it was the desire of the Corporation to have me continue, I should feel differently—but these conditions are not among the probabilities.

By return mail (June 1), Rogers offered his full support: "I am startled by your decision, though, from our previous conversations it is not wholly unexpected."

Runkle appeared before the Corporation on June 7 to announce his decision. Rogers did not attend but, sensing that some might like to see Runkle gone for good, backed him in a strongly worded letter: "I can not let him relinquish the position which he has filled so long and so disinterestedly without expressing my sense of the great value of his services to the Institute. Few persons know the labors and perplexities, which have been involved in carrying forward the plan of the Institute to its present widely expanded activity; but all who have marked its progress will, I am sure, agree with me in a most grateful recognition of the unflagging devotion to its welfare which Prest. Runkle has always shown, and will be assured that his zealous and disinterested labors as Prest. of the Institute must always have an honored place in its history." On June 17, a similar scenario played out before the assembled faculty. Runkle's tone at both meetings was humble, contrite, dispirited. The Corporation granted him paid leave for one year, a concession motivated not so much by appreciation for his loyalty or past service, but because Rogers insisted on it.

Rogers worked overtime to rehabilitate Runkle's image in the eyes of restive colleagues, Corporation members, even outside donors. One appeal in June 1878 painted events in deceptively rosy terms: "President Runkle has worked with the utmost zeal and disinterestedness for many years . . . often with inadequate means and appliances. He has now resigned the Presidency and will devote a year in Europe to the investigation of industrial education, then returning to his professorship in applied mathematics." Rogers, not a well man himself and now 73 years old, agreed to fill in pending a permanent replacement, but only on condition that the Corporation take action to raise $100,000, a first step toward pulling the Institute out of its financial hole.

The Runkle family—John, Catherine, children Kittie and John—set sail late in July 1878. So shattered was Runkle's health that by mid-August, in London, Catherine arranged a physician's consult. A diseased stomach lining, said the doctor, probably ulcers. Medications were prescribed and the patient was ordered "to avoid all violent exercise." "Mr. R.," Catherine wrote home to the Hodges sisters, Lucy and Emily (Catherine always referred to her husband in this formal way as Mr. R. or Mr. Runkle, he to her as "My dear wife"), "must not drink any beer, or coffee with his dinner—he might take it with his breakfast. He is to have a glass of light wine with his dinner and eat but few vegetables." Runkle improved on this (to modern ears) quirky prescription, and by early September had recovered enough that the family could cross the Channel to escape London's incessant smoke and fog. They spent more than a month in Paris. From there it was on to Cannstatt, just outside Stuttgart, site of mineral baths that Runkle hoped would speed his recovery. Cannstatt was the family's base for the next two years, with Runkle on a thrice-daily regimen of baths, Emser water, and warm milk.

His plan, to carry out a survey of select polytechnics on Tech's behalf, had to be put on hold more than once. "He has so little strength," Catherine wrote in February 1879, "that he cannot take as much exercise as he needs, or would like to take. His nerves are also weak, and I think that it would be impossible for him at present to attempt any work, but seven months may work great changes." By June it was clear that a return to teaching in the fall was out of the question. It would be a great risk, said Rogers, for Runkle to "expose himself to the Boston climate." The Corporation granted a further year's leave of absence—this time without salary, even though Runkle pleaded that "if the Inst. could afford a part of it, it would be very acceptable." Eventually, he made his way to several cities and towns—Stuttgart, Karlsruhe, Zurich, Berlin, Dresden, Freiberg, Vienna, Amsterdam, Delft, Komotau—where there were programs in industrial education, museums, technical exhibits that he wanted to study on site. He admired the schools he saw in Brunswick and Dresden. These, like Tech, were young, experimental, risky yet promising ventures. The polytechnic at Karlsruhe—the oldest in Germany, the one Rogers had used as a model for Tech—also impressed him, particularly its mechanical engineering program. Berlin's polytechnic, though handicapped by a crumbling building, boasted an innovative laboratory for testing metal strengths and a fine museum for mechanical models. Among other sites visited were Hermann von Helmholtz's physics lab in Berlin, the chemistry lab run by the Prussian Academy of Sciences, and the polytechnics at Munich and Vienna. A much-desired trip to Russia never materialized.

Runkle saw Tech measuring up well next to all the places he visited, and in some respects surpassing them. German institutions offered students at the undergraduate level few opportunities for lab work. Teaching methods remained didactic, rigid, lecture-bound rather than practice-oriented, leading Runkle to conclude: "As a teaching laboratory for the general student, there is nothing in Europe today which at all compares with that at the Institute, either in the method of instruction or in the capacity to accommodate large numbers of students." While he envied Europe's strong museum traditions, Tech students had no cause to feel deprived. There were good reasons, he said, to stay home: "It is undoubtedly better for the student to lay the solid foundation of his professional knowledge in the country where he will practice it. . . . The purely scientific knowledge is the same, but the details in application, methods in practice, and economic conditions are not the same; and it is important that the student should not be obliged in any sense to change his point of view as he enters upon his professional career. In the second place, the student is in a strange land, and free from all the social and educational restraints of his home life; and the fact is that too many cannot bear the change."

Runkle surveyed trade schools in Komotau, Stuttgart, and Freiberg, setups analogous to Tech's own mechanic arts program. "The whole morning was spent in the school," Catherine wrote from Komotau in November 1879, "for Mr. R. has seen none since he left home, that interested him so much, for it is quite like his work shops at home, having the Moscow school for its model." Runkle's faith in this type of education continued strong, in spite of an emerging consensus that Tech ought to focus on the professional and leave the paraprofessional for others. "There is a small, but growing, class of men abroad called *Technikers*," observed Runkle, who "do not claim to be, or rank with, engineers, but are simply skilled and well-educated mechanics. . . . If we could meet the demand which already exists, this kind of education would make rapid strides in America." Few disagreed on principle, but whether this could best be accomplished through Tech or through the high-school system was an open question.

Runkle returned to America in the fall of 1880 energized, his confidence regained, ready to get back to teaching. The news that Catherine was pregnant—"we are expecting an addition to our family some of these days"—boosted his spirits further. His welcome-back, however, was hardly warm. George Osborne, the other full professor in the math department, had insisted almost a year earlier that something be done about Runkle's title: Walker professor of higher mathematics. The word "higher," he said, should be removed, as it "gave the impression that certain advanced subjects in mathematics were taught by him." Rogers, as acting president, took Osborne's side and informed Runkle not only about the title change, but also that he and Osborne must serve from here on essentially as coequals.

Runkle hinted that he might resign over this. "Now you can see," he complained to Rogers, "with what pain & mortification I have read your letter, in which I foresee my doom. I am to be forced to sever all connection with the Inst. in order to preserve my self-respect. Is it possible that the Faculty, the Committee on the School & the Corporation will all consider this proper reward for the past? Do you consider this just? . . . It seems that Prof Osborne has forgotten who aided him to a place in the Inst, & prefers to treat me with disrespect. When I find that this feeling is shared by the Committee on the School & the Corporation I shall know that my usefulness is indeed at an end. Did the Corporation really mean when they voted to accept my resignation as President that I should retain my Professorship, as was also voted? Or is it true that doing the duties of the President for ten years to the best of my ability, & with all faithfulness, has proved my ruin?" Runkle felt stung that Osborne of all people, a long-time colleague from ephemeris work going back to the Peirce days at Harvard, then in 1866 recruited by Runkle from the U.S. Naval Academy, would turn on him in this way.

Rogers, surprised by his old friend's outburst, stood firm. But he was anxious to preserve a semblance of dignity for Runkle while "preparing him somewhat for the disappointment in store." As he pleaded with one Corporation member: "Considering [Prof. Runkle's] long services commencing with the formation of the School, might he not, in the event of his remaining be permitted for the present at least to retain the old title of 'Walker Prof^r of Math:' (not of 'higher math:' which was an addition to the original title & expressed nothing)." Rogers did not insist that Tech keep Runkle at all costs, simply that, if Runkle did stay, here was a mechanism that could ease friction between him and his colleagues. The result was an uneasy truce whereby Runkle retained his position minus the word "higher" in his title, negotiated a shared teaching arrangement with Osborne, and returned to the classroom—the best possible outcome under the circumstances. "I think the action in Prof^r Runkle's case," Rogers told Institute treasurer Lewis Tappan, "was just & kind to him & fair to the Inst: & has given me relief & satisfaction."

Runkle remained nominal department head, aided by his easygoing, nonconfrontational style. Rogers's successor as president, Francis Walker, relied on him, as Tech's most senior statesman—following Rogers's death in 1882—for historical insight that could inform current and future policy. But Runkle stayed to himself, mostly, mindful that without Rogers's intervention, he likely would have lost his job. In October 1884, he pleaded for a waiver of his son John's tuition fees. "I am sorry," he wrote to Lewis Tappan, "to feel obliged to ask the Corporation . . . to aid me by remitting my son's tuition. . . . If the grounds of my past service are not sufficient, will you pardon me for saying that in the days of the Institute's sore need in the past, I have contributed to this tuition many times over?" This referred to funds that Runkle had pulled from his own savings, a decade earlier, to help finance construction of the drill hall and annex. The Corporation may have denied the request—there is no record, anyway, that it was granted—and his son dropped out after a year. But he still had friends on the Corporation. Some who had subscribed to a fund ($600) to defray his expenses on the aborted mission to Russia in 1881 insisted he keep the money, which he did, gratefully, placing it in trust for his infant daughter Eleanor.

Runkle continued as Walker professor of mathematics and department head until May 1902, when, just short of eighty years old, he retired to the rank of emeritus. Harry Tyler ('84) recalled his "venerable but robust figure, the somewhat straggling locks of gray with a tawny tinge, the stimulating, luminous, unconventional exposition, the quick, incisive questioning, the surprising blackboard drawing, the inimitable touches of the confidential or the monitory, the constant substratum of uplifting earnestness and dignity. None of his students could fail to acquire admiring affection: very few could withstand the incentive to work. Which of them will not recall such characteristic expressions as this, 'Now, gentlemen, I am going to show you one of the most beautiful and interesting things you ever came across'?" By the time he departed in 1902, however, some of his colleagues had lost patience. "To deal with senescence and inefficiency was a difficult problem," biologist and Institute librarian Robert Bigelow recalled, and Runkle "should have retired years ago."

When he left, the Walker chair passed to Osborne—next in line for seniority—but Tyler became department head, part of a conscious effort to modernize the curriculum. Runkle and Osborne were relics. A newer generation, some with doctorates from German universities (Erlangen, Göttingen) and several up-and-coming American programs (Yale, Johns Hopkins, Cornell)—or, like Webster Wells who went to Lepizig (1879–82), with doctoral study but no degree earned—were poised to pull the program in new directions. It was a trend noted as early as 1890. "Our work in pure mathematics," observed the president's report for that year, "consists, on the one hand, of a limited range of subjects regarded as essential preparation for later technical work; on the other hand, of more advanced special subjects, open to students duly qualified." Tyler, a skilled administrator, helped nourish the scholarship of his contemporaries, Frederick Woods and Frederick Bailey, and went on to recruit a number of younger, creative mathematicians—Clarence Moore, Henry Bayard Phillips, Frank Hitchcock,

Norbert Wiener, Philip Franklin (Wiener's brother-in-law), Samuel Zeldin, Dirk Struik, and others. The department began to shed its service orientation and to value serious research, while a cadre of fine math teachers—Dana Bartlett, Nathan George, George Rutledge, and Leonard Passano—continued to meet obligations to the various engineering programs. In 1896, when Carl Friedrich Gauss—namesake and great-grandson of the German mathematician sometimes referred to as the greatest mathematician since antiquity—joined Tech's first-year class, the symbolism of this event was probably not lost on Runkle and his colleagues.

Runkle lived in Brookline until 1897, the year his wife Catherine died. Active in community affairs, he was a member of the Brookline Thursday Club (vice president in 1883, president in 1885) and the Brookline School Committee (1882–97, chairman for several terms). On school policy he remained a persistent advocate for the value of mechanic arts. Brookline acted on some of his ideas—courses in carpentry for boys, sewing for girls—and a school devoted to manual and technical instruction, at the corner of Druce Street and Clinton Road, was built and named in his honor. The John D. Runkle School still stands at 50 Druce Street. In 1897, he moved to Cambridge to live with his four children in a rented house at 15 Everett Street, near Harvard Square. The eldest, Kittie and John, worked locally—she as a schoolteacher, he as vice president of the National Coal Tar Company. The younger two, Eleanor and Gordon, were students, she at Radcliffe College and he at Harvard.

Runkle had just over a month to enjoy his retirement. The faculty sent him off on May 31, 1902, with "grateful appreciation for your eminent services to the Institute from its foundation, and your warm personal interest in your students and your associates." A few weeks later he fell ill and went to Southwest Harbor, Maine, to recover. "The coast of Maine, from Portland, east," a former student advised him on June 30, "is to my mind the finest part of the Atlantic coast, and I am sure that no place would aid so greatly in restoring your health, on account of the bracing air and healthful surroundings." Runkle died there on July 8, the last survivor of the Institute's original founders.

"All that we hold true and manly"

Francis Amasa Walker, 1840–1897

Tech's leadership vacuum was not easy to fill. An ailing William Rogers stepped in, guilt-ridden over having pushed John Runkle into a position that came close to destroying both himself and the Institute. Rogers expected to serve briefly, but this second term stretched to three years. His wife Emma and brother Robert urged him to bow out quickly. Tech's financial situation, meanwhile, kept deteriorating. The Institute was on edge, in crisis mode. While Rogers favored trying to get by with modest adjustments, the Corporation insisted on draconian measures: cuts in salaries, layoffs, departments merged or eliminated. He let them take the lead. In earlier days he might have pushed against the grain, but he wanted to devote whatever energy he had left to running the National Academy of Sciences.

It was up to Rogers to replace himself. The risk was too great, he felt, to recruit from the inside again. Besides, to pull from the ranks would leave a gaping hole in some part of the academic program. A graduate who combined a grasp of Tech's scope and mission with professional stature could have worked well, but the alumni body, like the Institute itself, was too young. Rogers would have to look outside.

One candidate emerged as the top, and only, prospect. A few years earlier Runkle had drawn Rogers's attention to Francis Amasa Walker, chief of the awards bureau for the Centennial Exposition, whom Runkle met in Philadelphia in 1876. Walker had served with distinction in the Civil War, been mustered out as brigadier-general, worked as a journalist and grammar-school teacher, joined the federal civil service, supervised the Ninth (1870) and Tenth (1880) U.S. Censuses, led the Bureau of Indian Affairs at a sensitive time in that agency's history, headed the department of political economy at Yale University's Sheffield Scientific School, and served as secretary of Sheffield's governing board. He had academic credentials, energy, experience, a proven record in overseeing complex organizations and projects. Add to this his relationship with Washington insiders, including Congressman (soon-to-be President) James Garfield, plus his commitment to the scientific method as vital to raising standards in higher education, and a better candidate would have been hard to find. Although educated at Amherst College when Latin and Greek were central to the curriculum—"The Classics," he said, "were always a hobby of mine"—he had moved with the times, become a strong advocate for science, engineering, and other progressive trends in education, and introduced quantitative methods into his own field, economics. He was outspoken, a challenger of tradition, at his best moving against the grain, pushing new ideas. He gravitated toward the outdoors, athletics, field work. As a military officer, he had led young men and possessed, according to one of his biographers, "a peculiar faculty for making himself, unconsciously, the object of their emulation." But while several presidential short lists—Williams College, for example, and his alma mater, Amherst—had already mentioned his name, no one had any idea if a position of this sort would appeal to him.

Rogers decided to sound him out. They met up in Washington, D.C., in mid-May 1880, when Walker happened to be there working on the census and Rogers was attending to National Academy business. Walker showed keen interest and Rogers came away impressed. George Brush, director of the Sheffield School, formerly an assistant to Rogers on the Virginia geological survey, recommended Walker without reservation, much as he hated the idea of his leaving Yale. William Niles, professor of geology at Tech and a one-time Sheffield student, went to see Walker in New Haven and paved the way for a formal offer, which went out under Rogers's signature on June 12, 1880.

But Walker's reply, on July 5, showed him having second thoughts. While he was confident that he could leave census matters in the hands of his deputy within a year or so, there were other issues to consider. How would his family react? He and his wife Exene, daughter of

Timothy Stoughton, a well-to-do businessman and sometime farmer in Gill, Massachusetts, had a substantial brood to raise, five boys and two girls. They had married on August 16, 1865, soon after his return from Civil War service. Stoughton, the eldest son, was born in 1867, followed by Lucy (1869), Ambrose (1870), Francis (1871), Evelyn (1876), Etheredge (1877), and Stuart (1879). The older four were well settled in New Haven schools. Walker worried about uprooting his household, comfortably ensconced at 68 Whitney Avenue near the heart of Yale's campus. He would have to consult with Exene—"my superb counselor and best friend," he told Rogers—before reaching any decision.

Other glitches emerged. Toward the end of 1880 President-elect Garfield, a close friend of Walker's since Civil War days and a key influence in persuading Congress to support a modernized federal census system with Walker in charge, considered him for an important cabinet post, secretary of the interior. In the end, Garfield passed him over for career politician Samuel Kirkwood, Republican senator and former governor of Iowa, but Walker remained in the running for several weeks. He had no special desire for the job anyway—his was a dutiful, but unmotivated candidacy—and he held deep reservations about life and culture in the nation's capital. "What I do consider the fatal plague of Washington," he wrote in his typically blunt, plainspoken way, "is the general spirit of hanging upon the government, seeking influence and patronage and support from without, and looking to public office as the chief end of man." His run-ins with career politicians were legendary. Once, on board a train, he happened to sit near two who irritated him with their crude gloating over a victory at the polls. When one turned to him and said—"Straws show which way the wind blows, eh?"—Walker shot back: "There's another proverb about straws, that drowning men catch at them." Most politicians had sense enough to stay out of range of his rapier-like wit.

The Garfield transition team wondered if he would consider going back to Indian affairs, but he said no. "I shall be out of the woods by April," he wrote to Rogers in February 1881, "and if your patience . . . shall hold till then, I shall not be obliged to ask further indulgence." In May, the Institute Corporation voted him in and he promised a start date of October 1. Yale took the news in stride. A disappointed but supportive George Brush wished him (and Tech) well. "I am aware," one Sheffield board member wrote, "that 'blood is thicker than water' and I can well understand how strongly you are attracted by the surroundings offered you at Boston." Walker's close friend, statesman and diplomat John Hay, Abraham Lincoln's former confidant and personal secretary, teased him about having deserted Washington and "waltzed away to aesthetic bean-pastures of Boston."

Walker reported for duty on November 1. The faculty met with him on November 5 and, on November 10, he was introduced to the Corporation and the Society of Arts. "I trust in him," Rogers said in handing over the reins, "as I would trust in myself were I of his years and had I his experience in administrative work. . . . I commend him to you knowing what are his sympathies and what are his capacities for usefulness." The additional delay was caused by turmoil following the assassination of President Garfield. Garfield, gunned down on July 2, clung to life (and office) till September 19. Walker refused to budge until the outcome was clear, as he had too much at stake in the census bureau—a decade of strenuous work—to leave it to the whim of a new cabinet secretary or to the unpredictable priorities of a new president. Then, as Tech held its collective breath, Walker hinted that he might stay on if Garfield's successor, vice president Chester Arthur, asked him to. "My plans are all knocked into *pi* by the President's death," Walker told his friend, publisher Henry Holt. "I hoped to be quietly on duty in Boston by this time . . . but can, of course, assume nothing respecting the wishes and purposes of the new President, at present." When no overtures came, Walker resigned and headed to Boston.

At Tech, reaction ranged from warm to ecstatic. The faculty voiced strong support: "We are much gratified to learn that the choice for succession has fallen on a gentleman of high reputation and eminent ability, who can carry forward the Institute in accordance with its original design and its established character." Some were thrilled to have in their midst a Civil War hero who still went by "General." Others, already unhappy with the hefty chunk of time given over to military instruction, worried about Walker's soldierly side but were optimistic that his reputation as a hard-nosed statistician would incline him to promote academic rigor at Tech. Still others suspected that, with his military background, he would prove a splendid organizer. A few harbored doubts as to whether an economist would make a good match for Tech. "To some of us it seemed strange," one faculty member recalled, "that a man, who had never made an experiment in a laboratory, should have such a ready knowledge and appreciation of the aims of the scientific research which forms the very life of this school, and which is the chief business of each one of us."

Walker's physical energy boosted confidence all around. He swept into Tech in grand style, an extroverted, take-charge sort: flamboyant, outspoken, larger than life—like "a great orchestra," said one colleague, "all its parts and instruments attuned to harmonious results [with] no discords under his magic leadership, the orchestra ever rendered obedient to his direction, the great symphony." The contrast with Rogers's and Runkle's subdued styles could not have been more pronounced.

Walker knew Boston well. Born there on July 2, 1840, the youngest of three children of Amasa and Hannah (Ambrose) Walker, he traced his paternal ancestry back to English immigrants who settled in Boston around 1630. His mother's side, originally French Huguenots, came from a line of wealthy merchants in Concord, New Hampshire. Francis's great-grandfather, Phineas Walker, served with distinction in the Revolutionary War under Ethan Allen and then-patriot Benedict Arnold, in the capture of Fort Ticonderoga in May 1775. Francis was proud of this heritage, particularly its survival for more than two centuries with "little or no admixture from other than British stock." The Walkers lived in a brownstone at Montgomery Place in Boston's South End, next door to poet and Harvard Medical School professor Oliver Wendell Holmes, whose son—Oliver Wendell Holmes Jr., later an associate justice on the U.S. Supreme Court—was a year younger than Francis.

Amasa Walker, a footwear manufacturer and one of Boston's leading merchants, retired from business the year Francis was born and in 1843 moved his family out to the same house where his father had raised him in rural North Brookfield, Massachusetts, about thirty miles outside of Amherst. He took up the study of economics, wrote on economic theory with a focus on monetary policy (his book *The*

Science of Wealth, published in 1866, was widely read and translated into several languages), taught on a voluntary basis at Oberlin College and Amherst College, and served terms in both branches of the Massachusetts legislature, as Massachusetts secretary of state, and then, in 1861–62, as U.S. congressman. A progressive, liberal thinker, he championed abolitionist Charles Sumner's bid for the Senate in 1851, was an avid partisan of William Lloyd Garrison's antislavery campaign and a key contact, too, on the underground railroad smuggling slaves into Canada.

Francis prepared for Amherst College at the Leicester Academy and New England Normal Institute in Lancaster. He excelled in school—"the only man I ever met," one of his Amherst classmates recalled, "to whom knowledge came without any effort"—and was a devoted athlete, especially in baseball and boxing. He struck his peers as a born leader: magnetic personality, outgoing, charming, an inspiring speaker and writer, able to put others at ease. A member of the Athenian Society, Delta Kappa Epsilon, and Alpha Sigma Phi, he had little patience for the sillier side of fraternity life, resigning from Alpha Sigma Phi "on account of rowdyism." He won prizes for ad-lib speaking and contributed to the college magazine, *Ichnolite*. In 1857, his writings began to lean toward economic themes. An essay of his, "Thoughts on the hard times," published in the popular journal *National Era*, shed light on the Panic of 1857 and the ensuing two-year, nationwide financial depression.

After graduation in 1860, Walker worked for a year in the law firm of Devens & Hoar in Worcester. On July 2, 1861, he enlisted for Civil War service and rose through the ranks from sergeant major in August 1861 to brigadier general in March 1865, his final promotion given for bravery in the Battle of Chancellorsville, May 1863. Walker's letters from the front evoked horrific scenes. The bones of his left hand were shattered by an artillery shell. In 1863, Governor John Andrew of Massachusetts recruited him to quell an assault on the state house by draft protesters. When Walker recounted the event in later years, as he often liked to do, "his eyes . . . flashed as he told of the posting of a battery of field-pieces behind the main doors of the building, and then, having thrown the doors wide open, inviting the mob to enter,—an invitation which was promptly and peacefully declined."

After the war, a brief, uneventful period teaching Latin and Greek at Williston Seminary in Easthampton, Massachusetts—too tranquil for his taste, probably—was followed by a volatile period in journalism starting early in 1868. With the real-life battlefield behind him, he took up literary warfare instead, sharpening and wielding verbal knives as a staff writer with the Springfield (Mass.) *Republican*. He "never winced," said one of his close friends, "whether he was facing men with guns in their hands or with pens, or with words of bitter and sharp controversy in their mouths." Walker churned out barbed op-eds on the federal election system, soldiers as politicians, processes for impeaching public officials, postwar reconstruction, the future of political parties, and economic subjects: taxes, fees (as hidden taxes), national revenue, debt reduction, retrenchment, and treasury and monetary reform.

Walker was recruited into the federal civil service by a family friend, David Ames Wells, U.S. special commissioner of internal revenue. Wells, himself a former staffer on the *Republican*, wanted Walker in Washington to revitalize the work of the statistics bureau, which had fallen on hard times owing to poor leadership. Wells was drawn by Walker's reputation for guts, energy, and organizational skills, and by his record of principled, reasoned perspectives on the economic issues of the day. But President Ulysses Grant's interior secretary, Jacob Cox, soon pulled him away to head the census bureau. Walker rid this corruption-riddled agency of its most blatant abuses—nepotism, bribery, the sense of "pushing forward our friends and relatives into good places" as an entitlement—but he found the challenge of raising statistical standards in a chaotic postwar environment more difficult. The 1870 census ended up deeply flawed, which Walker was quick to admit, particularly as regards undercounting in the South. Yet the innovative system of graphics that he introduced, including colored maps, atlases, graphs, and charts, helped make sense of random, raw numbers and clarified key demographic trends.

So pleased was the Grant administration with Walker's work that Cox's successor as interior secretary, Columbus Delano, added the Indian affairs bureau to his portfolio. Walker brought the same reformist vision to this agency, more crooked and incompetent than the other. A "task of Sisyphus," his old boss Cox wrote him, but if anyone could get it under control, Walker—with his by now famous "intolerance of rascality"—could. Walker found himself caught between sentimentalists who held the Indian-as-noble-savage view and an army convinced that the only good Indian was a dead Indian. He forged a perspective that had little in common with either. Indians, he concluded, were "like children . . . impressionable, susceptible, and capable of much good, but the precedent condition of doing anything for them is rigidly to control their attention and demand their presence." Conquer via tough love, in other words, a strategy of limited appeasement. Such views anticipated Walker's pragmatic, antiromantic outlook—in rebellion, partly, against his father's old-fashioned, quixotic liberalism—that would lead him to challenge economic orthodoxies, like laissez-faire, and to champion theories of Anglo-Saxon racial superiority. He became one of the original architects of the Indian reservation system.

Walker did not stay long at Indian affairs, just thirteen months (November 1871 to December 1872). He was anxious to get out of Washington, into some line of work where motives other than thievery, greed, and self-interest prevailed. Those who had his interests at heart urged him to move on. Senator Henry Dawes, Charles Sumner's successor as U.S. senator from Massachusetts, thought it best for him to "slough off Washington and breathe a pure atmosphere" somewhere else. Walker's brother-in-law, a member of the wealthy Batcheller family (shoe manufacturers) of Brookfield, Massachusetts, invited him to join the firm, which he would have done but for a last-minute offer of a professorship at Yale University's Sheffield Scientific School.

In 1872, one of Sheffield's star faculty members—Daniel Coit Gilman, professor of physical and political geography—resigned to become president of the University of California. Gilman, who had helped compile social statistics for Connecticut as part of the 1870 census, knew Walker well, professionally, and recommended him—"a pioneer in the sciences of economics and statistics, and a master of the scientific method"—to take his place. Sheffield offered Walker Gilman's professorship restructured so that political economy and history, not geography, were the focus.

Like Gilman, Walker was affiliated with Sheffield's "select course," an alternative to straight laboratory or engineering studies, "preparatory to other higher pursuits, to business, etc." Students gravitated to his unusual personality, this blunt, brash, straight-talking figure of a man decked out in elegant frock-coat and silk hat, defying the stereotype of anemic, withdrawn, frumpish erudition. Quickly, he became Sheffield's most popular teacher. His forceful, frenetic teaching style surprised some students. "I shall never forget," recalled freshman William Sedgwick, "my first sight of him when our class filed [in] for its first lecture in Political Economy. He was walking quickly back and forth behind the long demonstration table, glancing now out of the windows, now toward the freshmen, whom he was to teach. He was young, he was strong, he did not care to sit down, yet he was soberly and seriously in dead earnest." Sometimes he gave the impression of "a caged lion."

At Sheffield, Walker wrote prolifically for the popular press and churned out larger-scale works such as *The Indian Question* (1874), *The Wages Question: A Treatise on Wages and the Wages Class* (1876), *Money* (1878), *Money in Its Relations to Trade and Industry* (1879), *Land and Its Rent* (1883), and *Political Economy* (1883). The last two were completed mostly at Sheffield, even though their publication date coincides with Walker's second year as Tech president. Contemporary economists, including those who took issue with his conclusions, were impressed by Walker's blend of sound science and generous humanity. Besides his academic work, he ran as an independent candidate for Connecticut secretary of state against entrenched Republican and Democratic interests. While he lost the election, his stirring campaign speeches foreshadowed the mugwump movement that would help sweep Grover Cleveland into the White House. Walker ran for local office, too. He was elected to New Haven's school committee in September 1877 and held that seat until February 1880. The Democratic governor appointed him to the state board of education in July 1878, reappointed in succeeding years (through 1881) by Republican governors. In 1879, he served on the state's railroad commission.

Walker's reputation as exacting but fair led to his selection, in 1876, as head of the awards bureau at the Centennial Exposition in Philadelphia—a job that he carried off with remarkable tact even though, underneath, he seethed over a corrupt panel or two. At Philadelphia he met Tech's president, John Runkle, and reviewed the Institute's two award-winning exhibits—one for overall inventiveness, the other to the Lowell School of Design "for the beauty, and general merit of the drawings exhibited; for their useful effect on various important industries and especially for their bearing upon technical education in general."

Meanwhile, he was pulled in other directions. Daniel Gilman, who had left California to head the newly founded Johns Hopkins University in 1876, urged him to come to Baltimore to build Hopkins's program in economics and political science. Walker declined, but agreed to serve as nonresident lecturer there. In 1879, he gave a series of twelve lectures at the Lowell Institute of Boston, Tech's first opportunity for a look at the man who, unbeknownst to anyone, would become its next chief executive. President Rutherford Hayes appointed him as American representative to the Paris Exposition of 1878, and member of a three-man commission to argue the case against silver demonetization at an international conference, also in Paris. The silver negotiations ended, Walker groaned, with the American side "beaten horse, foot and dragoons."

He hated to lose, but such setbacks energized him. Still drawn to census work, in 1878 he drafted a bill that his friend Congressman Garfield edited slightly and brought to the House floor. Walker agreed to serve as superintendent once again, broadening the scope of the 1880 census beyond routine data on population, vital statistics, wealth, and industry to cover agriculture, transportation agencies, water power, forestation, shipbuilding, precious metals, wages, trade unions, strikes, mining laws, and urban social patterns. He was never one to complain—not about hard work, anyway—but this time the punishing schedule almost did him in. The 1870 census had fit into 4 volumes; the 1880 census filled 22 volumes plus a two-part compendium. Walker helped see the project through even after his departure, midstream, to become Tech president. He put in regular appearances at the census office until the final volume issued off the press in 1888, feeling fortunate just the same to be out of the pressure-cooker; the deaths of his successor, Charles Seaton, and three chief clerks he always attributed to overwork. The 1880 census, like that of 1870, found many critics and Walker himself called attention to weak spots so that future superintendents might learn from past mistakes. But the project was a success overall, reinforcing his stature as a first-rate statistician.

Tech had not seen a personality type like Walker's before, at least not as president. Rogers and Runkle were the archetypal quiet, gentle scholars. Most faculty followed this pattern, even in cases—Charles Eliot, for example—where cool, distant formality overshadowed gentleness. But Walker was aggressive, pugnacious, prone to volcanic outbursts. His alert, deep-set hazel eyes shone from behind a naturally florid complexion that grew brighter, sometimes beet-red, in the flush of excitement. "I was born," he reflected, "with an unfortunate disposition"—a short fuse, so to speak—that he was never able to control well, even late in life, and that some at the Institute recalled long after his death as "a fearsome thing" to behold. "It is fortunate for you," a mutual colleague once warned a politician who had suggested creative ways to line their pockets at public expense, "that you did not suggest this opportunity to General Walker [as] he is a man with a mighty temper and would have responded by throwing you downstairs." Walker carried himself ramrod straight, his broad shoulders drawn back, military-style. "I told him one time," said Boston mayor Charles Green, "that if he should shave off his mustache he would resemble Napoleon strikingly." No one doubted who was in charge. To colleagues his own age (or older) he was "a somewhat punctilious brother," to younger men and students "a rather austere father." Whether or not the Institute actively sought such a leader, to balance a decade of low-key, fumbling indecision by Runkle, followed by Rogers's valiant but increasingly frail efforts, here was a man ready to seize the proverbial bull by the horns.

Walker's language matched his mood. He was known to let loose with phrases—damn, damned, looked like the devil—deemed spicy by the standards of the time. Someone once asked him about such a tirade. "I don't remember" it, he said, "but the language is mine." His tongue landed him in enough trouble that he worked on self-control, choosing bits of text as mantras-in-waiting for the next explosion: "Nothing is so potent to clarify the judgment and sober the temper, in questions of right or wrong, as to know that a mistake will lead to a hard and long fight." While his temper simmered up, it rarely boiled over. Observers who saw it coming marveled at the way he kept hidden all but the rage in his eyes. But on those rare occasions when anger got the better of him, watch out.

Opening day one fall semester: the "cane rush" pitting sophomores against freshmen, a tradition established in 1880 or 1881. Freshmen grabbed the cane—an oak stick about four feet long—and sophomores rushed them to wrest control. The class that had the most hands on the cane after fifteen minutes would be declared the winner. Walker tolerated this bit of foolishness, barely, but this day his patience wore thin. A pile of yelling, wriggling bodies filled the main lobby. Walker, in his office to the left, sat buried in paperwork. Hearing the commotion, out he rushed and flung himself into the fray. As his fists flew—even his shattered left hand sprang to life—combatants retreated to walls and corners to nurse their wounds. Some were grateful he had not gotten hold of the cane. Walker stood his ground, center-stage, and with the boys now cowed, peppered them verbally. This was not a side of him they saw often and they had no desire for a repeat performance. Drills under their military instructor, General Hobart Moore, felt like an afternoon stroll by comparison.

For Walker, as he scattered these young fellows left and right just as years before he had sent the "skedaddling Rebels" on *their* way, the incident may have brought back a Civil War memory or two. But his bluster was occasional, fast-moving, temporary. Tech folks mostly saw his amiable, gentler side—a man who while "capable of truly Jovian wrath" had "a fun-loving temperament . . . and a personal charm which amounted to genuine magnetism." At the start of each semester he would make a point of camping out in his office, door ajar, available to anyone. He wanted to ease the transition for parents and new students, some frantic with separation anxiety, others nervous, intimidated, overawed by what lay ahead. Students who could not make it home at holiday-time received invitations, in Walker's semilegible scrawl, to dine at the Walker home, 237 Beacon Street. It helped that his sons were about their age. Four of the Walker boys went to Tech—Stoughton ('87), Ambrose ('91), Francis ('92), Etheredge ('99)—and were looked to as shields or buffers should their father grow too loud, or a mite physical, at any of these dinners. His nephew, Amasa Walker, came out from North Brookfield as a special student in the early 1890s. Even daughter Lucy studied at Tech, taking a class in German in 1887, and Walker's grandson—Francis Stoughton Walker, son of Francis Walker—would graduate much later, with the class of 1930.

Walker's accessibility made a lasting impression. Young Willis Whitney laughed when his father suggested he stop by the president's office for advice on what course to choose—"I thought that would be taking considerable liberty," Whitney recalled, "with a big and busy man"—but Walker took time with him, talked, listened, offered wise counsel. Roger Babson, on a similar mission as a second-year student, confided in Walker that he didn't much go for science or engineering. No problem, said Walker, do you like business? Definitely, said Babson, and Walker's reply—"You should be very thankful, because the graduates who succeed are those who can tell a quarter from a half-dollar, rather than those who can tell a strut from a tie." When he crossed the main lobby from his office to the Institute secretary's office, Walker would greet people, bow right and left to those coming through the Boylston Street entrance or down the steps of the inner stairwell, pausing now and then for a chat. In this he moved always against the grain, his path perpendicular to that followed by the stream of humanity struggling to get somewhere else, to another class or to lunch or to a meeting. He interrupted traffic flow seamlessly, without fuss, carving out gentle, pregnant pauses in a workplace dominated by intense effort, ambition, the obligation always to get from here to there.

Walker came in with no illusions. Edward Atkinson and John Cummings—the Corporation's voices of dismal pragmatism—saw to that, impressing on him their sense of the challenge ahead: how to pull the Institute out of a slump brought on by economic depression, compounded by mismanagement. But after his experience with the federal government, none of this struck Walker as overwhelming. Here he was in charge, with everyone around—Corporation members, faculty, students—falling in line, looking to him for guidance, leadership, a way forward. There would be few if any carping political hacks to contend with.

Walker sized up the situation quickly and set to work. Five problems, in his view, required most attention: physical plant, fundraising, curriculum reform, faculty recruitment, and student life. He began by appointing a committee "to inquire and report to the Corporation, respecting the immediate needs of the Institute." He chaired it; the other two members were Cummings and William Rogers (when Rogers died in May 1882, Edward Philbrick replaced him). The committee agreed on space as the most pressing priority, not just because the Boylston Street building and the two adjacent temporary structures (drill hall and annex) were bursting at the seams but because the organization of space had fallen into disarray. Chemical labs were scattered between the basement of the main building and the annex. Mechanical engineering had rooms on the fourth floor of the main building, with steam engineering experiments confined to an open basement hallway. Mechanic arts did foundry, forging, carpentry, and other practical work in the annex while occupying a room for recitations on the fourth floor of the main building. The Lowell design school ranged between the annex and the fifth floor of the main building. Everything else fit in between. Total floor space was 20–21,000 square feet. This had to be tripled to accommodate current programs adequately and leave room for expansion.

A new building, plans for which were announced by Walker in July 1882, went up in a few months. It was called simply that—New Building—until December 1891, when the Corporation renamed it the Walker Building in honor of Tech's president. Funds for construction came from the sale of the Institute's land in front of Trinity Church, on the other side of Boylston Street, granted by the state in 1873 but never developed. Trinity clergy and congregants, none too happy over the prospect of Tech's noisy students and smoky labs coming that close, had spearheaded a campaign to convert the area to a public park, the origin of what is now Copley Square. The Walker Building occupied most of the Clarendon Street side of Tech's block, necessitating removal or demolition of the gymnasium and annex, temporary structures erected in Runkle's time. The gymnasium was relocated to a lot on Exeter Street leased from the Boston & Albany Railroad Company on a two-year renewable basis, across the road from what was then the Harvard Medical School. The annex was torn down. Its primary occupants—Tech's auxiliary programs, the design school and the mechanic arts school—moved into new mechanics' shops on Garrison Street, beside the railroad, a half-mile distant.

In 1888, Tech bought two lots totaling almost 20,000 square feet on Trinity Place. These were followed in 1889 by an adjacent lot stretching southward and, in 1893, by more land on the Clarendon Street (eastern) side of Trinity Place formerly occupied by a skating rink, and a vacant field abutting railroad yards on the south side. Walker had to mollify fiscal conservatives on the Corporation who worried that snatching up land, with no clear means to build, was irresponsible. "I am, myself," he explained in May 1888, "disposed to feel that the Institute, in its present condition, can afford to pay the necessary price, as a sort of insurance or guarantee that it shall have room for expansion, if expansion shall be found to be its fortune, or fate, in the future." Several new Institute buildings went up on Trinity Place: engineering (1888), architecture (1892), and a boiler house (1893) to serve the Institute's growing energy needs.

Under Walker, Tech embarked on a building program limited by space shortages in a neighborhood where demand for residential and commercial development was on the rise. The Walker Building was useful, practical, serviceable. It was not an eyesore, exactly, but it had a factory-like look, rectangular, clunky, dismal, in contrast with the elegant, classic lines of its neighbors: the Main Building—called the Rogers Building, officially, as of May 1883—and the Natural History Building. About the same ground dimensions as Rogers (90 × 156 ft.), it consisted of four stories and a basement. The chemistry department occupied the two upper floors and some of the basement, for industrial chemistry. The pure chemistry labs were named for oil baron Jerome Kidder, who died in 1882 leaving Tech a substantial bequest for chemical education and research. Physics occupied some of the basement, the entire first floor, and much of the second floor, except six classrooms used for modern languages and mathematics. Over on Trinity Place, the Engineering Building (52 × 148 ft.) comprised five stories plus basement and was devoted to civil and mechanical, while mining stayed in Rogers. Labs occupied the basement and ground floors; the upper floors held drawing, recitation, and lecture rooms. Engineering, like the Walker Building, was functional, boxy, unattractive. But the Architectural Building (58 × 66 ft.) combined functionality and exterior charm. "The architects," said Walker, "are going to be as proud as peacocks." The mechanic arts shops on Garrison Street, while just a single story high, sprawled across 24,000 square feet, more than twice the ground area of the engineering and architectural buildings combined.

The original goal, to triple the Institute's physical plant, was quickly met then as quickly exceeded. By the mid-1890s, Tech boasted eight times what it occupied when Walker became president. The new space was not simply larger but better utilized as well, tied to state-of-the-art developments in equipment, apparatus, and technical know-how. Walker returned from a trip to France and England in the summer of 1885 impressed not so much by what he found there, as by his sense of how well Tech had managed with scarcer resources. But Tech's expansion never kept pace with exploding demand created by growth in the student body, new fields and disciplines, new frameworks and methods within established fields and disciplines. As the campus spread out, so too did the Institute's debt burden and, with it, the Corporation's fiscally hawkish mood.

Tech's money woes, its indebtedness, its chronic teetering on the brink of insolvency were underscored, for Walker, by what he saw on his visit to Europe. Unlike the universities and technical institutions of Britain and France, Tech had no endowment or public subsidies. It ran on fumes, so to speak, hand-to-mouth, year by year if not day by day, dependent on tuition fees and the occasional kindnesses of friends, acquaintances, sometimes strangers. Walker laid out some hard truths. The Institute must build an adequate endowment, either that or face chronic uncertainty and possible collapse; it must protect itself, he said, in a not-so-kind allusion to the failures of the Runkle administration, "against the possibilities of temporary internal mismanagement." An endowment would help Tech "enlarge its means of present usefulness."

Walker's sense of urgency rose as pressures on the Institute grew—student enrollment tripled in less than a decade, new fields and subdisciplines emerged—with no additional sources of funding in sight. He pointed to a vicious cycle that saw resources flowing to established, wealthy institutions like Harvard, as newer, poorer ones languished. A half million dollars would keep Tech afloat, he estimated, while a million would "place us in as good a financial condition as the poorest school of our rank in the United States. . . .The needs of the Institute are so great because the Institute itself is so much needed." In January 1887, he petitioned the state legislature for $200,000. That June, the state offered half the amount payable in two annual installments and subject to two conditions: that Tech match the proffered grant from other sources and provide twenty free scholarships distributed fairly across the state's senatorial districts, with preference to needy students. Corporation member William Endicott spearheaded a private subscription drive that pulled together matching funds by November. Another petition was drawn up in January 1888 asking for an additional $100,000. This time the state granted the full request.

But as economic conditions deteriorated following the nationwide Panic of 1893, triggered by railroad overbuilding and bank collapses, Walker pushed even harder. "It will soon be necessary," he wrote in December that year, "if indeed the time has not already come, for the friends of the Institute and the people of Boston and Massachusetts to decide whether this great school of industrial science . . . shall be allowed to suffer serious and enduring injury, and possibly irreparable disaster, from lack of pecuniary means." Hoping to rouse state and local officials to action, he pointed once again to the relative generosity of Europeans toward *their* educational institutions—the Swiss government's subsidies, for example, to the polytechnic institute at Zurich. In the meantime, he continued to scout out resources both private and public. In 1893, he tried without success to persuade trustees of the Franklin Fund, originating in cash bequeathed to Boston by Benjamin Franklin (whose name, Walker noted, was carved in stone on Tech's Rogers Building), to support the work of the Institute's physics department. The trustees decided, instead, to create an independent technical institute, the Franklin Institute, which would open its doors in 1908. This failure was offset by the legislature's decision, in April 1895, to award Tech $25,000 a year for six years plus $2,000 a year to support 10 scholarships over and above the 20 scholarships already established under the acts of 1887 and 1888.

When Walker arrived at Tech in 1881, the course structure was relatively straightforward:

I	Civil and topographical engineering
II	Mechanical engineering
III	Mining engineering, or geology and mining
IV	Building and architecture
V	Chemistry
VI	Metallurgy
VII	Natural history
VIII	Physics
IX	Elective
X	Science and literature

New layers and sublayers were added over the years so that by 1896, at the end of his presidency, the structure looked like this:

I-1	Civil engineering (general)
I-2	Civil engineering (highways, railroads, and railroad management)
I-3	Civil engineering (geodesy and topography)
II-1	Mechanical engineering (general)
II-2	Mechanical engineering (marine)
II-3	Mechanical engineering (locomotive construction)
II-4	Mechanical engineering (mill)
III-1	Mining and metallurgy (general)
III-2	Mining and metallurgy (metallurgy)
IV	Architecture
V	Chemistry
VI	Electrical engineering
VII	Biology
VIII	Physics
IX	General studies
X	Chemical engineering
XI	Sanitary engineering
XII	Geology
XIII	Naval architecture

The options within each course remained fluid. Some were dropped, others retained, new options added then discarded, existing options tweaked, then tweaked some more. This framework, a decade and a half in the making, mirrored Walker's curricular push on several fronts: expanding, deepening, in some instances fragmenting subject areas to meet the demand for specialized skills in engineering fields. The tension between widening and subdividing did not concern him as the best moves, he held, never proceeded in tidy, predictable strings.

In 1882, the physics department introduced a course in electrical engineering (first known as VIII-B, then, beginning in 1884, as VI). This step responded to new technical requirements and the needs of modern commerce. The natural history department adopted a new identity, too, one year later, with VII-B as an option for students who wanted basic knowledge across varied disciplines—biology, chemistry, physics, mathematics—useful for medical studies. The first two students, Frank Pickernell and Richard Pierce, graduated in electrical engineering in 1885. Pickernell went on to become chief engineer for American Telephone and Telegraph in New York, while Pierce was a principal in the Chicago firm of Pierce & Richardson, Electrical and Mechanical Engineers. The number of students opting for Course VI exploded, and within a few years electrical supplanted civil and mechanical as Tech's most popular course. In 1894, it had 137 full-time students to mechanical's 111 and civil's 88, with graduation rates following a similar pattern: 33 bachelor's degrees were awarded that year in electrical, 31 in mechanical, 21 in civil. Together, the electrical, mechanical, and civil programs registered over 60 percent of full-time students.

Tech's graduate-degree programs, on the books since 1872 but never active, drew renewed attention. While scores of graduate students registered at Tech, however, very few advanced degrees were awarded during Walker's administration. In his first year as president, there were 15 graduate students, none of whom aimed for a higher degree. Four of these were Tech graduates (Frank Stantial '79, James Atkinson '81, Harry Cutler '81, and George Mower '81), along with three Harvard graduates, two from Smith College, two from Dartmouth College, and one each from the U.S. Naval Academy, Massachusetts Agricultural College, Brown University, and Cornell University. The first master's degree (S.M.) was awarded to Tech graduate Frederick Fox Jr. (chemistry) in 1886. Other S.M.'s followed in 1887 (Arthur Noyes in chemistry), 1890 (William Thurber in general science), 1893 (Prescott Hopkins in architecture), 1894 (Frederic Fay in civil engineering), 1895 (Fred Mann in architecture, Walter Scott in chemistry, Charles Abbot in physics), and 1896 (Frank Bourne and Herbert Chamberlain in architecture, George Defren in chemistry). Eighty graduate students were registered in 1896, just nine of these as candidates for advanced degrees. There would be no doctorates until 1907, when three were awarded in chemistry.

The focus in biology, called natural history before 1889, moved away from its old-fashioned roots in descriptive morphology toward sanitary science, bacteriology, epidemiology, and industrial biology. Sewage disposal, air and water quality, plant and human diseases, food preservation, and fermentation emerged as topics of special interest. In 1889, a new program—Course XI, sanitary engineering—was established as an offshoot of Course I (civil) and Course VII (biology). It stimulated connections between chemistry (Course V), civil engineering, and biology. Walker underscored Course XI's cross-disciplinary potential, "the increasing necessity for co-operation between the engineer, the chemist, and the biologist, in dealing with questions affecting the public health." In 1892, the first six sanitary engineers graduated: Richard Chase, William Locke, Elmer Manahan, George Merrill, Frank Shepherd, and Joseph Warren.

By the mid-1880s, chemistry offered two options: organic and industrial. The industrial option broke off in 1888 as Course X—chemical engineering—to satisfy growing demand for graduates with specialized knowledge and experience in fuels, lubricants, dyes, textiles, metallurgy, fertilizers, and other industrial lines. The decision to recycle course numbers—X had been used for the by-now defunct philosophy course, and at different points for the "science and literature" and "elective" courses—caused some confusion over the years, but seemed the best way to avoid proliferation. (The numbering scheme eventually reached XXV, when "interdisciplinary science" was created in 1971.) Course XII, geology, appeared in 1890 merging studies from civil, chemistry, mining, and economics into a program organized largely around surveying and mapmaking. Mechanical's marine option, founded in 1892, became naval architecture—Course XIII—the following year. "It would be difficult," wrote Walker, "to find any other direction in which the Institute could develop which would have a greater promise of usefulness," referring to the rise of shipbuilding programs in the navy and merchant marine.

Walker never conceived of specialization as an end in itself, or even as a primary focus. His own interest lay in broadening, which required new vitality in Tech's general education program. This, he realized, would be difficult for an institution whose raison d'être lay elsewhere. "As the primary object of the Institute has been known to be technical education," he observed in 1883, "it has not been found easy to attract the required degree of attention to the projected branches of study, on the part of pupils or their parents." Students often treated subjects in the humanistic disciplines as moments of relaxation—with an opportunity, perhaps, for a fast-and-easy cultural veneer—in a curriculum crowded with studies that were more useful, hence viewed as more valuable.

Walker set out to modify this outlook. He began by revamping the science and literature and elective courses, which had become a refuge for the fragile or the unmotivated, students who shied away from the rigors of the professional courses or who had trouble deciding what they wanted to do in life. But instead of emphasizing intrinsic value, Walker sought first to bridge the general and the professional. Course IX, renamed "general studies" in December 1881, began offering three options in 1882, each "of a nature to enlarge the views and enrich the life of the man of business." One option emphasized physics, another chemistry, a third geology, botany, and zoology. The idea was not to give students free range to dabble, but to shape a coherent program that, while general, had a professional objective: to prepare students for the commercial side of science and technology, a combination of "practical and liberal tendencies." In effect, Course IX was Tech's first attempt at a modern-day business school, with a technological twist. Walker recommended it for "young men whose purpose it is to become merchants, manufacturers, or bankers, and who desire a preparation for active life, liberalizing in its tendencies, but without any influence to alienate the student from the ideas, tastes, and habits which are appropriate to practical business pursuits."

Walker insisted that Course IX be no watered-down version of the curriculum offered by older colleges like Harvard. While denying any wish to malign such institutions, he found a way to paint them in rather grim, pathetic terms. "It is a familiar feature of classical colleges," he lamented in 1883, "that large numbers of students, who are by nature neither vicious nor idle . . . relinquish at a very early date in their collegiate career, all scholarly ambition, and come . . . to accept the part of doing nothing well, contented to be known as poor scholars. . . . Surely this is a poor preparation for life, at least for practical business life, where success is to be gained far less by talent or acquirement than by promptitude, by punctuality, by industry, by self-respect and by strict attention to duty." Tech students, he bragged, were cut from a different cloth: "It is making no unseemly boast to say that students of that class are exceedingly rare in this school." Elsewhere, in stronger terms, he implied that Tech's culture as compared with that of Harvard and other such colleges was "repugnant to foppery, extravagance, triviality, and indolence." The Institute's objective was not to "polish the surface," but to "build up the substance of mind and character."

While Course IX never attracted students in a big way, numbers grew steadily after 1890: 11 graduates during its first nine years (1881–89) and 36 during its next seven (1890–96). But Walker aimed to expose every Tech student, not just Course IX majors, to new levels of substance and rigor in the humanistic disciplines. In 1889, he proposed reforms in the teaching of English, "the difficult and delicate duty of giving to the students of a school of industrial science that instruction and practice in the use of language and in the art of composition which shall be most for their benefit, professionally and socially." This was not to suggest that technical students were at a disadvantage compared to their classical counterparts. The technical student, claimed Walker, was a sharper thinker who needed merely to harness cognitive power to enhanced communication skills. "Trained, day by day and year by year, in the objective study of concrete things, he sees nothing vaguely; the images he forms are definite and distinct; what he knows, he knows perfectly. If fine writing be the end in view, these mental characteristics may or may not be advantageous; but for the purposes of simple, straightforward, manly expression, whether in description, in exposition, in narrative, in argument, or in business correspondence, they are a source of great power." Walker pressed the faculty, too, to sustain ample offerings in history, literature, political economy, philosophy, modern languages, anthropology, sociology, government, and business law. A statistician himself, he wanted to introduce students to standards for quantitative analysis in the humanities and social sciences as well as in technical disciplines.

This period also saw renewed attention to the summer school. Begun in Runkle's time, it was placed on a more systematic basis under Walker. Field trips occurred almost every summer between 1872 and 1887. Groups of mining and metallurgy students led by Robert Richards, with other faculty and Ellen Richards along to help out, went to Colorado, Michigan (Lake Superior), Virginia, Vermont, Pennsylvania, New York (Lake Champlain), and Canada (New Brunswick and Nova Scotia). Beginning in 1887, civil engineering introduced a four-week required program in topography, geodesy, geology, and hydraulics, the first in South Deerfield, Massachusetts. Subsequent summer sessions were held in Schoharie (N.Y.), South Deerfield, Delaware Water Gap, Keeseville (N.Y.), and East Machias (Maine). East Machias became a long-term site for programs requiring field work. Mining and metallurgy, however, continued to roam. In June–July 1888, 15 students traveled to the Eustis copper mine in Capelton, Quebec, divided up into squads, and alternated work assignments—one week down in the mine, one week above ground. According to Walker, "the students all took a four days' turn at drilling and blasting; and . . . quite won the hearts of the miners by the readiness with which they acquired both the skill and the judgment required for economical mining . . . setting timbers and laying track." The party camped out in tents, under the stars.

A third summer school—in architecture—began in 1893, with a visit to the Chicago World's Fair. In the summers of 1894 and 1895, students explored colonial architecture in Salem and Portsmouth. They spent nearly two months bicycling around England and France in 1896. In 1894, the Institute first offered summer classes on campus in mathematics, physics, chemistry, biology, French, German, and a few other subjects, but registration was spotty. Walker was convinced that numbers would soon grow, so he kept the program going.

The instructing staff more than quadrupled under Walker, from 37 in 1881 to 153 in 1896. The most rapid expansion occurred within the lower ranks of instructor and assistant, while the upper ranks—full, associate, and assistant professors—grew more gradually, a pattern shaped by Walker's desire to build a permanent faculty from the ground up. Among Walker's recruits in the 1880s were William Sedgwick for biology; Lewis Norton, Thomas Pope, Thomas Drown, Augustus Gill, Arthur Noyes, and Henry Talbot for chemistry; George Swain and Dwight Porter for civil engineering; Alfred Burton for topographical engineering; Cecil Peabody for applied mechanics, later naval architecture; Jerome Sondericker for applied mechanics; Webster Wells, Harry Tyler, and Dana Bartlett for mathematics; Theodore Dippold and Alphonse van Daell for modern languages; Davis Dewey for history and political science, later economics and statistics; Charles Levermore for history; George Barton and Heinrich Hofman for mineralogy; C. Frank Allen for railroad engineering; Harry Clifford for physics; Francis Chandler for architecture; and Peter Schwamb, Herman Hollerith, and Edward Miller for mechanical engineering. Frederick Fox Jr., Tech's first recipient of a master's degree, was hired in 1886 to help Ellen Richards teach lab techniques in sanitary chemistry. After 1890 came George Carpenter and Arlo Bates in English; Charles Currier in history and political science; Frank Laws and Harry Goodwin in physics; Willis Whitney, William Walker, F. Jewett Moore, Samuel Mulliken, James Norris, Joseph Phelan, and James Crafts in chemistry, the latter a returnee after two decades living and working in France; George Haven in mechanical engineering; Frederick Woods and Leonard Passano in mathematics; Désiré Despradelle in architecture; Charles Norton, William Drisko, George Wendell, and William Coolidge in physics; Samuel Prescott in biology; Charles Spofford in civil engineering; and William Ripley in economics and sociology—"the first teacher in this department," Walker proudly announced, "of our own breeding."

Walker's first two recruits as special lecturers, a category introduced in 1883, were both in architecture: Arthur Rotch on theory of decorative painting and E. P. Treadwell on practice of decorative painting.

Lecturers reinforced very specific knowledge. Walker came up with the idea partly from Johns Hopkins, partly from Harvard. He had served at Hopkins as a special lecturer in political economy during the 1870s, and at Harvard he was appointed university lecturer on resources of the United States, on a one-year renewable basis beginning in 1882. Walker adapted the system for Tech's unique purposes, bringing in luminaries from various fields on flexible terms—some regularly, once or twice a week, others two or three times a semester depending on need. The aim was to cover certain key topics, chiefly recent developments in engineering that could only be touched on in the regular curriculum.

Walker took special care with his lectureship appointments in the humanities and social sciences. John Fiske, preeminent historian of his day, taught at Tech in 1893–94, and legendary constitutional scholar John Jameson came up regularly from Brown University to lecture that same year. John Chipman Gray, Harvard's Royall professor of law, taught business law from 1889 to 1892 and was followed for a couple of years by the brilliant young attorney, Louis Brandeis, soon to become one of America's most distinguished jurists. "Mr. Brandeis has to a remarkable extent," Walker observed in 1895, "overcome the difficulties attending a course in such a subject given in a scientific school, where the time at command is closely limited. He has made use to the utmost of the special aptitude which students of technology and science possess for dealing with concrete cases." The benefits went both ways. "Those talks at Tech," Brandeis said later, "marked an epoch in my own career," referring to the framework of intellectual cross-linkages between law, business, and politics that would occupy a pivotal place in his two-and-a-half-decade career on the U.S. Supreme Court. Walker also brought in lawyer and historian A. Lawrence Lowell to teach political history in 1893–94. Lowell, a member of the Tech Corporation since 1890, would succeed Charles Eliot as Harvard president in 1909. Walker was an informal special lecturer himself, introducing a generation of sophomores to principles of political economy. Paul Litchfield ('96) recalled him whipping off his spectacles and shaking them at students to stress a point: "Desire will always exceed supply, but it is not important until *desire* becomes *demand* on the part of people willing and able to pay."

At times, Walker worried about attrition within the faculty ranks. Some of the brightest and best kept being drawn away by offers elsewhere—testimony to Tech's growing reputation, but a challenge, too, for replacement strategy, how to pull in faculty of equal or higher ability. John Ordway (industrial chemistry and metallurgy) left for Tulane University in 1884. Charles Wing, originally hired by Runkle for chemistry, went to private industry in 1885. Tech lost talented physicist William Pickering ('79) to Harvard in 1887—not a surprise, exactly, as Pickering's older brother, Edward, having gone the same way a decade earlier, offered him an appointment at the Harvard College Observatory where the research and teaching programs in astronomy were more in line with his own interests than anything Tech had to offer. Jules Luquiens, a fixture in modern languages since 1874, accepted a professorship in Romance languages at Yale in 1892. Charles Levermore and George Carpenter both left in 1893, Levermore to head Adelphi Academy in Brooklyn, New York (later Adelphi University) and Carpenter to become professor of literature at Columbia. When Thomas Drown left in 1895 to become president of Lehigh University, Walker felt crushed: "The history of the Institute of Technology is full of painful losses . . . sustained for the enrichment of other institutions. Harvard, Yale, Columbia, and now Lehigh, have in succession carried away from us some of our most valued instructors."

Illness and death also took their toll. Chemistry professor William Nichols died on July 14, 1886, after undergoing surgery in Hamburg, Germany, the first Tech faculty member to die in harness, so to speak, while on the active faculty roster. Nichols had founded one of Tech's earliest, most widely recognized research programs—sanitary chemistry, tied to the testing of city water supplies—and was a dedicated teacher. Nichols's sanitation work was carried on by William Sedgwick, although more from the bacteriological side. Also in 1886 Tech lost an instructor in modern languages, William Cook, who accidentally shot himself while hunting. Charles Otis, professor of modern languages hired under Runkle, died in 1888.

Walker grew increasingly concerned about the faculty's well-being; Holman and Schwamb went on leave in 1890 with Walker's "earnest hope that both these admirable teachers and highly accomplished scholars may be restored to us another year, in the full enjoyment of health." Eugène Létang (architecture), away on sick leave in 1889, died in 1892. Walker sent Sedgwick to Europe for a precautionary rest in 1891. A compulsive worker, Sedgwick instead devoted much of the time to surveying the latest developments in sanitation research. Walker called the unexpected death of Lewis Norton, in 1893, "one of the severest blows" ever sustained by the Institute. Chemical engineering, as a course distinct from chemistry, had been Norton's brainchild, evolving into one of Tech's most popular programs.

Through all this, Walker remained confident that the Institute's reputation for serious scholarship—as a place, he liked to say, where students came to work, not to play—made it a natural magnet for faculty of the highest caliber. "There is something in the atmosphere of the place, in the character of the studies and exercises pursued; in the distinctness with which the students see the purpose and object of those studies and exercises, as bearing upon their own success and happiness in life; in the close and familiar contact of teacher and student, at the bench, across the table, around the machine; doubtless, also, in the class of students who are attracted to the school—which makes the work of instruction here singularly agreeable and inspiring." Charles Wing told him that in ten years at Tech, dealing with hundreds of students, he had encountered very few "black sheep," less than a dozen all told.

One trend that Walker wanted to keep in check had to do with inbreeding, the perception that Tech's faculty was drawn primarily from the ranks of its own graduates. There was some truth to this; in 1895, for example, more than three-quarters of the instructing staff held Tech bachelor's degrees. For some, this weighting was sensible, even desirable, because of the uniqueness of the curriculum, particularly in technical disciplines, which outsiders might find difficulty connecting with. But for others, it raised warning flags about provincialism, narrowness, a lack of healthy, diverse influences. Walker advocated the widest possible tent, going to considerable lengths "to secure liberality and breadth of character by bringing in men who represent the training and culture of other institutions." Devoted statistician that he was, he marshaled figures to prove his point. In 1895, Tech's instructing staff included 12 Harvard men, as well as graduates of Yale (4), Amherst (4), Johns Hopkins (3), and Bowdoin (2); 1 each from Boston University, Columbia, Iowa State, Wesleyan,

Williams, University of Virginia, West Point, Lafayette, University of Illinois, University of Vermont, Pennsylvania State, University of Minnesota, and New Hampshire College of Agriculture and the Mechanic Arts; and 6 with degrees from foreign universities. In other words, 25 outside colleges and universities were represented among 44 (of 147) instructors on Tech's staff. "I think this is doing pretty well," Walker remarked, "so far as preventing the recognized evil influences of in-and-in-breeding is concerned."

He pointed, too, to the growing number of Institute-trained instructors eager for exposure to different environments. By his count, nine held higher degrees from other institutions, chiefly abroad. Harry Tyler set a trend in 1887, the first in a string of credential-seekers whom Walker granted leaves of absence for this purpose. Tyler earned a doctorate in mathematics at Erlangen in 1889 under algebraists Paul Gordan and Max Noether. Noyes, Gill, and Talbot followed, all in chemistry at Leipzig in 1890, then Frederick Woods in mathematics at Göttingen in 1894 and Willis Whitney in chemistry at Leipzig in 1896. Sedgwick and Dewey both came to Tech with Ph.D.'s from Johns Hopkins in 1881 and 1886, respectively. After graduating at Tech, Ripley earned his at Columbia in 1893. The chemists, in particular, returned home gung-ho about pushing Tech toward innovative research.

Aside from biology under Sedgwick, chemistry was Tech's most research-oriented field at the time, a trend begun under Nichols and then carried on by Noyes and others in later years. An indexing project that Noyes began in 1895—"Review of American chemical research," published first in *Technology Quarterly*—grew by 1907 into *Chemical Abstracts*, the international bible of information in the field. The most prolific member of the chemistry department, Noyes authored or coauthored eight scholarly articles in 1895 alone. His coauthors were sometimes students (Charles Abbot '94, William Hall '95, John Dorrance '95, Willard Watkins '95), sometimes colleagues (Willis Whitney '90).

Physics also moved on the research front. Charles Cross, Charles Norton, and Harry Goodwin grew intrigued, for example, by Wilhelm Roentgen's discovery of the x-ray in 1895. Cross and Norton compiled a trove of photographic plates, mostly images of hands—normal hands, damaged hands, deformed hands, arthritic hands, hands with broken fingers, hands with cracked knuckles. Walker was among the eager volunteers. Norton x-rayed his left hand, the one shattered during Civil War action, showing four smashed wrist-bones that had re-fused and pushed fingers forward and closer together. The machine malfunctioned—a pitch change, a whining sound, a blotch on the resulting negative. "The General," Norton recalled, "disturbed one of our wires in some way and it fell down on the back of his hand causing considerable discharge from the induction coil. . . . He was making rather wry faces about it but held his hand absolutely still, in spite of the repeated shocks, without complaint." Cross attributed Walker's stoicism to his years of military discipline.

Student numbers increased in every year of Walker's presidency, from 302 in 1881 to 1,198 in 1896. The momentous 1,000-mark was broken for the first time in 1891, when 1,011 students entered. The largest rates of increase occurred in his early years, 1881–85, with smaller, but steady, gains thereafter. The number of degree-earners climbed, too, with 24 in 1882 and 188 in 1896.

As with faculty, Walker aimed to broaden, deepen, and vary the student body. While the population remained primarily New England based—between 55 and 65 percent of each entering class came from Massachusetts—its reach gradually extended to students from a swath of the American continent, and from more foreign countries. Students came from 13 states (excluding New England) and 4 foreign countries in 1881, and in 1896 from 35 states and 9 foreign countries. The foreign countries represented in 1881 were Panama, Mexico, Dutch Guiana, and Nicaragua; in 1896, Chile, Cuba, England, France, Japan, Mexico, Canada, Turkey, and Venezuela. Walker pointed, too, to signs that Europe was starting to look to Tech, rather than the other way around. The "fruits of the training of this Institute are, so far as I have seen, not equaled on the Continent," said one British politician in a speech delivered in 1895.

The earliest documented black students arrived during Walker's administration, when Tech first included race—the label "colored," for example—as an identifier in the academic record. Because the practice was never systematic, it is difficult to know for sure how many black students attended, or even who the first one was. The earliest student tagged as "colored" was Robert Taylor, of Wilmington, North Carolina, who entered in the fall of 1888. Following him was William Johnson, of Jamestown, Rhode Island, in 1890. Then came Richard Lewis, of West Roxbury, Massachusetts, in 1892, Frederick Hemmings of Boston in 1893, the Dixon brothers—Charles Sumner Dixon and John Brown Dixon, of Washington, D.C.—in 1894, and William Smith of Altoona, Pennsylvania, in 1896. By this time, ironically, Walker had become an outspoken racial theorist, promoting ideas of Anglo-Saxon superiority in forums around the country. "There is nothing here, aside from a few kinds of personal service," he wrote in 1891 about prospects for a massive wave of blacks migrating from south to north, "which the negro can do, which the white man cannot do as well, or perhaps better." When someone asked his opinion on the nation's capital as a site for a national university, Walker replied that Washington, D.C., was "no better and no worse than other cities, though the presence of a large colored population is always a source of great danger."

His view of Jews, Slavs, and other non-Anglo, non-Nordic Europeans was equally negative. In 1892, he lectured widely on "immigration and its evils," urging fellow Americans to feel "great apprehension and alarm" about unwashed hordes banging on the nation's gates. "We are now draining off great stagnant pools of population," he told a convention in Philadelphia in 1895. "Hundreds of thousands, representing the lowest stage of degradation to which human beings can be reduced . . . are found among the new citizens whom the last decade has brought into the Republic. . . . They are beaten men from beaten races. . . . It is idle to think that in any short time we can make these people as good citizens as if they came from Teutonic stock, representatives of the men who met under the oak trees in the forests of old Germany." Walker's views stoked anti-immigration sentiment, fed into the nation's growing obsession over waves of inferior, foreign races inundating America's shores, and helped inspire the early twentieth-century eugenics movement, focused around racial hygiene and racial betterment, that led to abuses—forced sterilization programs, human experimentation, even genocidal urges.

Jewish students could be found at Tech, albeit—like blacks—in smallish numbers. Gerard Swope, later General Electric president and a stalwart of the Tech Corporation, entered as a freshman in 1891, from St. Louis, Missouri, and thrived there. His father, a German Jewish immigrant, ran a merchandise trading business. But with respect to race and ethnicity, Walker was less concerned about individuals or local academic policy than about the grander scheme of things—population shifts, birth rates, immigration patterns—that he had observed and analyzed as federal census director. His ideas deeply influenced some students. William Ripley, who joined Tech's economics faculty three years after graduating in 1890, authored a book called *The Races of Europe* (1899) that echoed Walker in significant ways. "This great Polish swamp of miserable human beings," wrote Ripley, referring to Jews, "threatens to drain itself off into our country . . . unless we restrict its ingress." In one of his projects, with Tech students as subjects, he used old-fashioned, discredited methods of craniometry—measurement of heads—to assess how shape and size related to ethnic origin, intelligence, and other factors. He opined that it was "impossible for white people" to settle in tropical climates, like the Philippines, with much hope for survival and that native populations should be handled with force, as "familiarity breeds audacity." None of this appears, however, to have influenced Tech's policy—its absence of policy, actually—toward racial and ethnic minorities.

Walker's presidency mirrored another rich irony. While he was the epitome, said one admirer (Charles-Edward Winslow '98), of "all that we hold true and manly," he also proved attentive to women's needs. Early on he took stock of how women had fared at Tech: the role of the coed Lowell Institute programs; the women's laboratory, which had provided opportunities for 102 women, mostly science teachers, since its founding in 1876; chemical labs in the new building to replace the inadequate women's laboratory; several women—Ellen Swallow Richards, Marie Glover, Evelyn Walton, Clara Ames, and Carrie Rice—awarded the Institute's regular degree, all in chemistry, alongside their male classmates. In 1883, he counted 11 women students at Tech in a population of 443, a minuscule proportion but noteworthy, he felt, in light of women's relative lack of interest in scientific and technical work.

Walker once questioned whether women as a group had the strength and resilience for Tech's rigorous curriculum: the number "is never likely to be large, considering the nature of the professions to which our courses lead, and the severity of our requirements for admission and for graduation." But in an effort to promote women's comfort levels at Tech, he announced in 1883 that a special space—the Margaret Cheney Reading Room, named for a deceased Tech student and daughter of writer, philanthropist, and women's-rights advocate Ednah Dow Cheney, who donated funds for this purpose—would be fitted up for their exclusive use on the third floor of the New Building. Walker considered the room a sanctuary where "students of this sex" could find "adequate facilities for retirement and rest in the intervals of recitation and laboratory work." As years went by, he discovered that gender-specific lines of success and failure, and of tolerance for hard work, were blurrier than he expected. Many men and women did well, while some men and women fell by the wayside.

Walker pushed the cause of women's education at Tech, with mild to moderate success. By 1887, the number of women students almost tripled from the time he took office in 1881, a pattern that continued into the 1890s. There were 25 women in 1887–88, 41 in 1892–93, 58 in 1894–95, and 75 in 1895–96. A substantial fall-off in the early 1890s—from 33 in 1889–90 to 23 in 1890–91—disappointed him. But women graduated in every year of his presidency except one (1883). Few in number, they were confined to a small group of courses and none ventured into engineering. Most (16) earned bachelor's degrees in chemistry, while some graduated in biology (7), architecture (6), physics (3), geology (3), and general studies (3). The first to graduate in general studies, Marcella O'Grady ('85), taught biology at Bryn Mawr College and Vassar College, went to Europe for further study in 1894, earned her doctorate at the University of Freiburg in 1895, studied at Würzburg under the eminent cytologist Theodor Boveri, whom she married in 1897, worked as his assistant until his death in 1916, then returned to America to teach at Albertus Magnus College in New Haven, Connecticut. The first in physics, Annie Sabine ('88), did graduate work at Ohio State University; the second, Margaret Maltby ('91), earned a doctorate at the University of Göttingen and taught physics at Wellesley College, afterward at Barnard College. The first in biology was Caroline Woodman ('89), in architecture Sophia Hayden ('90), and in geology Dixie Lee Bryant ('91). Hayden was chosen to design the women's building for the World's Fair in Chicago in 1893. Of 6 graduates in biology in 1892, 3 were women—a fact that Walker took special pride in, pointing out that they "began their professional work by conducting with marked success large vacation classes in natural history at the seashore." The precedent set by Ellen Richards—of Tech women marrying Tech men, or Tech faculty—took hold with a vengeance during Walker's presidency, when there were no less than thirty such nuptials.

While Walker never considered women prime candidates for technical careers, if a woman showed desire and talent, and if some commercial or industrial enterprise was open to hiring her, he helped match them up. In 1896 he recommended Harriet Gallup, Tech's lone woman chemistry graduate in 1894, to George Eastman for the post of works chemist at Kodak in Rochester, New York. Eastman had told Walker that he would be "favorably disposed toward a woman for the position." Walker, a distant cousin of Eastman's (Roger Eastman, early American colonist, was a common ancestor with marital ties to the Ambrose family), persuaded him that Gallup would fit the bill well. Tech's promotional pamphlet for the 1893 World's Fair, however, had noted that the Institute was "by the nature of the case, essentially a man's college, though the Corporation and Faculty have seen no reason why any person who wishes to do the work of the school, and is qualified for it, should be excluded by reason of sex." Hardly a warm invitation, but it did suggest a flexible policy.

The dominant class of students was white and male, yet diverse in several respects. Old-time Boston Brahmins rubbed elbows with sons and daughters of working-class Irish. Some came from wealthy families, others were poor, some struggled in the middle—a hodge-podge, unlike Harvard's relatively homogeneous group. The Du Pont family, industrialists of Delaware, sent its sons—William ('76), T. Coleman ('84), Alfred ('86), Maurice ('88), Pierre ('90), Henry ('94), Irenée ('97)—and a host of their descendants would follow, making Tech a family tradition. There were Appletons, Cabots, Conants, Cushings, Bigelows, Lorings, and Forbeses as well, all from prominent New England families, but also folks from modest circumstances: Duffs,

O'Gradys, Collinses, Rooneys, Dolans, and O'Learys. To accommodate this varied group, Tech allowed students who could not afford to pay for a full course, or who had to work to support themselves or families, to enter as "specials" to take one, two, three, or more subjects as time and means permitted. Specials met the same academic standards as regular students and their goal, sometimes, was to move eventually into degree-granting programs. Walker resisted pressure to abolish the system. He considered it a reasonable goal, however, to reduce the proportion of specials to regulars, and this he did, from a high of 46 percent in 1881 to 28 percent in 1896.

Tech's graduates could be found in positions all over, from the smallest, most remote outposts to the densest urban centers—not just in the United States, but also in Europe, the Far East, South America, and other places overseas. Many worked in mining, others in transportation (particularly roads and railroads), power production and distribution, chemical manufacture, public health, sewage, and sanitation. Some lived as itinerants, moving from spot to spot, from one specialty to another, as industries shifted locale or focus, or as new settlements emerged. Tech graduates were known as adventurers, risk-takers, explorers inspired as much by a thrill or a challenge as by a safe income or job security.

During Walker's time, career choices grew more daring as well, reaching beyond the applied sciences to include management, education, even research. While many graduates still went into engineering, more and more could be found as company executives, professors, or research specialists and independent consultants—chemists, physicists, biologists, astronomers—in the public and private sectors. Besides Gill, Noyes, Tyler, Talbot, Mulliken, Goodwin, Prescott, Whitney, Wendell, Coolidge, and others who joined the Tech faculty, those who went on to prominent academic and research careers included biologists Edwin O. Jordan ('88), George Whipple ('89), Gary Calkins ('90), and Albert Mathews ('92); physicist George Burgess ('96); economist Francis Walker ('92), the president's son; meteorologist A. Lawrence Rotch ('84); and astronomers George Ellery Hale ('90) and Charles Abbot ('94). Hale, unhappy that Tech's research spirit was not more vigorous than he found it, stayed for four years but spent much of his serious work-and-study time with the Pickering brothers at the

Harvard College Observatory. But it was clear that research opportunities were emerging, even if too slowly for some.

Among the group that would become leaders in American commerce and industry were Arthur D. Little ('85), chemical engineer and pioneer in industrial research; the brothers Morss—Everett ('85) and Henry ('93)—of Simplex Electrical; Alexander Rice McKim ('86), consulting architectural engineer; classmates Charles Stone and Edwin Webster ('88), principals of Stone & Webster of Boston, "electrical experts and engineers"; banker Charles Hayden ('90); manufacturer Albert Bemis ('93); Frank Lovejoy ('94), of Eastman Kodak; Gerard Swope ('95), of General Electric; Alfred Sloan ('95), of General Motors; Paul Litchfield ('96), of Goodyear Tire & Rubber; and many more. Tech also produced graduates who went into lines of work that had little if anything to do with science or technology: stock raiser (cattle, not Wall Street), farmer, coffee planter, rancher, decorator, real estate agent, druggist, attorney, artist, journalist, literary editor, creative writer, bookkeeper, traveler (implying exploration), librarian, banker, voice coach, music teacher, salesman, military officer, missionary—an eclectic range that underscored Walker's conviction that Tech's scope was essentially limitless. He even raised the possibility that the Institute might get into the business of training philologists. "I see no reason," he wrote to one prospective candidate, a woman, in December 1896, "why a person who has received the bachelor's degree . . . should not attain the Ph.D. with us, in German and French." While he advised that Yale might better suit her goals, and although nothing came of the idea, there was a chance that the first doctorate awarded at Tech might have gone to a woman in the humanities.

The most controversial aspect of Tech's curriculum was neither range nor emphasis, but what some on the outside perceived as indifference to students' welfare. Complaints about overwork, occasionally heard in Runkle's time, increased under Walker. He found himself having to reassure a restive public that the Institute did not and would not push students to the brink. Still, the course load was jam-packed and the dropout rate continued high, particularly among freshmen. Those left standing after the first year breathed a sigh of relief, but some went away struggling to recover lost confidence.

This last group became enough of a concern that Walker, in 1887, felt compelled to survey recent graduates. "Every now and then," one of his cover letters stated, "the question of *overwork* arises in this school, on the complaint of some parents. We know that such complaints are often unreasonable, just as we know that they may, at times, be well founded. . . . With a view to giving light on this question, will you kindly inform me, whether . . . it appears to you, that, in order to maintain a fair standing in your class, you were obliged to study harder than was consistent with your health and with your own proper interests?" In another letter he left little doubt as to where *he* stood: "The grumblers are not infrequently among those who are fully able to keep up with the class." Walker's tone suggested conclusions he expected to draw and replies tended to follow his lead.

Most responders confirmed that Tech's curriculum had not worn them out, that it had made them stronger, if anything, better able to handle whatever the workplace threw at them, and that it should not be watered down for the sake of the weak, the lazy, or the undisciplined few. Very few responses complained that "more work and study was required of me . . . than was consistent with my health or best interests." But the image of Tech as a slave-driving machine did not disappear. It grew, in fact, into an identifying emblem that insiders sported with pride and outsiders observed with awe, sometimes alarm, from a distance. On December 7, 1890, the *Boston Herald* slammed Tech for its alleged cruelty to students. "The Institute of Technology is not a place for boys to play, but for men to work," Walker wrote back. "This is the point we start from. We expect those who come to us, asking for our degree, to take up the work of their lives then and there, definitely and seriously, and to labor thereafter as they will have to do in business, if they are to succeed." Such rebuttals made little impact, but Walker kept trying—not so much, after a while, in the popular press, as one-on-one with students, their parents, and educators at other institutions. At least one Corporation member, the formidable Edward Atkinson, resigned over the issue. "I am of opinion," he wrote Walker in March 1890, "that the impressions which I have derived myself from my own observation and unhappy experience, as to the injudicious methods of instruction and the over-work of students in some departments, are thoroughly well grounded."

Walker helped adjust the curriculum so that Tech did not seem quite so daunting. One strategy, the five-year option, allowed students to stretch studies over a longer period. This option had been in place since 1878, but it assumed new meaning in the early 1890s when Walker recommended it for those who "from considerations of health, lack of thorough preparation or other cause, have reason to anticipate exceptional difficulty." Very few took advantage of it, less than 5 percent of all registered full-time students. There may well have been a stigma attached. Pressures to compete, to work hard, to get on with it—all these combined to make the five-year option look like a haven for the frail, the lazy, and the unmotivated. Walker was never fond of it. He took far more interest in students at the other extreme, those who worked double-time: Samuel Hunt ('95) and Charles Parmelee ('95), for example, Tech's first students to walk away with two degrees earned simultaneously—Hunt in electrical and chemical engineering, Parmelee in civil and sanitary. Walker did, however, caution against a rush to earn dual degrees: "Since it is naturally regarded as a feather in a man's cap . . . there will always be some ground to fear that a student may attempt this, not for the sake of a larger knowledge and a broader culture, but for the distinction attending it, with the result that either he may fail in both courses, or . . . that he will get through . . . both . . . without doing any part of the work very well." In 1894, a system of official advisers went on the books. Each student was assigned to a member of the instructing staff, who would offer friendly advice as needed. But Walker pointed out that this in no way implied either a promise by Tech to serve in loco parentis or an abrogation of what was expected of students in the way of "responsibility for their own lives [and] accountability as ordinary citizens of the community."

Even as he played down concerns about student life, Walker recognized the need for healthy social outlets and activities. Life for the typical serious Tech student was limited, cramped, confined. He rented a room for nine dollars a week in a brownstone walkup, attended school all day, spent evenings alone buried in homework, took a stroll around midnight over to Washington Street, say, for coffee and a wheat cake at Sennet's, came home fortified for a bit more work, then slept for three or four hours before the start of a new school day. These were Tech's so-called grinds. Others needed more than a midnight coffee break. Alfred du Pont ('86) gadded about at seedy night spots, played (for pay) in a burlesque orchestra, and became a camp follower and close confidant to the frequently inebriated, famous prizefighter John L. Sullivan.

Some students got into deeper trouble, and the local media loved nothing more—except, perhaps, crime stories—than to write up Tech boys' high jinks. One episode in the mid-1880s provoked widespread amusement and, for the Institute, no little embarrassment. A Tech student, known as "a darling masher," rushed from a nearby poolroom into his barber's Columbus Avenue shop one Saturday morning waving a lady's diamond ring, inscribed "From Clement to Annie," proof of his latest conquest. The man in the barber's chair leapt up shouting, "That's my wife's ring—where did you get it?" The student stammered "I-I-I f-f-found it" and rushed out. Clement started after him, before realizing that a man racing down the street with a lathered face and wrapped in a neck sheet would draw too much attention. Meanwhile, the German Adonis—as the *Boston Globe* dubbed him—banged on the front door of Clement and Annie's townhouse, blurting to an astonished Annie what had happened and how they needed to match up their stories. In rushed Clement, pistol drawn. The student ran for dear life, Clement laid into Annie, and divorce papers went on file the next day. The lesson learned by the student, according to the *Globe*, was not that he should quit philandering, but that next time he confided a scandalous bit of news to his barber, he should first know whose whiskers were hiding under the lather.

Students, mostly, were just restless. Buried in stuffy labs, workshops, and classrooms much of the day, with little open space to run wild in after hours and few organized sports at the level that students elsewhere took for granted, they found other channels for youthful energy. The steps of Rogers fronting on Boylston Street became the preferred spot to congregate between classes, at lunchtime, and in the late afternoons. Students went there to chat, smoke, share stories (mostly apocryphal) of romantic conquests, make plans for the evening. Passersby would stop in their tracks and guests at the Hotel Brunswick across the way would rush to adjacent windows to watch as students jostled and cuffed each other. The appearance of Walker or some other authority figure signaled the denouement of these mini-dramas, but there were days when witnesses were treated to a grand climax: a barrage of water-filled, paper missiles, say, dropped from inside Rogers, thirty feet up, sending folks below scampering for cover, a towel, or perhaps a change of clothing. The missile-makers would defend themselves, facetiously, as part of Tech's most serious student group, intent on testing Newton's laws of gravity.

Walker set out to transform this culture, which shifted wildly—often unpredictably—between grimness and frivolity. The lack of a cohesive campus made the problem more difficult. As Tech offered no housing, a good portion of the student body rented rooms in the South End, within easy walking distance of Boylston Street. Unlike the Back Bay neighborhood to the north and east, settled mostly by old-money Brahmins ensconced in roomy, elegant townhouses passed from generation to generation, the South End was densely populated with Irish, Polish, and Italian immigrants, blacks, students, young professionals, the unemployed and the semiemployed, a community of diverse, transient tenants all competing for inexpensive lodging. Rooming houses lay between, sometimes atop, small businesses—mom-and-pop grocery stores, billiard and pool rooms, dance halls, pubs—along teeming thoroughfares, Columbus Avenue and Tremont Street. Tech students relished the freedom of this lifestyle, guarded it jealously, and boasted of its benefits compared with what students at regular colleges like Harvard had to live with: invasive rules, constant monitoring, at times a stifling level of surveillance. But it had its downside: a chaotic, unstructured feel.

When Walker surveyed student residential patterns in 1893, he identified at least five groups defined by relative proximity to, or distance from, campus. About 54 percent lived close enough to walk, but the remainder faced a crushing daily commute, some from more than sixty miles away. Walker grew convinced that improved, Tech-based meals and exercise would boost morale, inspire fellowship, enhance overall quality-of-life. Jones's Lunch, introduced in Runkle's time, was closed down in 1889—its standard had begun low, and gone downhill ever since—and a new eatery, run by Ellen King with overall supervision by Tech instructor, chemist, and nutritionist Ellen Richards, appeared in the basement of the Rogers Building in the fall of 1890. Mrs. King's café served a wholesome lunchtime meal, at 20 cents each, to three or four hundred students a day. "How much difference it makes," Walker observed in 1893, "to the present and future health of the student whether he eats a cold lunch in the corner of the laboratory, or takes something warm and palatable, under pleasant surroundings and in company with his fellows." The gymnasium also came in for increased usage with the hiring of instructors Albert Whitehouse and Herman Boos to put students through their paces in a systematic, controlled way. New locker and bathing facilities helped, too.

Walker saw athletics as a key not only to physical health but also to building moral fiber, manly strength, self-discipline, a responsible mindset overall. His own exploits in boxing and baseball at Amherst had helped prepare him, he believed, for service on the battlefield, for a career, for life; and he wanted to duplicate this experience at Tech. But in his view sports crazes like football that threatened to overwhelm college life should be kept in their place, on a rational level. Walker promoted organized sports consistent with students' primary obligations in the lab and classroom. Tech's Athletic Club, formed on October 13, 1894, joined the New England Intercollegiate Athletic Association, competing against ten other colleges and universities: Amherst, Bowdoin, Brown, Dartmouth, Trinity, Tufts, Vermont, Wesleyan, Williams, and Worcester Polytechnic. While Amherst and Dartmouth tended to dominate, Tech came out winners in 1894. The Football Association, organized on March 6, 1895, played in a league with Dartmouth, Brown, Amherst, Tufts, and the prep schools Exeter and Andover. Tech's record here was often pitiful (2–6 in 1896), but no matter—spirit, said Walker, always trumped outcome. The Institute also fielded teams in tennis and baseball; the first tennis club drew up a constitution and elected officers in April 1886. The Technology Yacht Club was organized on November 27, 1894. Cycling enthusiasts created the Technology Wheelmen in May 1896. The Gun Club, which went in for shoots at suburban ranges, and the Hare and Hounds Club, combining outdoor swims with cross-country running, also appeared in 1896. But Walker warned away all sports fanatics. "Opportunities for championship athletics . . . are not large," he wrote to one applicant in November 1896, and "the standard of scholarship at this school is incompatible with that degree of attention to athletics which in these days is essential to 'winning form.' Our students are expected to attend to their scholarly work. . . . I should advise no man to come hither with a view to an athletic career."

Walker saw structured, campus-based pastimes as both rest-and-relaxation opportunities for serious students and ways to keep less serious students out of trouble. He encouraged the establishment of professional societies. The 2G Society, for mining engineers, was formed in the fall of 1881. Mechanical engineers followed in 1882 with Sigma M.E. Also that year, civil engineers started Gamma Sigma Upsilon. In 1885, the Biological Club appeared in March and K_2S, for chemistry students, in the fall. Some of these groups were short-lived. Gamma Sigma died in the mid-1880s, succeeded in February 1889 by the Civil Engineering Society. The Sketch Club was founded by architecture students on October 15, 1886, at a meeting where instructor Thomas O'Grady Jr. proposed a student-run periodical that finally saw light in 1891, as *Architectural Review*. In 1889 came the Tech Electrical Club and the Geological Club, around the same time as the Biological Journal Club. The Electrical Engineering Society (1893), Naval Architectural Society (1896), and Mechanical Engineering Society (1897) followed. The earliest of these groups were secret societies open only to majors in their respective disciplines, crosses between fraternities and professional groups. They were called "local societies" for several years and generally grouped with fraternities and social clubs. When restrictions on membership seemed likely to limit intellectual goals, the preferred term became "professional societies."

Fraternity chapters sprang up under Walker, himself a fraternity man going back to his Amherst days. The first, Sigma Chi, appeared on March 22, 1882, followed by Alpha Tau Omega and Theta Xi in 1885, then Delta Psi, Delta Tau Delta, and Phi Gamma Delta in 1889. Tech chapters of Chi Phi, Delta Kappa Epsilon, Theta Delta

Chi, and Phi Beta Epsilon were inaugurated in 1890. Delta Upsilon appeared in 1891, Sigma Alpha Epsilon in 1892. Tech chapters had no permanent residences to begin with. They met when and where they could—saloons, cafés, park benches, or empty Tech classrooms after hours. In 1896, 192 Tech students were frat members, just over 16 percent of the total student population.

Other groups had more ambiguous identities—hybrids, some of them, between a fraternity, a professional society, a guild, and a social club. The Cooperative Society was formed in April 1886 to help students economize on academic and living expenses. There was Hammer and Tongs, founded in the mid-1880s, and Delta Sigma. The Mandaman Club and Sherwood Club enjoyed role-playing of an exotic type. Mandaman centered on American Indians (fantasies, anyway, about Indian life), while Sherwood, founded in February 1895, evoked characters from English lore: Robin Hood, Little John, Friar Tuck, outlaws, and yeomen. Tech women formed a club of their own around 1890. Originally a secret society, Eta Sigma Mu, it came to be known as Cleofan in 1895 and was dubbed by male classmates "a sweet society of fair ones." The Walker Club, founded on December 4, 1894, and named for Tech's president, produced plays often satirically based on Institute events or characters. The first of these productions—"Professor Jones, Instructor" and "The X-Ray Machine"—were staged in 1896. The Institute also supported a branch of the Young Men's Christian Association, formed on January 17, 1895. By the mid-1890s, students were coalescing in groups with regional interests at heart: the Southern Club, Cincinnati Club, Chicago Club, Washington Club, Maine Club, St. Louis Club, Philadelphia Club. Some had ties to preparatory schools, like the Exeter Club and Andover Club, both founded in February 1894. The Technology Club, established in rented quarters at 71 Newbury Street, across the street from the Rogers Building, was intended, Walker said, to foster "opportunities for social intercourse" between Tech faculty, alumni, and upperclassmen. The club formally opened on October 7, 1896.

Walker encouraged a wide range of extracurricular activities. The first issue of a student newspaper, *The Tech*, rolled off the press on November 16, 1881. It began as a biweekly, then turned to weekly publication in 1892. In 1885, the class of '87 introduced a yearbook,

The Technique, which—like *The Tech*—became a permanent fixture in campus life. *The Senior Portfolio*, an annual begun by the class of 1894, published formal, portrait-style, studio-quality photographs of graduates. *Technology Quarterly* started in 1887 as a science-oriented magazine edited by members of the senior and junior classes, but by 1892 it had turned into an Institute-wide organ publishing minutes of the Society of Arts, scholarly articles by faculty and students, notable news and commentary. The Institute Committee, established in 1893, aimed to nurture a spirit of constructive self-governance among students and consisted of the four class presidents plus two additional representatives from each class. The Glee Club gave its inaugural concert at Berkeley Hall in the South End, corner of Warren Avenue and Berkeley Street, on December 28, 1883. The Tech Banjo and Guitar Club was formed in October 1884, but early attempts to sustain an orchestra met with limited success. Often the programs planned were too large, ambitious, or demanding. "We are sorry," *The Tech* editorialized in November 1887, "to know that the music selected for this year is to be all of a heavy character. The 'Largo' [by Handel] is all right in its place, but what the orchestra needs to do first is to 'feel' each other . . . . Light, popular dance music will accomplish this quicker than anything else, and it would be folly not to give it prominent place." A mandolin club began in 1890, riding a craze triggered in Boston by a group of traveling Spanish musicians. The Glee Club, Banjo Club, and Mandolin Club merged in 1890. The inaugural concert of the Technology Minstrels, a song-and-dance act in blackface, helped raise money for the school's baseball team in December 1895; a similar group, the V.L. Club, founded in April 1883, had not survived long. Several smaller ensembles came and went in the 1890s—the Quintet Club, the Senior String Quartet, a Septet. Other performing arts groups included l'Avenir and Der deutsche Verein, which staged plays in the original French and German.

No mere technical institute; a university, rather, polarized around science and engineering. Tech would come to embrace this identity as uniquely its own, and Walker was the first to give it conscious form, active guidance. But his life outside was equally busy, rich, and varied. Someone once counted his eleven distinct occupations, without saying what these were. Local clubs vied for his membership; he joined several when he moved to Boston—notably Oliver Wendell Holmes's famous Saturday Club—and stuck with them. With a reputation as provocateur-extraordinaire, Walker could be relied on to stir things up when meetings turned deadly dull. Once, at Trinity College, Dublin, a paean that he delivered on Irish valor inspired glee, shock, or outrage depending on which end of the political (and sectarian) spectrum his listeners happened to stand. In his speeches, articles, and books, he tackled the most controversial topics of the day, and relished doing so. The positions he took on foreigners, immigration restriction, socialism, labor unions, taxation, international bimetallism, military intervention overseas, and reform in high-school mathematics teaching, to name a few, triggered debate that dogged him wherever he went. He drew energy from it all, or seemed to. In 1885, he became the first president of the American Economic Association, a post he held for seven years—"an imposing figure," wrote one member, "as he presided over the meetings." He was also, for a time, president of the American Statistical Association. As a member of the Boston School Committee (1885–87), the Massachusetts State Board of Education (1882–90), and various civic groups, he threw himself into the business at hand with little concern for niceties, even less for the toll that such commitments might take on his health. One Boston city councilor recalled his "quick, decisive, energetic way . . . which some people thought was brusque, [but] he was one of the most kind-hearted of men, and any one who wished to go to him for information, for advice in any way, always found him most warm-hearted and sincere in offering advice and help where it was asked."

December 1896 brought a heftier than usual wave of invitations. Walker fended off a bunch of them, then thought half-seriously about writing an article, "Killing a man," that would describe "the manner and ways in which decent and well-meaning people . . . surround the poor victim on every side until he is fain to surrender and give up the last chance he has of getting a little rest or a little pleasure during the next two weeks, all for the purpose of delivering an address for some infernal society." Even so, at month's end he rushed off to meetings of the American Historical Society in New York and the American Statistical Association in Washington, D.C. He felt weighed down, physically unwell; irritated, too, by naggers who insisted he join a U.S. delegation heading to Brussels for an international conference on bimetallism. On January 4, 1897, eager to put this matter to rest, he told a *Boston Herald* reporter that the rumor he would go to Europe "was absolutely without foundation, and that he had not countenanced it in any way."

It was a typical day, otherwise. After finishing up at the Institute, Walker spent the evening at home with his family. Mrs. Walker and their daughters Lucy and Evelyn were there. Walker worked before dinner on a series of ten lectures promised for the Lowell Institute that winter and spring. The family played a card game or two after dinner, his favorite way of unwinding at the end of a hard day. Everyone retired to bed at the usual time, soon after ten. Around midnight, Mrs. Walker heard him mutter something—a casual comment about what he must do next day—and begin breathing heavily, then struggle for air. She sent for Dr. Elbridge Cutler, their physician neighbor across the street. Cutler arrived just as Walker took his last breath. "Without a word and without pain," Cutler said, "he folded his arms across his breast and passed away. There was no struggle, no suffering." Cutler diagnosed apoplexy.

Next morning, Ellen King arrived at work around nine. As she stepped off the streetcar on Boylston Street, laden with supplies—fresh produce, meats—acquired at her weekly visit to the Faneuil Hall and Quincy markets, she did not anticipate anything out of the ordinary. Outside, the usual bedlam: anxious passengers, driver and conductor maneuvering around ice patches and snow drifts, a fallen horse slowing traffic further on. She expected to see her kitchen staff, inside, hard at work preparing for the lunch-hour crowd of ravenous Tech students. "I opened the door and stood upon the top step bewildered and speechless," she recalled. "The room, usually bright with light and activity, steam rising, preparations going on at long tables, was gray in the half light which came in through the glass in the door. . . . The people who always gave me a cheerful greeting . . . were like strangers. They had on their outside garments, sat in groups and did not stir or speak. Probably the time was not more than a minute when I had a sense of a man coming from the dim room beyond. He held a paper in his hand which he gave to me. No word was spoken, as one in a dream I read the lines—'Massachusetts Institute of Technology, January 5th 1897, President Walker is dead.'" Mrs. King dropped her parcels, sank into her chair—frozen, like her staffers, into disbelief. She kept looking at the headline, hoping she had somehow misread it.

Upstairs, in the main foyer, students gathered in small groups whispering. They looked ready for work—textbooks and drawing boards tucked under their arms—but no one could focus, and by 9:15 all classes were suspended until January 11. Faculty members drifted in and out of the secretary's office, a few stopping to answer students' questions, giving (some seeking) comfort. Outside, Tech's wide stone steps—scene of many a rowdy display—filled with quiet, dejected students: a contrast that puzzled passersby until, according to the *Herald*, the flag was lowered to half staff and "the traveling public in the Back Bay had an inkling of the calamity that had come upon the city, and upon the Institute especially."

It had been assumed that Walker, just 56 years old, would remain at Tech's helm for at least another decade. His family came from a line that lived to grand old ages, eighty, ninety years; no one could put a name to the last Walker who had died under age sixty. Walker's friend and next-door neighbor, Henry Haynes, a fellow member of the American Antiquarian Society, blamed overwork. He told of strolls down Beacon Street, then glimpses of Walker through a window at all hours, at his writing table, "throwing off sheet after sheet with marvelous rapidity." Walker, said Haynes, was "literally worked to death in the endeavor to do more than one man's strength could possibly accomplish, and from his never having learned to say 'No' to any request for service to the community."

The funeral, at Trinity Church on January 8, was as close to official as Boston could go for someone who was not a mayor, a governor, or a congressman. City Hall and the Public Library shut down. In bitter cold, the family led the hearse down Boylston Street followed by faculty, students, representatives of city and state government, colleagues from all over: the National Academy, Yale, the military. George Brush, John Runkle, and Charles Eliot joined the dozen or so honorary pallbearers. Following interment at Mount Auburn Cemetery in Cambridge, the Walker family—Exene and her seven children—returned to 237 Beacon Street, enveloped now in gloomy quiet. They would not stay long. Within a few years, Mrs. Walker sold the place to Tech's chapter of Phi Beta Epsilon, which converted it to a fraternity residence. She moved with her daughters Lucy and Evelyn to Brookline and, like Emma Rogers before her, lived to a grand old age, ninety-three. Two of her sons—Ambrose and Etheredge—also lived well into their nineties. Stoughton Walker, despondent over the state of his health, leaped to his death from a transatlantic ocean liner in April 1912, aged 46. Francis, who followed in his father's footsteps as an economist, died on January 15, 1950, in his late seventies.

"Uneasy lies the head"

James Mason Crafts, 1839–1917

Francis Walker's death could not have come at a worse time, with the spring semester just about to open. The faculty met in emergency session on January 5, 1897, elder statesman John Runkle in the chair. They shared out duties. The department heads, Runkle, and faculty secretary Harry Tyler would take charge for a week. Davis Dewey, Arlo Bates, and William Sedgwick would prepare memorials and resolutions. James Crafts, Gaetano Lanza, and Tyler would handle all official correspondence, while Sedgwick and Tyler attended to everything else. The faculty resolved to appoint a chairman pro tem whom the executive committee could look to for advice as the search for a new president got underway.

All eyes turned to Crafts, an organic chemist and probably Tech's most distinguished scholar at the time. But he resisted. "I feel rather used up today," he wrote to physics professor Charles Cross the morning before the faculty met again on January 9, "and shall not go to the faculty meeting. I write to say that the night has not brought change of counsel and I do not see how I can perform the duties of chairman of the faculty and think you can make a selection which will be more advantageous to the Institute." This looked definite, but Crafts, as it turned out, was coaxable. On January 13, the faculty offered him the appointment and he accepted.

Crafts had little administrative experience to show, and even less of a desire to test his leadership skills. Walker had pushed the chemistry department headship on him when Thomas Drown left to become president of Lehigh University in 1895. But Crafts had agreed only to act as head and, as far as he was concerned, this new commitment would not last long either. He would serve as emergency placeholder—a figurehead—then get back to his research the minute Walker's replacement was named. In his mind, he would always be more useful in his lab, at his workbench.

Crafts's lineage, like many among Boston's social elite, traced back to seventeenth-century immigrants to the New World. One of his ancestors, Colonel Thomas Crafts, a noted patriot in the Revolutionary War, had read aloud the declaration of independence from the balcony of the state house, on July 25, 1776, a moment of "great joy expressed by three huzzas from a great concourse of people, assembled on the occasion." James's father, Royal Altamont Crafts, said to have been the first manufacturer of wool muslin in the United States, was a wealthy merchant whose empire stretched from Massachusetts to Louisiana. The family of Marianne (Mason) Crafts, James's mother, came originally from Portsmouth, New Hampshire. Her father, Jeremiah Mason, was that state's most famous lawyer, apart from Daniel Webster, and a sometime public official who served as state attorney general beginning in 1802, U.S. senator (1813–17, as a member of the Federalist Party), several terms in the state legislature (1820s), and president of the Portsmouth branch of the United States Bank (1828) before moving to Boston in 1832.

Born on March 8, 1839, James Crafts grew up on an exclusive stretch of Tremont Street near the Boston Common. He had one sister, Mary Elizabeth, born on April 1, 1840. The Crafts home lay a stone's throw from Temple Place, where William Rogers lived with his wife Emma, and her father and brother. Rogers, a regular visitor to the Crafts home, introduced young James to science through simple, improvised experiments. With Rogers's encouragement, Royal and Marianne Crafts allowed their son to convert attic space into a laboratory. One night he found himself "surrounded by glowing eyes, which effectually banished sleep, until it was discovered that they had their origin in numerous little pieces of phosphorous which had become scattered around his bedroom."

James attended private elementary schools in downtown Boston—Mr. Kidder's School in Bowdoin Square and the Sullivan School near the Park Street Church, a block or two from his home—followed by college prep at Boston Latin School, then a year under the tutelage of Charles Eliot's cousin, Dr. Samuel Eliot, a rigorous scholar and ardent anglophile who modeled himself after the famous Dr. Thomas Arnold, of Rugby. James intended to go to Harvard College, but, drawn to science, opted instead for the Lawrence Scientific School, Harvard's applied science division. He entered Lawrence in 1856 and studied chemistry under Eben Horsford, graduating in 1858. Simon Newcomb, later to become one of America's foremost astronomers, was in his class. James felt pulled initially toward both geology and analytical chemistry, fields tied to mining and mineralogy. He stayed on at Lawrence for postgraduate work, but left in frustration after a year because of Louis Agassiz's dominance there. Agassiz favored pure science over applied fields like mineralogy and was preoccupied with building his visionary museum of natural history, what became the Museum of Comparative Zoology. Horsford, who years before had studied under Justus von Liebig in Giessen, Germany, advised James to seek greener pastures in Europe.

He set sail in 1859, accompanied by a friend and former Lawrence student, Francis Washburn. It was the first of two extended stays abroad. This one lasted five years, the second more than a decade. James studied at the Bergakademie in Freiberg, Saxony, famous for its curriculum in mining engineering. Karl Friedrich Plattner had died the year before, but James worked in his lab and gained experience with the innovative blowpipe techniques he had developed. After a year, with his interests shifting from mining to chemistry, he went to work in Robert Bunsen's lab at the University of Heidelberg. Bunsen and Gustav Kirchhoff were in the early stages of their research on spectrum analysis, which James helped with.

In the early 1860s, Crafts moved to Paris to work under Adolphe Wurtz of the Ecole de Médecine, later holder of the inaugural chair in organic chemistry at the Sorbonne. He grew close to a fellow Wurtz student, Charles Friedel, of the Ecole des Mines. Crafts and Friedel shared similar backgrounds. Both were from wealthy, cultivated families, pursued science at an early age, and became independent, original thinkers. Their personalities—straightforward, quiet, reflective—were alike, too. With Crafts by now focused on organic chemistry and Friedel an expert in metallurgy, they began a long, fruitful, cross-disciplinary collaboration on crystalline structures and synthetic mineralogy. The first of their many joint papers appeared in 1863, in the *Comptes rendus de l'Académie des Sciences*, on vapor densities of silicon compounds. They worked under primitive conditions. According to Crafts, Friedel "never had a well-equipped laboratory provided with labor-saving contrivances . . . most of the work . . . was done in a small room in the fine palace on the Luxembourg Gardens which is used for the School of Mines, whilst a dark, vaulted cellar served for the artificial production of minerals at high temperatures and pressures." But the stimulating environment more than made up for this. Crafts enjoyed Friedel's company, benefited from his ideas, and became part of his wide circle of friends and colleagues from the worlds of science, music, art, and politics. Crafts's father, meanwhile, who retired when the Civil War broke out, fearing collapse of the textile market, followed him to Europe, spent his last years there, and died at La Tour, near Vevey, Switzerland, on May 25, 1864. His mother, who had stayed behind, remarried seven years later and continued to reside in Boston. Marianne Crafts's second husband, Francis Bleeker Ellison, of Brooklyn, New York, was a commodore in the U.S. Navy.

With few options open to him in Europe, Crafts returned to America in 1865. Friedel might have welcomed him as a long-term guest in his cramped workspace, but no American would have been considered for permanent appointment at a French university, *école*, or research institute. Options in Boston were limited, too, so Crafts headed west and spent more than a year as inspector of silver mines in California and Mexico. He said so little of this experience in later years that associates were left to speculate. One called his western travels thrilling, another hazardous, another pleasant.

In 1866, Crafts moved back to Boston, set up a private laboratory in the home of his old professor, Eben Horsford, and began a series of studies on arsenious and arsenic acids. Soon he cast about for an academic position. Cornell University hired him as professor of chemistry in 1867, partly on William Rogers's say-so, a year before the school officially opened its doors. Rogers would have liked nothing better than to bring Crafts to Tech, but with Charles Eliot, Francis Storer, and Cyrus Warren in place, and successful, there was no room for another chemist, not even one of Crafts's credentials and outstanding promise. Rogers had no way of knowing that within three years Eliot, Storer, and Warren would all be gone. A short-term opportunity at Tech opened up in the spring of 1868, but fell through. Warren, Tech's professor of organic chemistry, fell ill and needed someone to fill in. He recommended Crafts, even offering to pay him out of his own pocket. His one worry, about Crafts's tendency to lecture over students' heads, would not pose a problem, he assured Rogers, because Crafts knew this was something he had to fix. Why Rogers did not carry through with the plan is unclear, possibly because Crafts was on his way to Ithaca or because of Warren's warning that Crafts might not connect well with beginning students.

Some took Crafts for the perennial bachelor, wedded to his work, but on June 13, 1868, just before taking up his Cornell appointment, he married Clémence Haggerty, daughter of Ogden and Elizabeth (Kneeland) Haggerty, of New York. The couple moved to Ithaca that fall. They would have four daughters. Anna Kneeland Crafts (born 1869) would marry Boston realtor Russell Sturgis Codman, and Marianne Mason Crafts (born 1871) would marry Gordon Knox Bell of New York. The younger two—Elizabeth Sedgwick Crafts (born 1879) and Clémence Crafts (born 1880)—remained single and resided permanently with their parents.

Crafts was Cornell's first professor of chemistry and dean of science. With chemical texts either unavailable, scattered, or written in German and French—languages that most American students did not handle well, if they knew any at all—he put together a textbook, a little work, as he self-deprecatingly called it, for the use of beginners in chemistry. Entitled *A Short Course in Qualitative Analysis, with the New Notation*, it appeared in 1869. Crafts shared the philosophy that had inspired Eliot and Storer to produce a text of their own, *Compendious Manual of Qualitative Chemical Analysis*, the year before. Both works aimed to reduce reliance on lectures and recitations, to increase hands-on work, to spur the experimental spirit, and to guide observation in a laboratory setting. Crafts's book, dedicated to his mentor Wurtz, went through at least six editions, not as many as Eliot and Storer's but enough to keep it in use for many years. Later editions were revised by Crafts's Cornell colleague, Charles Schaeffer.

Crafts's stay at Cornell was brief, just two years. In 1870, John Runkle recruited him to replace Francis Storer, who had followed his brother-in-law Charles Eliot to Harvard when Eliot became Harvard president. At Tech Crafts pushed to modernize lab facilities. In 1872, he introduced a new invention by Bunsen, filter devices that "have proved so easy to manage, so inexpensive, and of such value . . . that it is proposed to provide each desk . . . with them next year." Other apparatus came from Europe, all on Crafts's initiative: a Steinheil spectroscope from Munich, a Soleil saccharometer from Paris, and a Lingke balance from Freiberg. A Becker balance came from New York, although Christian Becker, a German immigrant in the 1850s, had learned his craft in his native Hannover. To some, especially those who had begun to think of Tech as America's answer to Europe's great polytechnics, Crafts seemed way too taken by French and German traditions. But Europe, he held, was where the great advances in chemistry were happening, and America had a long way to go to catch up. He often assigned German texts. A work by Friedrich Kekulé daunted his students, but they managed with the help of Tech's professor of modern languages, Charles Otis.

Crafts's debt to the European academic ethos showed in his emphasis on research, discovery, and publication. "It is to be hoped," he wrote in 1872, "that in the future a larger number of the students may be able to go over untrodden ground before finishing their course, and also that the laboratory may offer sufficient inducements to attract more experienced workers to make investigations there." This was a tall order for the typical Tech undergraduate, whose focus was on mastering elementary concepts useful in engineering. But to pull students to a higher level, Crafts campaigned for coursework leading to advanced degrees. In 1872, the faculty adopted his proposal for the Doctor of Science (D.S., S.D., D.Sc., or Sc.D.—they could not quite decide what initials to use), a plan well ahead of its time. Tech did not award doctorates until 1907, preferring to offer none rather than to cheapen the concept by falling short of European standards. Yale, meanwhile, had awarded its first doctorate in 1861, followed by the University of Pennsylvania in 1871, Cornell in 1872, Harvard in 1873, and Princeton in 1879. Tech wanted to avoid the ridicule thrown at Harvard first for awarding master's degrees to students who paid five dollars after staying "out of jail for five years," and secondly for handing out doctorates at a rate that led one faculty member to call Harvard "the premier American Ph.D. mill."

Crafts yearned for Europe's more lively research climate. His own original work had ground to a halt in America. Aside from his textbook, the only three publications to his name during this period—one of them coauthored with Friedel—were articles in French journals, work completed before his return from Paris. In 1874, when illness forced him to take a leave from teaching, he escaped to Paris once again. This time he stayed more than a decade, and his research output soared. He wrote dozens of papers, some coauthored with Friedel, in the *Comptes rendus* and in the *Bulletin* of the Chemical Society of Paris. In 1877, Crafts outlined with Friedel a process of synthesis that became known as the Friedel-Crafts reaction, with broad applications in organic chemistry. The French Academy awarded him the coveted Jecker prize (2,000 francs) in March 1880, and in April 1885 he was named Chevalier de la Légion d'honneur by the French government. His close ties to French science and French scientists led some to assume that, in spite of his Anglo-sounding name, he was a Frenchman born and bred.

Reluctant to lose Crafts, Tech kept him on the books as a nonresident professor. Sometimes he was listed as "absent in Europe," meant to suggest that here was a man with Institute connections working alongside the world's most illustrious scientists. Tech badly wanted him back. Then in 1880, Rogers wrote to say that patience had worn thin: "You have been absent for six years, residing abroad and occupied, most of the time in scientific work . . . and . . . not signified to the Institute any wish to resume the duties of the Chair with which your name was associated." Charles Wing, who had taught Crafts's subjects since 1874, was awarded "organic" as part of his title in order to make, in Rogers's words, his "Prof. . . . commensurate with the subjects which it actually embraced."

Rogers's letter appears to have gone astray, for two years later Crafts expressed surprise that his name had been omitted from the 1881–82 catalog (it was left out in 1880–81, too, on Rogers's instructions). Wing learned of Crafts's pique through chemistry professor William Nichols, to whom Crafts had hinted that he would like to return at some point. Crafts now insisted that all references to his name be purged from Nichols's periodic bibliography of work done by those connected with the Institute (eleven scholarly articles by Crafts in 1882 alone). Wing, conscious of Crafts's stature, urged Rogers and Walker to repair the damage. "Aside from any question of Ethics," he wrote to Rogers in January 1882, "it seemed to me of importance in view of the unsettled state of the Chemical Department, that Crafts' connections with the Institute should continue. . . . I believe that in the field of Organic Chemistry his name today outranks that of any other American Chemist, and that if the Chemical Department is to have a development corresponding to the growth of Chemical science his cooperation will greatly strengthen us." Neither Rogers nor Walker appears to have made much if any effort to bring Crafts back to Boston—perhaps they saw it as a lost cause—but Tech continued to lay claim to him as a former faculty member, as when news reached Boston in 1882 of two impressive presentations that he had made before the Chemical Society of Paris.

Crafts returned in 1888, but for several years moved back and forth between Paris and Boston. In Boston, he settled with his family at what became their long-term residence—59 Marlborough Street—in the heart of Back Bay. Walker, sensing a change of heart, lured him first as a Corporation member (elected 1890) and then, in 1892, as professor of organic chemistry. "The accession of a chemist of Professor Crafts' reputation," wrote Walker at the time, "a teacher of his experience and exceptional powers of inspiring interest and enthusiasm on the part of students, marks an era in the history of the Institute." He promised Crafts wide latitude for research and asked him to continue on the Corporation. Walker worried only that Tech, with its limited research personnel and facilities, compared to what Crafts had grown accustomed to in Paris, might not be able to hold him.

Crafts continued to press for a higher level of scholarship than many of his undergraduate students—fledgling engineers, primarily—considered useful. In January 1894, he asked instructor in physics Harry Goodwin to map out thesis topics that would require not only original chemical research, but sophisticated overlap with physics and biology as well. He motivated Goodwin and other talented junior faculty to think long, hard, and creatively about research topics and methods. Crafts used the thesis requirement as a strategy to build a cadre of scholars, and it was partly due to him that chemistry emerged as Tech's first solid research program, with Henry Talbot, Arthur Noyes, Augustus Gill, Samuel Mulliken, William Walker, F. Jewett Moore, and others carving out projects for themselves and their students. Some physics faculty, eager to cultivate a similar climate, worried about Crafts drawing energy away from them. As Charles Cross warned Goodwin: "The Chem. Dept., through Professor Crafts want a room for gas-analysis & that kind of chemical physics and I think we had better keep the electro-chemistry in our own hands."

A soft-spoken, genteel man, Crafts was known also as a purist on matters of academic integrity and rigor. Students caught cheating on exams or falsifying lab data could expect harsh consequences. "We have no use for lying or any misrepresentation," he once said, "and, more than that, our chief endeavor is to learn the difficult art of seeing and proclaiming the exact truth with no reference whatever to consequences." The phrase "somewhat severe," that Walker had once used to describe Crafts's junior colleague, William Nichols, applied equally to Crafts. But neither man had a cruel streak. Their aversive style, more mannerism than disposition, signified their way of pushing students, colleagues, everyone around them to high standards.

Crafts inspired awe but little affection. He took after his maternal grandfather, Jeremiah Mason, whose awkward, lumbering, six-foot-six frame had terrorized courtroom adversaries, even the thunderous orator Daniel Webster. Crafts was not, one student recalled, "a man to have much hold on the students." He could come across as impatient, intimidating, and dismissive. When absorbed in a problem, "he was so bent on its solution that his actions became almost gruff toward those working with him in the laboratory. At times a certain reticence gave him a superficial air of aloofness. He had the peculiar mannerism of suddenly wrinkling his face, making his glasses drop from his nose." But while distant and occasionally caustic, he was always fair, objective, and polite—an odd blend, captured by one student in 1899:

> He is a scholar, and a ripe and good one;
> Exceeding wise, fair-spoken, and persuading;
> Lofty and sour to them that love him not;
> But to those men that seek him, sweet as summer.

The popular media joked about his desultory speeches—"of a serious and scholarly tendency," in implied contrast with Walker, the caged lion, who conveyed more visceral energy. Crafts, in turn, had little respect for the press. In 1897 he poked fun at journalists for writing up silly inventions—a machine, for example, to prevent snoring in railway sleeping cars—while important developments in science went unnoticed. He was famous for brushing off reporters as know-nothings. But he was refined, even at his most cutting. Most of all, he shied away from the limelight.

Crafts never bargained for the level of publicity that the faculty chairmanship brought him, even though it came, initially, from a not-unexpected source. Charles Eliot, aware of Walker's misgivings about Harvard, had stayed off the topic of Tech–Harvard cooperation, consolidation, or merger all through Walker's presidency. With Walker gone, however, and Tech rudderless, a number of new, Harvard-inspired feelers went out as the dust settled on Walker's grave. On January 23, 1897, a Harvard literature professor, Barrett Wendell, rose to speak before the Beacon Society of Boston. Wendell, famous for his lively discourses mixing Shakespeare, Homer, and Dante with local politics and gossip, touched on Tech–Harvard relations. "We have here in New England two institutions," the *Boston Advertiser* paraphrased him, "each striving to develop the best qualities of its students but . . . instead of being allies, as they should be, are really rivals in certain branches." A "waste of energy," Wendell said. "No institutions could surpass, and few equal, the Institute of Technology in its line, and even Harvard . . . has to admit this superiority. Recognizing this fact, and also . . . that Harvard is unapproachable in its wider field . . . combined they would do better work than either could alone." Wendell urged "harmony of action" to replace "mutual distrust."

Whether or not he put Wendell up to this, Eliot was more than ready to revisit the issue. An official overture reached Tech in mid-April, just as a new presidential search was getting under way. Harvard, "being convinced that the great public interest of professional scientific education would be promoted by an alliance between the Schools of applied science in Boston and Cambridge respectfully invites the government of the Massachusetts Institute of Technology to consider whether some plan cannot now be devised for the accomplishment of such an alliance. To bring it about the President and Fellows will cordially enter into any practicable arrangement not inconsistent with their legal obligations." The offer went not to Crafts, as faculty chair, but to Tech executive committee member Augustus Lowell.

Crafts took charge as discussions moved forward. Unlike Walker, he supported the idea of Tech–Harvard cooperation in principle, although what precise form this should take, he was not sure. The two sides met for the first time on April 27. Tech was represented by Crafts, treasurer George Wigglesworth, and Corporation members Augustus Lowell, Howard Carson ('69), and Thomas Lothrop. Harvard brought Henry Walcott and Francis Lowell to the table. Eliot assumed that this overlap of Harvard men—Carson was the only representative on Tech's side without a Harvard connection of some sort, either as an alumnus or board member—would not only smooth discussions but promote a result favorable to Harvard. The Tech group, however, led by Crafts, drove a hard bargain, insisting that while it would gladly work out a range of cooperative strategies, even some form of alliance, any talk of union was out of the question and autonomy was nonnegotiable.

Contrary to what Eliot expected, Crafts's loyalty to his alma mater Lawrence and his devotion to pure over applied science played little if any part in shaping his attitude toward the merger. Crafts refused to concede the point argued by Eliot, Nathaniel Shaler, and others at Harvard that independent technical schools could never thrive except within the orbit of a larger university. The reverse, he held, was often the case: "It is quite true that technical education is successfully carried on by universities, but the success, as at Yale, has sometimes been little due to the sympathy of the collegiate government." A good argument could be made that schools like Sheffield and Lawrence had prospered despite, not because of, their affiliation with universities. Crafts argued further that "divergent habits of thought between academical & technical faculties and their governments have retarded the growth of schools of applied science under the university system, notwithstanding that the men in a university are bound together by old traditions, and by a fine esprit de corps." Harvard's merger proposal, seen in this light, presented Tech with as much handicap as opportunity, and Crafts urged caution.

Eliot pressed forward anyway and, by early August, the Harvard side proposed a joint school to be known as "the Massachusetts Institute of Technology and Lawrence Scientific School." The Tech side agreed, again in principle, that alliance, association, consolidation, cooperation—whatever it might be called, short of merger—was possible, even desirable, primarily to eliminate duplication of coursework and, in Crafts's words, to "diminish undue rivalry." But Crafts and company objected to a number of items: board membership crossovers, shared faculty, joint salary scales, and, most of all, to the name "Massachusetts Institute of Technology and Lawrence Scientific School" defined as "the department of applied science of Harvard University." Harvard agreed, then, to eliminate all reference to Lawrence, to accept "Massachusetts Institute of Technology" alone, and to consider the consolidated school not merely "the school of applied science in connection with said University," but "the only school of applied science connected with said University."

Crafts led Tech's negotiating team as faculty chair, no president yet in place but with several candidates under review. The first offer went to Elihu Thomson, mainstay of General Electric's research program, part-time teacher at Tech (Walker had him appointed, in 1894, as special lecturer on recent developments in applied electricity), and a powerful advocate of basic research. He declined, choosing the quiet of his laboratory over the irritations, intrigues, and busywork of administration, just as he had refused to take on executive responsibilities at GE. Also in the mix was Walker's old friend from Yale and Johns Hopkins, Daniel Gilman. A firm believer in science for its own sake, Gilman had resigned as president of the University of California when state legislators tried to transform it into something that resembled, in his view, a glorified agricultural school. Then, as Hopkins's first president, he had modeled the university after the German system with doctoral-level studies, research, scholarship, and publication demands of a very high order. Augustus Lowell approached him on Tech's behalf in July. Gilman, still president of Hopkins, fiercely loyal to that institution, and, at 66 years old, in the twilight of his career, also declined. There was talk, too, of Roger Wolcott, governor of Massachusetts, coming on board at the end of his current term. "No, I am not the man," Wolcott told the press in August, "but I have heard a prominent gentleman named for the place. I am not at liberty to state his name." Wolcott likely meant Gilman, already in the no column himself. Other prospects included Worcester Polytechnic's president Thomas Mendenhall, even an insider candidate like Tech biology professor William Sedgwick.

When none of these worked out, Tech turned to Crafts as the default option. Here was a man not quite of Walker's mettle, but with the backbone, now proven in negotiations with Harvard, to keep Eliot at bay, to guard Tech's autonomy, to protect its identity and vital interests. "Prof Crafts was not entirely willing to accept the position principally on account of the condition of his health," declared the *Boston Globe* on October 21, 1897, the day after his election by the Corporation, "but his mild objection has now been overruled and . . . he will accept the position." Health, however, had little if anything to do with it. Aside from Crafts's main worry—did he really want to give up organic chemistry?—the achievements of his predecessors, Rogers and Walker, left him feeling insecure. How, he asked himself, could he live up to their record much less make a mark of his own? As he stepped into the presidency—"this perilous seat," he called it—he took some comfort in the knowledge that his predecessors had put in place such fine systems that everything would work out well as long as he kept the administrative machinery moving in the same general direction.

Self-doubt aside, Crafts continued to hold his ground with Eliot through the fall of 1897, and a loose arrangement more along lines of association than consolidation was agreed to by both sides in November. First as faculty chair, then as Institute president, Crafts showed a level of political astuteness that Eliot never expected from this lab-immersed fellow chemist whose scientific reputation far exceeded his own. "The chief thing," Crafts warned Augustus Lowell, recommending that Eliot be kept off any board appointed to oversee the associated schools, "is what seems likely to be the ultimatum of Harvard concerning share in our government. . . . The more I think of it, the more I think, that if we exclude Eliot, such a man will do us more good than harm [and] would be more likely to carry back our notions to the University than give us theirs." Eliot, meanwhile, began to wonder about

Harvard losing the upper hand. With Crafts having marshaled Tech to a position of negotiating strength, Eliot considered pulling out of the deal altogether. He asked to see Crafts urgently in December.

The two met on Christmas Eve to hash out various details. Tech and Harvard would not give joint degrees, for example, or issue a common catalog. But Eliot, on Christmas Day, hinted that it was time to call off the formal agreement. "Our Corporation," he wrote to Crafts, "contemplated a 'consolidation,' as appears in their first resolution on the subject. That issue being apparently impossible, it seems to me that you and I can personally effect all the avoidance of duplications and the accentuation of differences which are practicable. The proposed form of 'association' does not offer any important advantages in my judgment over an avowed relation of good-will and mutual support. . . . The memorandum as a whole seems to me . . . almost completely ineffectual. I am very glad that you and I had that frank talk yesterday. It assured me that you and I can bring about a happier relation between our two institutions than had heretofore existed." A frustrated Eliot revived negotiations one week later, hoping to bring parts of the original, Harvard-weighted proposal back into play. But Crafts insisted that this ground had been covered and that a return to square one made no sense. A polite note of closure, drafted by Crafts and signed by him and other committee members, went to Eliot on January 10, 1898.

At Tech, the outcome was met with gratitude for Crafts's success at fending off a menace to the independence of the Institute while preserving cordial relations with its powerful neighbor. "The union," wrote physics professor Silas Holman, "seems not to be analogous to a business partnership, but rather to a marriage. Better no wedding than one in which the possibility of divorce must be weighed in advance. But as there is to be no wedding, let us hope for a fruitful celibacy." One Tech alumnus warned against further negotiations, calling this round "the first step of the avaricious desire of Harvard to extend its resources." Some in the media, however, took Tech to task for making unreasonable demands. "The Technology boys," opined the *Globe*, "would not oppose having Harvard absorbed by the Institute of Technology, but they do not seem to like the idea of having Harvard absorb the Tech."

Few remarkable changes occurred during Crafts's brief two-and-a-half-year stint as president. At the outset he stressed that while he would do his best to carry on Walker's policies, he was no Walker. "No radical changes have occurred . . ."; "No changes have been made . . ."; "No new work has been undertaken . . ."; "The work has progressed along the usual lines"—these were typical preambles in Crafts's annual reports, understated next to Walker's histrionic style. Tech's practice of allowing faculty leeway in self-governance made the job, Crafts said, "as little burdensome as possible." He relied on faculty secretary Harry Tyler to handle administrative details other than finance and faculty appointments. Without Tyler, noted for his "unerring memory, tireless energy, and outstanding ability, who . . . did work enough for four men with tact and wisdom," Crafts could well have found himself lost in a sea of paperwork.

The student population remained stable, overall, showing a small decline from 1,198 in Walker's last year (1896–97) to 1,178 in Crafts's last (1899–1900). As under Walker, about three-fifths hailed from the New England area (nine-tenths of these from Massachusetts), with a scattered inflow from other regions—typically, 35–38 states were represented—and a small overseas contingent. Nationals of a dozen foreign countries came to study at Tech during Crafts's time: Canada, Cuba, Denmark, Dutch Guiana, England, France, Germany, Jamaica, Japan, Mexico, Russia, and Turkey.

The teaching staff grew slightly, 153 (1896–97) to 172 (1899–1900), almost all of the increase accounted for in the lower ranks (instructors, assistants) and special-lecturer category, with the professorial group remaining flat. In 1899–1900, there were 24 professors, 9 associate professors, and 21 assistant professors. Crafts pushed through several promotions, but made few hires from the outside. One noteworthy recruit was Swiss scholar Adolph Rambeau, who came from Johns Hopkins in 1899 to replace Alphonse van Daell (deceased) as professor of modern languages. Key promoted faculty included Arthur Noyes (chemistry), Harry Goodwin (physics), and Frank Laws (physics) in 1897; Heinrich Hofman (metallurgy), Henry Talbot (chemistry), Dana Bartlett (mathematics), and Henry Pearson (English) in 1898;

Jerome Sondericker, Allyne Merrill, and Edward Miller (mechanical engineering) in 1899. A few of Crafts's appointees to the lower ranks were destined for promotions later on: Charles Spofford and George Hosmer (civil engineering) in 1897; Arthur Blanchard and Miles Sherrill (chemistry), William Coolidge (physics), and Charles-Edward Winslow (biology) in 1899. In 1898, Alice Loring (architecture) joined Ellen Richards as the only other woman on the instructing staff. All except for Loring were Tech graduates. Unlike Walker, Crafts actively recruited from within and never allowed either himself or the Institute to be put on the defensive about inbreeding. A couple of significant losses resulted from retirements and departures. Physics professor Silas Holman withdrew in 1897 on account of poor health, and long-term special lecturer on metallurgy, Henry Howe (son of civil-rights activist Julia Ward Howe), left in 1897 for a professorship at Columbia University.

No new courses appeared. The same 13 that had ended Walker's term concluded Crafts's, the only difference being that a few new options—heating and ventilation, architectural engineering, and landscape architecture—were added: the first to Course II (mechanical engineering), the second and third to Course IV (architecture). Most students graduated in mechanical engineering, followed by electrical engineering, civil engineering, architecture, chemistry, mining engineering, chemical engineering, naval architecture, general course, physics, sanitary engineering, biology, and geology (in that order). No new first degrees were offered. The number of graduate students remained constant, around 70 to 80 each year. Very few of these, less than ten at any one time, worked toward higher degrees. Four master's degrees were awarded in 1897 (two in architecture, one in physics, one in chemical engineering); 5 in 1898 (one in mechanical engineering, one in architecture, one in physics, two in chemical engineering); and 3 in 1899 (one in architecture, one in chemistry, one in biology). Tech's first master's degrees in chemical engineering (George Bixby, 1897), mechanical engineering (Albert Smyser, 1898), and biology (Charles-Edward Winslow, 1899) were awarded on Crafts's watch. Special students, those not registered for a degree course, made up 25–30 percent of the student body, a proportion that Walker thought optimal. Also following Walker's lead, Crafts took note of women students as a group. He counted 69 in 1897 (16 regulars, 53 specials), 47 in 1898 (17 regulars, 30 specials), 53 in 1899 (14 regulars, 31 specials), still with none venturing into an engineering discipline.

One small innovation—required summer reading for students heading into their second year—came out of stepped-up efforts, begun under Walker, to reinforce general studies, to expose students to nonprofessional subject matter, to broaden the program without losing professional focus. Crafts outlined the goal this way: "On account of the very large portion of the student's time which in a technical school is necessarily occupied by his purely professional studies, it is difficult to introduce into the course as large an amount of instruction in literature, history, and general science as would be desirable. And it is believed that the plan of required summer reading is the only feasible one by which the difficulty can be overcome. In the preparation of the list of books, great care was used to select only those which are readable and attractive in character, as well as instructive, in order not to make the requirement an onerous one; for it was thought even more important to awaken interest in reading than to impart information." Crafts hoped, too, that such an exercise would diminish a trend, observed by faculty and employers, of students falling into "slovenly and incorrect habits of expression, so that their later written work is inexcusably bad." But Tech's primary mission remained unchanged. "It is our business," Crafts emphasized, "to test the capacity of young men on the practical as well as the theoretical side, and to make our diploma as far as possible a guarantee of fitness to enter immediately on a career of professional usefulness." Tech could never join Eliot's influential campaign for free electives—broad subject choices, regardless of major—and expect to maintain professional benchmarks.

These values were widely shared by students and faculty alike. Most students viewed Tech's inherent difference from the traditional college as a mark of distinction. "A college education does not mean quite the same to a Tech man that it does to the average college man," beamed one class day speaker in June 1899. "A college education may be a mere adjunct to a social future—a sort of boutonniere added to the dress of a gentleman. . . . But such an education is not to be had at Tech. Tech does not have in her gift flowers; she has seeds. A man must sow and reap his harvest, the mother always by his side to direct, but never to do the work for him. Men go to Tech not to have their alma mater seal them gentlemen, but to make them workers." To underscore the gravity of the occasion, Tech's commencement ceremony remained spare if not grim—"shorn of decorations," the *Globe* wrote in June 1899, an air of "absolute simplicity" and "businesslike plainness" in the style preferred by William Rogers, the edge taken off only by Mrs. Crafts's warm, cheery greeting to graduates at the reception afterward. In 1899, a Dartmouth College professor—F. Parker Emery, who had taught English and history at Tech from 1887 to 1893—concluded that Tech students worked twice as hard as Dartmouth men, and that the Institute operated on the "law of the survival of the fittest." He intended the comparison as a tribute to Tech's high standards.

In one respect, Crafts inherited a vastly different environment than the one Walker found when he arrived at Tech in the early 1880s. The late 1890s saw benefactors, new and old, give more generously than before, so Crafts had fewer financial worries. The John and Belinda Randall Charities Corporation donated $50,000 in 1897 and $20,000 more in 1899. According to Crafts, the intervening year, 1898, was a veritable annus mirabilis—a miracle year, with "embarrassing evidences of prosperity"—as Tech's coffers swelled by more than a million dollars. Thc largest bequest ($750,000) came from former Corporation member Henry Pierce, who had died in 1896, with substantial contributions also from the estates of Edward Austin ($400,000), Julia Huntington James ($140,500), Ann Dickinson ($40,000), John Foster ($10,000), John Carter ($6,250), and Willard Perkins ($6,000), along with a gift from George Gardner ($20,000) as a scholarship fund. In 1899, Augustus Lowell gave $50,000 for a trust—the Teachers' Fund—to benefit Tech's instructing staff in case of illness, retirement, disability, and survivors' needs. Lowell's generosity struck a special chord with Crafts, grateful for his loyalty to Tech despite his strong Harvard ties as an alumnus, trustee, and father of Harvard government professor A. Lawrence Lowell, who would succeed Charles Eliot as president in 1909. "Mr. Lowell's admirable letter," Crafts told Tech treasurer George Wigglesworth in July 1899, "is an additional proof of the large minded, judicious interest that he takes in the Institute. . . . a gentleman [who does] a large hearted thing with simplicity & good taste."

Crafts's spirits rose as Tech's balance books showed more black than red. He began his presidency skeptical, cautious, downbeat. In December 1897, he referenced the quandaries—space, for one—that had haunted his predecessors, and that now confronted him. "It will be long before we can leave behind us the anxieties of such problems." But within a year his tone had shifted. The Pierce Building, funded by Henry Pierce's generous bequest, opened in 1898 next to Engineering B, the former architecture building, at Trinity Place. This increased Tech's floor space by about 25 percent, a much-relieved Crafts noted in December 1898. It was all very modern—fireproof, steel beams, finest equipment for the labs in biology and industrial chemistry, incandescent lighting for the architectural and mechanical drawing rooms. Two-thirds of the basement stored mechanical's heavy machinery; the remaining third went for a new cafeteria, Ellen King's establishment having moved out of the Rogers Building to make more room for mining engineering. Biology and geology were also transferred from Rogers, with biology occupying most of Pierce's third floor and geology sharing space on the second floor with the Margaret Cheney Room, moved over from the Walker Building. Architecture took the fourth floor, industrial chemistry (chemical engineering) the fifth floor; general lecture and recitation rooms were on the first floor. Rogers, no longer such a tight squeeze, now had space "to meet the want, long felt, of a suitable library for general purposes, and a reading room which can be kept open at all times for the use of students." Up to this point the growing book, journal, and report collections had been scattered between various laboratories and professors' offices.

As Crafts's spirits rose, so did his engagement with the Tech community. He drew energy not only from the Institute's new wealth, but also from its standing as a Boston landmark. "Farther up Boylston st the great square structures of the Massachusetts Institute of Technology loom up to the right," trumpeted the *Boston Globe* in November 1897, pointing out-of-towners to must-see city sights. Tourists paused to gawk, while others came with more serious purposes in mind. Scientific associations liked to hold conferences at Tech, attracted both by what went on there and by its location, at the heart of a busy, thriving hub, convenient to hotels, restaurants, historic landmarks. The American Public Health Association and American Institute of Architects had chosen Tech for their conventions in the 1870s, while the American Dental Association, Massachusetts Medical Society, American Association of Mechanical Engineers, American Economic Association, and American Historical Society gathered there in the 1880s. In August 1898, no less than three groups—the American Association for the Advancement of Science, American Chemical Society, and Society for the Promotion of Engineering Education—met at Tech separately, the AAAS to commemorate its fiftieth anniversary. The American Statistical Association and American Institute of Electrical Engineers both held conventions there in 1899.

Crafts's administrative inexperience cost Tech in just one instance that mattered. In 1898, during the Spanish-American War, the Institute had an opportunity to consolidate and expand its innovative naval architecture program begun under Walker. A rough tally taken by Crafts that year showed a number of alumni (59) and current students (16) on active duty with the U.S. Navy in the Caribbean and Pacific theaters, many on engineering and ordnance assignments. Secretary of the Navy John Long, aware of Tech's expertise in this area, asked Captain F. W. Dickens to assess the Institute's ability to train U.S. Naval Academy graduates in building warships. Dickens went to Boston in September 1898. He interviewed naval architecture professor Cecil Peabody and would have liked to talk with Crafts, but Crafts was unavailable. In his report to Long, which recommended against Tech and in favor of reinforcing the Naval Academy's own program, Dickens implied that Crafts's absence from the discussions may have hurt Tech's chances. Long accepted Dickens's recommendation, despite a protest letter from Crafts, and Tech lost out. The Navy soon reconsidered, and an advanced course in naval construction—under warship design expert William Hovgaard, of the Danish Navy—was established at the Institute in 1901. But Crafts was thought to have dropped the ball, which drew unfavorable comparisons between himself and the ever-alert Walker.

Few faulted Crafts, however, on his sensitivity to student life outside the classroom. In this, he took his cue from Walker even though their styles differed. Crafts's reaction to the near-hooliganism of the annual freshman-sophomore cane rush was muted next to Walker's. ("Lost a tooth," one student grumbled, "and most of the clothes I had on.")

While Walker threw himself physically into the fray—quelling chaos by force—Crafts stood on the sidelines in quiet, clear disapproval, then graciously invited the ringleaders to meet him in his office. As one freshman recalled:

> Never shall we forget the day when Rogers ran with gore,
> And the Sophomores celebrated by destroying furniture.
> It was a stirring sight indeed to see that furious fray;
> The President and Faculty were deeply moved they say,
> And anyone upon the floor could very plainly see
> That President Crafts was greatly moved by the charge of the Freshman V.

Crafts, like Walker, hoped to channel this youthful energy in more productive ways. Dozens of extracurricular activities that had come to life in Walker's time were sustained under Crafts, with some new ones added. For sports fans, the Hockey Club and Bowling Club both emerged in 1899. Crafts warned that Tech could not seriously compete in athletics—"sports can never," he said, "receive the same attention in this Institute that they have . . . in the more leisurely college life"—but that opportunities for healthy, restful, pleasant exercise should be encouraged. An advisory council on athletics met for the first time on January 1, 1898. A Chess Club and Camera Club were mentioned in 1899, as was Die Gesellschaft, an all-German conversation group with two membership categories: *ehren Mitgleider* (honorary, mostly faculty) and *wirkliche Mitgleider* (regular, students only). The Mining Engineering Society was formed on November 30, 1897, a new Naval Architectural Society in March 1900. In 1899, the Chauncy Hall Club—Chauncy Hall, like Andover and Exeter, was a feeder prep school for Tech—joined the growing list of groups whose members shared backgrounds and relationships that predated their Institute careers. No new fraternities appeared, but existing ones thrived with Crafts's qualified blessing. "Students dwelling together have established clubs which have nothing classic about them except their Greek-letter names, and . . . when not too elaborate they form pleasant associations of young men with tastes and occupations in common . . . small centers in which good fellowship and attachment to their institution have active sway." In Crafts's view, fraternities also promised "to lessen the temptations incident upon residence in a large city." The YMCA unit, meanwhile, published a handbook with helpful hints on rooms-for-rent, affordable supplies, even coursework. The Society of Arts, Tech's public-outreach arm, continued to host sessions where students and others heard distinguished scientists and engineers discuss inventions, research results, and ideas in progress.

For all his exposure to Parisian culture, Crafts was very much a proper, puritan Bostonian. At the time, groups like the Watch and Ward Society fought hard to protect Boston's moral standards. The Public Library kept a locked, limited-access room for housing objectionable books such as Walt Whitman's *Leaves of Grass*. Show business also came under close scrutiny, hence the famous phrase "banned in Boston." But theater at Tech thrived because Crafts, conservative though he was, shared little of the Watch and Ward's extremist outlook. The Tech Show, a student musical review first staged in May 1899 to raise money for the Institute's athletic association, managed to escape the self-appointed guardians of community values. Later it evolved into light or comic opera and then, after World War I, into musical comedy with ballet thrown in as a show of solidarity with wartime ally France. But at first it was firmly minstrel, with an all-white, all-male cast in blackface and drag. Even without a good script or catchy musical numbers, the producers could count on howls of laughter from the audience whenever the leading lady or chorus girls appeared on stage with flat chests, hairy legs peeping out from under dainty petticoats, and voices in the baritone or bass ranges. The humor was always crude. Harvard's equivalent, the Hasty Pudding Theatricals, appealed to more refined tastes, with vulgarity offset by social satire and literary parody.

The Technology Club, founded in Walker's time, gained new vitality under Crafts. While Walker had intended it as a place for faculty and upperclassmen to meet informally, it served as a rallying ground for alumni groups beginning in the late 1890s. Plans for an alumni magazine gathered momentum, meanwhile, with *Technology Review* appearing for the first time on January 1, 1899. As regional alumni groups formed in increasing numbers—New York City in 1895, for example, followed by Philadelphia and the Connecticut Valley (both in 1896), Pittsburgh and Western New York (both in 1898), and Washington, D.C. (1899)—the Technology Club fostered a spirit of institutional loyalty that Crafts considered essential.

On February 3, 1899, the club tied in telephonically to banquet halls filled with Tech alumni in seven cities: Chicago, Milwaukee, New York, Philadelphia, Washington, Pittsburgh, and St. Louis. The link to Chicago worked fine, to other cities not so well. Crafts, a better sport than the technophobic Walker (who never ventured near a telephone, if he could help it), seized a transmitter as alumni gathered around receivers. "Aesop's fable," he said, "tells us how once upon a time the crane invited the fox to dinner and put the food in a vessel with a very long and narrow neck, so that the fox could not get a taste of it. It seems to me that you have gone a great many steps farther, inviting us to dinner with a thousand miles . . . between us. Let me suggest that next time you should at least give us a selenium plate, if not a dinner plate, so that we can see you eat by telegraph and enjoy your company in that way." Crafts made three attempts at the first sentence. When no one could hear him, he passed his notes to someone else to deliver. The event, technically a mixed success, turned into an emotional triumph, galvanizing alumni support for the new century ahead.

Crafts was soon ready to move on. On October 24, 1899, he surprised everyone by resigning effective the end of the academic year. The executive committee urged him to reconsider, but he stood firm. Oliver Wendell Holmes's famous couplet, about the cares of academic administration, kept ringing in his ears:

> Uneasy lies the head that's born to rule,
> And most of all whose kingdom is a school.

Funds were up, but with Tech constantly outgrowing its space—and with few options for expansion downtown, even after 55,000 more square feet were purchased at Trinity Place in April 1900—talk of a new home was already in play, somewhere outside the city, maybe Newton Highlands or Roxbury or Cambridge, at a cost of $2,000,000 or more. Meanwhile, the six-year state grant negotiated by Walker in 1895 would soon expire. Crafts, on edge about this, appealed to state legislators in May 1900 for "an emergency measure to put things on a sound footing." He was guardedly optimistic that Tech would have no need to tap into the public treasury much longer, a prediction based on his sense of alumni about to reach an age and levels of success where they could be "expected to help very largely."

But the cycle of worry began to wear on him. He was only 61 years old, with plenty of productive research energy left, just no time for it. Since his return from Paris, all he had to show was a handful of articles, two coauthored with Charles Friedel, in *Comptes rendus* and *Bulletin* of the Chemical Society of Paris. The only other piece—"A lecture upon acetylene," printed in *Science*, March 13, 1896—was a transcript of remarks he had made before the Society of Arts earlier that year. "My reasons for taking this step at this time," Crafts wrote in his letter of resignation, "are founded upon my desire to return to purely scientific occupations. My term of office has shown me the wide field of educational problems, both within and outside the Institute, which should be studied; and I have found that such studies and the performance of administrative duties, although not in themselves burdensome, leave little freedom for the pursuit of experimental science. A choice must be made between administrative and scientific occupations, and it is the latter which I choose. . . . All these ties are severed with great reluctance, to return to a field which aroused my early enthusiasm and which still claims my most active interest."

Crafts longed for the flexibility and security that European academic scientists enjoyed. "I hope that in years, perhaps far distant," he told an alumni gathering on December 29, 1899, "you will witness a change in the direction of our studies and a progress toward those which are generally considered in Europe as belonging to the highest forms of university education. Much that we teach in our universities would be considered by European critics as belonging to preparatory schools. What distinguishes above all things the highest education supported by public funds in European states is the freedom of the position made for the professor. A man who has invented a new science or a new philosophy has a chair created for him. He becomes a pensioner of the state for life, with no onerous obligations to teach, and with the confident assurance that his own way of bringing his knowledge into use will be the best way." Crafts's goal, now, was to simulate this lifestyle.

For reasons that remain unclear, Tech did not ask him to stay on as a Corporation or faculty member. Some years after he died, one of his daughters signaled that the family was "still much aggrieved at what they believe to be ill-treatment of their father after his resignation from the Presidency. He expected, of course, to be re-elected a member of the Corporation, or of the Faculty, or of both; but no such vote was taken, and during his remaining years he was, in large measure, ignored." Yet following his resignation, Crafts was more keen to rent lab space as a private researcher than to take on teaching or administrative duties. "As to the matter of a room for laboratory purposes," he advised treasurer George Wigglesworth in March 1900, "my intention was to offer the Institute a sum based upon the floor space occupied. If for instance 500 sq ft it would be 1/100 part of the H. L. Pierce Building which has I believe on all its stories somewhat more than 50,000 ft and cost with its land less than $250,000. On these terms the Institute could build a corresponding room for any purpose when the next building is put up. As my occupancy will not be for a great many years and after that my laboratory would fall again into the general use again, the Institute would be the gainer." Tech gave him the space refurbished, well equipped, free of charge. "It has been a source of pleasure," his successor as president, Henry Pritchett, would observe in December 1900, "that he still remains with us as an adviser and friend, and continues to pursue his chemical researches in a private laboratory in the Walker Building." For Crafts, the arrangement came with advantages besides freedom and flexibility: a familiar workplace within easy reach of his home on Marlborough Street, as well as proximity to colleagues—especially Arthur Noyes and others in chemistry—whose work he found stimulating. He felt warmly enough about Tech to donate two thousand dollars for general purposes in 1901.

Crafts rigged up a private lab, too, at his summer home, known as Griffin Croft, in picturesque Ridgefield, Connecticut, near the Berkshire foothills. For the next decade and a half, he traveled back and forth to Boston. Much of the time he spent at his workbench. His retirement proved less productive than he hoped, partly because of failing health. He never fully recovered from a severe bout of neuritis, which laid him low in 1911. Still, a half dozen substantial papers emerged from his labs between 1901 and 1915. One, on catalysis in concentrated solutions, appeared in the *Journal of the American Chemical Society* (1901). The rest—mostly on thermometry and vapor densities—shifted from organic toward physical chemistry and were published in Crafts's preferred French journals: the *Comptes rendus*, *Bulletin* of the Chemical Society of Paris, and *Journal de chimie physique*. After Mrs. Crafts died in 1912, their two single daughters—Elizabeth and Clémence—kept house for him until his death in Ridgefield, on June 20, 1917.

"Into touch with the world at large"

Henry Smith Pritchett, 1857–1939

Tech wanted a resonant, charismatic figure to replace James Crafts, the quiet, withdrawn scholar. Word went out that there was no hurry, that the Institute would bide its time, but the rumor mill churned just the same. Outgoing Massachusetts governor Roger Wolcott removed himself from consideration early—"the duties," he said, "would be too confining, and the position would not be congenial . . . in other respects." Columbia University president Seth Low's name also came up. Low's experience in public life as mayor of Brooklyn, before he went to Columbia, and his skill in orchestrating the university's move from midtown Manhattan to Morningside Heights, were viewed as highly relevant with Tech facing a possible move of its own. Also among the prospects were three celebrated naval officers—William Sampson, John Long, and Alfred Mahan—a consequence of Tech's growing interest in naval architecture after the Spanish-American War. But Sampson, hero of the Battle of Santiago de Cuba and commandant of the Boston Navy Yard, preferred the Navy and Long and Mahan also balked. "It hardly seems to be the proper thing," said Long, "to decline in advance what has not been offered you. . . . Besides my tastes do not lie in that direction." The only internal Tech candidate, William Sedgwick, wanted to remain a faculty member, content with his research and teaching in bacteriology, industrial biology, and sanitary engineering.

On March 15, 1900, the Corporation chose Henry Pritchett—an ideal candidate, they felt, combining in one person the scientific brilliance of Crafts with the administrative energy of Francis Walker. Pritchett, an astronomer, had transformed the politically troubled U.S. Coast and Geodetic Survey into a model of efficiency since taking over as superintendent in 1897. He had shown vision, too, an inclination to think deeply about science and technology. Some Corporation members recalled a talk he gave at Worcester Polytechnic Institute's commencement in 1898. The topic, "The astronomer as engineer," held no particular resonance for Tech, which had as good as abandoned astronomy to Harvard when Edward Pickering and, later, his brother William went over to the Harvard College Observatory. But Pritchett addressed larger problems in the pure and applied sciences. American engineers had begun to lose touch with science, he warned, and this had put their creative, innovative drive at risk. Besides intellectual breadth, Pritchett brought the outsider's perspective. He was a Midwesterner born and bred. "It is a good omen for the future of this Institute," said one Corporation member, citing the dangers of inbreeding in an increasingly competitive, interdependent world, "that its new president comes from the great valley of the Mississippi."

Henry Pritchett's ancestry traced back to English-Welsh settlers who arrived in Virginia in the early eighteenth century. His grandfather Henry Pritchett pushed west in the mid-1830s, taking with him slaves as well as family and settling near Fayette, Missouri. His father, Carr Waller Pritchett, was a sometime farmer, Methodist minister, teacher, and astronomer; his mother, Betty Susan (Smith), also had ties to Virginia. Carr Pritchett's academic interests led him to Harvard in 1858 for a year of study under astronomer Asaph Hall. While there, he almost certainly came in contact with John Runkle, Simon Newcomb, and others active with the ephemeris work overseen by Benjamin Peirce. Back in Fayette by the end of 1859, Pritchett taught at Central College until the Civil War forced its temporary closure. Missouri, a border state, saw friends and families deeply divided. Carr Pritchett joined the antislavery camp and spent the war years mostly in Washington, D.C., working for the U.S. Sanitary Commission, while one of his brothers fought on the Confederate side. After the war, the

Pritchetts moved to Glasgow, Missouri, not far from Fayette, where Carr opened a school—the Pritchett Institute—and founded an astronomical observatory with funds from local tobacco heiress Bernice Morrison. The institute ran by a strict moral and religious code, but it also offered a rigorous, expansive curriculum that included music, modern languages, and science at the secondary and post-secondary levels, even awarding bachelor's degrees.

Born in Fayette on April 16, 1857, Henry Pritchett grew up in Glasgow and attended his father's school for nine years. After graduating in 1875 he headed east, as his father had done, to study under Asaph Hall, who had since moved from Harvard to the Naval Observatory in Washington. There Pritchett met Simon Newcomb and other distinguished astronomers, some of whom, like Hall and Newcomb, were also Harvard transplants. He worked two years as one of Hall's assistant astronomers, then one year as astronomer at the Morrison Observatory back in Glasgow. In 1881, he joined the faculty of Washington University, St. Louis, where he helped build an astronomy department along the observational-descriptive lines laid by Peirce, Newcomb, and Hall. Pritchett's professional writings during this period included a report on the total solar eclipse of July 29, 1878. This was followed by a study of Jupiter's rotation, in *Proceedings of the American Association for the Advancement of Science*, 1881, and papers on the satellites of Mars and Saturn, Mars's diameter, comets, the transit of Mercury, longitudes, double stars, and the solar corona of 1889 in the *American Naturalist*, *Publications of the Astronomical Society of the Pacific*, and other periodicals. As astronomer to the transit-of-Venus expedition in 1882–83, he carried out orbital calculations in New Zealand, Australia, India, China, and Japan.

While Pritchett stayed abreast of developments in his field, his interests were broad, eclectic, and never quite predictable. He read widely in philosophy and became an advocate of ethical culture, the movement founded by transcendentalist Felix Adler in 1876 to promote meaningful living outside organized religion. Pritchett grew close to scholars in fields other than his own and to professionals outside the world of academics: attorneys, judges, physicians, businessmen, and politicians. He wrote on nonastronomical subjects. In a paper appearing in *Publications of the American Statistical Association* in 1891, he proposed a predictive formula for population growth and migration that would have caught the attention of Francis Walker, who as director of two U.S. censuses thought long and hard about the application of mathematical models to population trends.

Pritchett was sociable, versatile, and flexible. Science and technology excited him even when the subject matter was unfamiliar. A year out of college, in the summer of 1876, he wandered the grounds of the Centennial Exposition in Philadelphia, marveling at the creativity and inventiveness on display. "An American," he wrote, "cannot but feel proud of American machinery. . . . Passing through this wonderful exhibit of human ingenuity and seeing hundreds of machines acting with perfect precision and almost human intelligence, the thought arises: how far will human invention go?" Philadelphia was Pritchett's introduction to Tech. He had never been to Boston, but almost certainly he looked in on the Institute's award-winning exhibits. Four years later, while in Boston to give a paper before the American Association for the Advancement of Science, he attended a reception for delegates hosted by William Rogers at the Hotel Vendome. "I went as a young beginner in science to see a man long since grown famous."

A generalist at heart, Pritchett was committed enough to his own specialty that in 1894 he headed to Europe for doctoral studies. But he set out as much to learn about foreign lands and peoples as to seek new knowledge, or to acquire a credential. Pritchett, in his mid-thirties and already a widower (his wife Ida had died in 1891, shortly after the birth of their fourth child, also named Ida), swept from place to place with two small sons, Harry and Edwin (Ted), in tow; the younger children, Leonard and Ida, had stayed home in the care of Pritchett's sister, Sara. The threesome began in Germany and moved through Holland, Switzerland, Austria, France, Hungary, Greece, Bulgaria, Serbia, and Turkey. Then it was on to the University of Munich, where Pritchett studied under the distinguished astronomer, Hugo von Seeliger. Seeliger's specialty was Saturn—he had confirmed James Clerk Maxwell's theories on the composition of Saturn's rings—and Pritchett worked on Saturn, too. Each night, long after he tucked his sons into bed, he would sit up scribbling in English, then German.

Pritchett earned his Ph.D. in a few months. "I have just come through the *Examen Rigorosum* with flying colors," he wrote to his father in May 1895, "and in fact surprised myself by carrying off a *summa cum laude*, which is the highest honor a man can get in a German university. In consequence I am filled, as the old Negro said, 'Wid' pride and vanity.' It does not mean so much in America, but over here a *summa cum* only comes about once in ten years, and to a foreigner very, very seldom and has never before been obtained after one semester's work."

Aside from its intellectual rigor, German academic life impressed him for its rich social traditions. He grew attached to *Kommers*, the custom of professors, docents (lecturers), and students gathering after-hours in favorite cafés and beer houses to talk academics and politics while swilling beer and occasionally breaking into song. *Kommers* could be formal, requiring membership as in a fraternity. Pritchett joined such a club, a mathematics *Verein*, which met from time to time at a local drinking establishment. At the outset of each session participants would discuss problems they had read about or worked on. They would then spend hours generating *Gemütlichkeit*—a warm, friendly ambience—through song, talk, drink, and ad-lib speechifying. A marvelous tradition, Pritchett observed, "in many ways good for the young fellows." Once he visited the *Kommers* of a history *Verein* in Berlin, where the great classicist Theodor Mommsen, his enormous head of straggly white hair and fierce black eyes set behind tiny, thick spectacles, inspired the gathering—an experience similar, Pritchett told his father, to "what the Methodists call a means of grace."

After this, Pritchett found it hard to imagine himself back at Washington University teaching freshmen. He stayed in St. Louis just two more years. In 1897, U.S. treasury secretary Lyman Gage recruited him to superintend the U.S. Coast and Geodetic Survey. Earlier superintendents had been scientists with few political instincts—Alexander Bache, 1843–67; Benjamin Peirce, 1867–74; Thomas Mendenhall, 1889–94—or politicians like William Ward Duffield, Pritchett's immediate predecessor, who understood little about science. Pritchett, on the other hand, balanced science and politics with remarkable skill. A Democrat in a Republican administration, he bridged party lines seamlessly, helped create the U.S. Bureau of Standards, and oversaw the production of high-quality maps and charts during the naval crisis provoked by the Spanish-American War. A trusted confidant to President William McKinley, he joined a negotiating team that settled longstanding differences over the Alaska-Canada boundary. His was a familiar, congenial presence at the Cosmos Club and other venues on the city's hectic social circuit. Washington always felt to him like a delightful, hospitable town, unlike Francis Walker, who loathed its culture of sycophancy, patronage, and corruption. Besides President McKinley, Pritchett grew close to political stalwarts like vice president Theodore Roosevelt, Gage and his assistant Frank Vanderlip, interior secretary Ethan Hitchcock, and House member Joseph (Uncle Joe) Cannon, known for his colorful, pugnacious style.

No matter which circle he wandered into, Pritchett was admired for his charm as well as for his scientific and administrative know-how. Rarely did he quarrel and he always kept his temper in check, a resolution made as a child after witnessing a brutal argument. It took a lot to rattle him. Only in rare instances, as when betrayed by an ally, did his confrontational side emerge. He relied on fellow scientists to act sensibly, and when they did not he told them so directly. "A bureau officer who undertakes to administer the affairs of government work," he wrote to one Harvard professor who had carped about a staff reshuffling at the Coast Survey in 1898, "finds himself at once under the fire of politicians, and this he expects, but it is a little disheartening to be compelled to defend one's self from the attacks of professors in institutions like Harvard, when one is simply trying to carry out honestly the plan of efficient administration which might naturally be expected to command the support of all scientific men."

Such a high-profile choice came with one drawback: he would have much to tie up in Washington before he could get started at Tech. But the Institute was used to such delays; Walker, after all, had taken a year and a half to reach Boston. Pritchett's acceptance letter to Institute treasurer George Wigglesworth, dated April 2, 1900, expressed mild relief—not the flight instinct Walker had conveyed—to leave Washington for the relative calm of Boston. "After three years of struggle with Congress (pleasant enough in many ways) I am sure you cannot appreciate how pleasant it is to find such a spirit among the men of your Executive Committee. I look forward with the keenest anticipations to my acquaintance—I hope my friendship—with you, and whatsoever of devotion I can command shall go into the work." Before heading to Boston, he helped ease the transition for his successor Otto Tittman, nephew of Julius Hilgard, whom William Rogers had helped persuade President Garfield to appoint as Coast Survey superintendent in 1881.

Washington was sorry to see him go. McKinley begged him to stay, and when Pritchett declined, sent him off with a challenge. "I hope that some way will be found to teach the young men in our schools a better estimate of the dignity and honor of serving one's country well, that they might understand they are factors in the country's upbuilding and must learn to take upon themselves its responsibilities." Pritchett, like McKinley, saw public service as a worthy goal of American higher education. So when he took the stage for his inauguration as Tech president, on October 24, 1900, he laid out an expansive vision. "Is education to have for its object the training of the intellect," he asked, "or is it to aim at the development of character, or is it to undertake both objects?" Still, he acknowledged Tech's primary mission—the education of engineers—and vowed no dramatic changes. "I come to you with no new message and as the herald of no new gospel." He addressed the student body as if they were all engineers. The next quarter century, he said, belonged primarily to the well-trained engineer. "May I hope that in your preparation you may bear in mind as your ideal of an engineer, not only one who works in steel and brick and timber, but one who by the quality of his manliness works also in the hearts of men; one who is great enough to appreciate his duty to his profession, but likewise, and in a deeper sense, his duty to a common country and to a common civilization." A humanized professional, in other words; technical skills, alone, were not enough.

The inauguration—Tech's first—marked a radical departure in style, a desire to jettison assumptions about the Institute's proper sphere. Simple, spare dignity went by the boards in favor of grandeur, opulence, and showy display. Previous incumbents, in line with Rogers's preference for minimum fuss, had simply shown up for work—no ceremony, no speechmaking, maybe a few words of introduction to the faculty behind closed doors. But Tech, seeking to recast its image for the twentieth century, hoped to project an aura less austere, more extroverted.

By early afternoon, Boston's just-built Symphony Hall filled with students, faculty, and invited guests. A processional followed, led by Corporation and faculty marshals. Behind them walked Pritchett, trailed by former Tech presidents Runkle and Crafts, U.S. senator Henry Cabot Lodge, the president of the state Senate and speaker of the House, Massachusetts governor Winthrop Crane, Boston mayor Thomas Hart, chief justice Oliver Wendell Holmes Jr., U.S. circuit judge Francis Cabot Lowell, colleagues of Pritchett's from Washington (Vanderlip, Newcomb, Tittman), and representatives of colleges and universities nationwide—Charles Eliot from Harvard, Nicholas Murray Butler from Columbia, and a score more. These dignitaries covered the stage, several rows deep. Crafts spoke on the faculty's behalf, predicting for Pritchett a smooth path compared to the tortuous byways of Washington, and "precious opportunities for companionship and friendship." Lodge underscored Pritchett's commitment to science and public service. "He will not forget that the little world he guides and rules is part of the greater world of the United States, borne on the mighty current of the national life as the tides of ocean bear the ship, and that he who serves the country best, in training her sons, best serves the noble institution committed to his care." A reception followed at the Rogers Building. It was relatively simple, organized around a library table with tea poured by faculty wives. That evening a thousand or more students gathered on Exeter and Blagden Streets, then marched with torchlights to Pritchett's residence at 337 Marlborough Street. A band played "Hail to the chief" as Pritchett emerged onto his front steps, beaming, along with his wife and dinner guests.

It was an auspicious start. Pritchett bonded with students, and they with him. Crafts had been stodgy, physically imposing, aloof, gruff, and businesslike, while Pritchett projected youthful energy, sociability, and warmth. Both were cultivated and well traveled—men of the world, so to speak—but polar opposites otherwise. "I wish to be your friend and to be with you in every way," Pritchett introduced himself to the student body. "I believe in the open door policy, and with the door of my office always open, I shall be glad to see you whenever you will, for talk, consultation or advice. I ask for your friendship, as I hope to give you mine." The offer was unrealistic—he would have been overwhelmed had even a tiny portion of the student body taken him up on it—but it signaled something about what he wanted to build at Tech, a culture with softer edges. Students who packed Huntington Hall that afternoon cheered him loud and long.

The Pritchetts were the first presidential family to remind students of what their own home lives looked like, or for some, what they wished for in a home life. Pritchett was 43 years old when he took office, about the age of the typical Tech student's father; Crafts, at 60, had been more like a grandfather, but with little hint of grandfatherly affection. Pritchett's new wife, Eva, eagerly took to the role of surrogate mother. When Ernst Henne ('02)—"a slight flower among a patch of lusty weeds," according to one classmate—died of typhoid fever during the winter of 1900–1, her attentiveness was much commented on. The memorial, led by a dour, unsympathetic Lutheran minister, took place one miserable, sleety day in undertaker's rooms next to a burlesque theater downtown. Mrs. Pritchett sloshed over, sat through the service, comforted the grieving parents, then sloshed back with the few Tech students who had shown up to pay their respects.

The Pritchetts, newlyweds when they reached Boston, had become engaged just a few days after his appointment as Tech president was announced. Mrs. Pritchett, daughter of San Francisco lawyer Hall McAllister, was also sister-in-law of Francis Newlands, the Nevada congressman best remembered for authoring the 1898 resolution annexing the Republic of Hawaii as a U.S. territory. The four children from Pritchett's first marriage—Harry, Ted, Leonard, and Ida—lived with their father and stepmother in Boston, joined in due course by their half-sister, Edith, born in 1901. Pritchett, according to his eldest son Harry, "possessed that rare quality that caused us to feel great devotion for him." Tech students felt much the same way, drawn also by his occasional boyishness. Frederick Hunter ('02) happened to be in the president's office one day when a snowball fight raged outside. "Mr.

Hunter," said Mrs. Pritchett, "I should think that you would enjoy being out there," with her husband interjecting: "Why, I would enjoy being out there myself." Mrs. Pritchett, always on the lookout for ways to connect with students, organized a series of receptions twice a week, Tuesdays and Saturdays, 5–6:30 p.m., in the palatial, multiroomed mansion at 147 Bay State Road where the president's family moved within a year or two of their arrival in Boston. The residence (now home to Boston University's Elie Wiesel Center for Judaic Studies) was lent to the Pritchetts by Corporation member Charles Weld, for whom it was built in 1900. Mrs. Pritchett hosted her first open house there on November 14, 1903, drawing an eager crowd of youngsters in search of homemade refreshments, a glowing hearth, the comforts of family life.

When tragedy struck early on, the Tech family drew closer. In the thick of the twentieth annual cane rush, held on November 15, 1900, eighteen-year-old Hugh Moore ('04) ended up dead—a victim of suffocation—at the bottom of a pile of several hundred wriggling freshmen and sophomores. Pritchett, a few weeks into his fledgling presidency, took prompt action not only to halt these violence-filled skirmishes once and for all, but also to reshape Tech's extracurricular life. He began by proposing "a rational system of physical culture" in which students would be required to exercise under careful supervision. Football, he argued, ought to be abolished. "Only such competitive sports should be encouraged as may secure out-door exercise, and particularly those, such as track athletics, which afford opportunity to the individual, rather than those which depend on expensive and time-consuming team work. In general, all such sports should be conducted in a way not only consistent with good scholarship, but subordinate to it." A mass meeting called in October 1901 to discuss the future of football at Tech drew 231 students, who voted narrowly—119 to 112—to do away with intercollegiate competition; intramural games were allowed to stay. Pritchett appointed an advisory council on athletics with Frank Briggs ('81) as chairman. Franklin White ('90) was hired as medical adviser and lecturer on personal hygiene. Field Day, first held on November 19, 1901, replaced cane rush with a tug-of-war and other controlled tests of strength. Fencing and golf joined the roster of approved sports. A rifle club was formed in 1906.

These changes signaled a concerted effort on Pritchett's part to enrich the nonacademic lives of Tech students, who were far more likely to fail, he believed, "by reason of lack of ability to deal with men than by lack of technical knowledge." Pritchett boosted interest in fraternities, professional societies, musical and dramatic clubs, and other groups. A new fraternity, Psi Alpha Kappa, appeared in 1901, followed by chapters of Theta Chi and Phi Sigma Kappa in 1902, then Phi Kappa Sigma and Alpha Epsilon in 1903, Delta Sigma Phi in 1905, and Lambda Phi in 1906. Frats already in existence began to buy, or rent, Back Bay townhouses for use as residences. In 1901, Phi Beta Epsilon purchased and moved into 237 Beacon Street, Francis Walker's old house. Most frats stayed in the neighborhood. Chi Phi, Delta Kappa Epsilon, Delta Upsilon, and Sigma Alpha Epsilon were all on Newbury Street, Phi Gamma Delta and Delta Tau Delta on Marlborough Street. Alpha Epsilon was on Gloucester Street, Phi Sigma Kappa on Hereford Street. Sigma Chi set up house in Brookline, while the brothers of Delta Psi resided at Louisburg Square, on Beacon Hill.

Some students worried about the growing sway held by fraternities—"Down with fraternities at Tech! Abolish clannishness!" read one flier from 1901—but frat and nonfrat types settled into peaceful coexistence, for the most part. While 18 percent of Tech students were fraternity members in 1900, by 1906 that proportion had jumped to 26 percent. A Wisconsin Club joined the roster of regional social groups in 1903. A Missouri Club, with native-son Pritchett as honorary president, appeared in 1904 along with a Pennsylvania Club, followed a year or two later by an Ohio Club, a California Club, and a Texas Club. A Civic Club grew active around 1905, organizing mock parliamentary debates. In 1905, students from current and former British colonies, including Canada, formed the British Empire Association of Technology. Mexican students formed El Circulo Mexicano by 1907. The Mechanic Arts High School Club and Brookline High School Club were formed in 1905 and 1906, respectively, to promote Tech secondary-school ties. An all-purpose discussion group, Osiris, first mentioned in 1903, counted Pritchett among its patrons. A new student newspaper, *The Institute*, appeared in October 1904 and a reconstituted Naval Architectural Society on November 13, 1901. The Chemical Society, organized in 1903, restricted membership to students majoring in chemistry, chemical engineering, and physics. A Catholic Club appeared in 1905 to serve the fellowship needs of Tech's Roman Catholic students and "to prevent them from drifting from the ideas nurtured in their homes."

Pritchett also laid plans in 1901 to accommodate about 175 students in a six-story apartment building, to be called Technology Chambers, at Irvington and Saint Botolph Streets, just off Huntington Avenue. The building, while not owned by Tech, was to have been collaboratively managed with its owners and modeled on the English university house system, a place where students would live, eat, study, and develop a sense of community under the watchful eye of assigned tutors. Students moved in by the fall of 1902, but the college atmosphere never quite materialized. Overall residential patterns remained diffuse, scattered, and unpredictable. In 1904, roughly 40 percent of students lived at home (with parents, typically), 30 percent in Boston boardinghouses or bachelor apartments, 10 percent in suburban boardinghouses, and 20 percent in fraternities. A residence known as Tech House, at 138 Eustis Street, accommodated eight students in 1905.

In Pritchett's view, the lack of a cohesive campus did not mean that Tech students had to forgo a meaningful social life. The German system impressed him as a model in this respect. In urban areas like Munich and Berlin, the setup was similar to Tech's: a group of buildings served core academic functions, with no residential quarters except official ones for administrators and key faculty. Students bunked where they could, each according to his means. There were drawbacks to this lifestyle, but advantages as well—exposure, for example, to a broader range of events, settings, and personalities than cloistered campuses could offer, plus social and academic values fostered through the *Kommers* and *Verein* traditions. Pritchett aimed to replicate this atmosphere at Tech. The Technology Club moved in 1901 to larger, better furnished quarters on Newbury Street. The Technology Union, formally opened on December 6, 1902, occupied space once used by the Lowell design school above the mechanical shops on Garrison Street. The Club and Union turned into community social centers. "In both of these," wrote Pritchett, "the idea is constantly developed that the student is asked into a life of moral and intellectual liberty, but a life which carries with it at the same time certain responsibilities, and always the responsibility to be a gentleman."

Pritchett's approach, particularly its libertine European side, did not always sit well with conservative, puritan Bostonians. On December 13, 1901, a group of students gathered for a smoker at the Tech Union. Smokers were like *Kommers*, but less frequent, less animated, and less intellectually stimulating. Pritchett, hoping to inject a bit of German-style vitality, organized for beer to be served. Local clergy got wind of this through an embellished newspaper account. A hundred students, the report said, gathered in a Tech classroom and drank to their hearts' content—"every one present had all the beer and tobacco he desired [and] there is no doubt that this style of assembly will hereafter be a feature of Technology life." Pritchett found himself on the defensive: "I regret that I have been put forward as an advocate of a beer cultus." While he would never force alcohol on anyone, he argued that there were "right and proper" places where moderate alcohol consumption promoted "temperate and clean living." The campus smoker, he suggested, was one such place: a controlled environment under faculty supervision, beer consumed in limited quantities (56 pints for 97 men on this occasion, about one small stein each). Better this, he suggested, than uncontrolled carousing at hotels or restaurants off-campus. A slippery slope, the clergymen replied, denouncing Tech for bringing "the German drinking habit among its pupils . . . decadence in sentiment and practice." The *Boston Globe* cattily advised Pritchett to invite Britain's visiting Prince Henry in "as he passes by and offer him a glass of beer."

Pritchett did not abandon the practice, even as he realized that it would have to be watered down for local consumption. *Kommers* stayed, albeit a drier *Kommers*. Alcohol, while never expressly forbidden, was no longer a central feature. Pritchett encouraged Frederic Field Bullard ('87) to compile a book of Tech songs not just for use at smokers but to promote all-around school spirit, camaraderie, community values. The first edition of Bullard's *Tech Songs: The M.I.T. Kommers Book* appeared in June 1903. A number of the entries—rollicking, occasionally crude—referenced drinking and smoking. Bullard's own "A stein song" captured what had so appealed to Pritchett as a student in Germany:

For it's always fair weather
When good fellows get together
With a stein on the table and a good song ringing clear.

"Bingo (A marching or street song)" began with the lyric, "Here's to M.I.T. drink it down, drink it down." "On Rogers steps," by Thomas Estabrook ('05), fondly referenced a pastime that some Bostonians viewed as a nuisance—Tech students hanging out for hours in front of the Boylston Street entrance, "troubles smok'd away, with merry comrades near." One student, asked his opinion of *Kommers* in 1907, said it was "one of the few things that stand between me and insanity."

Events at the Tech Union often closed with students gathered around the piano singing from the *Kommers* book. A nostalgic Pritchett sometimes joined in. Students, grateful for his efforts, saw this year, 1903, as a watershed in Tech's social life. "A change . . . has taken place," editors of the yearbook *Technique* observed,

> in the character, in the mode of living, to the point of view of the average Tech undergraduate. . . . Tech is known pretty generally throughout the country as a place of grind; and in the sense that it is a place of hard and steady work, it is a grind. But its reputation is even worse than this. A great many people imagine that when a man enters Tech, with the intention of graduating, he must form a resolution to abandon all idea of sociability and amusement; that he must read over the grim doorways of the Institute buildings: "Abandon ye all the human nature ye ever had, who enter here," "Become a steam engine as soon as possible, or get out." It is toward the elimination of this feeling that the change . . . is showing itself. . . . The grind is gradually coming out of his shell and discovering that good scholarship and good fellowship are perfectly compatible. He is learning that the faculty of making friends will be worth far more to him in future life than a little extra knowledge in the theory of least squares.

Meanwhile, Pritchett organized periodic convocations—mass meetings, as the students called them—where invited guests shared rousing, motivational words of wisdom. Author, clergyman, and newspaperman Edward Everett Hale spoke in 1901 on "rules of life for young men to follow": accept the universe as it is, live in the open air, rub elbows with everyone, create avenues for dialog with powerbrokers. Hale painted the typical Tech man as "an apostle . . . whose duty it is to bring civilization into the wilderness, and who is by means of his engineering knowledge, to help in building up the vast unknown regions of the world." Lucius Tuttle, Corporation member and president of the Boston & Maine Railroad, urged students in 1902 to "read good books and be manly men."

Pritchett aimed for a more cosmopolitan student body. While aggregate numbers fell off somewhat during his presidency—from a high of 1,608 in 1902–3 to a low of 1,397 in 1906–7—the range and diversity widened, geographically. The proportion of students from Massachusetts dropped five to ten percent from Crafts's and Walker's time, when it had hovered consistently at around 60 percent. Pritchett saw this as a sign of diminished provincialism. Numbers increased from all regions of the country but one, the southeast, with western states, California and Colorado in particular, showing the largest growth. Overseas numbers were dramatic, too, more than doubling between the time Pritchett took office and his departure. In 1900, 33 foreign students entered from 11 countries, while in 1906 76 entered from 26 countries. In response to some who worried that this trend risked pushing Tech away from its roots, Pritchett argued that expansion—never constriction—promoted local, regional, national, and global interests: "The attendance of students from abroad is the best barometer we have of our own alertness and our own fitness." He was particularly proud of the sharp rise in Chinese students (2 in 1903, 8 in 1904). The *Globe* in October 1904 printed a feature article—"Six Chinamen now at Tech"—based on an interview with these students, two of whom were called "bright, intelligent looking boys, rather above the height of the usual Chinaman . . . fitting for mining engineers." Two additional Chinese students entered the following semester.

The number of women students, however, declined. A moderate increase between 1900 and 1902, from 44 to 63, was followed by a precipitous drop-off: 20-odd registered between 1903 and 1905, then only 13 in 1906. The environment, never overly hospitable to women, grew less so during this period. At Pritchett's Symphony Hall inauguration women students had been segregated from their male classmates, pushed off to a block of seats on the edge of the auditorium, while wives, mothers, and sisters sat in the balconies. To build solidarity, the Association of the Women of the Institute organized and met on December 29, 1900. In 1906, when the first woman (Ida Ryan, in architecture) earned a master's degree at Tech, Ellen Richards carried out a survey under the association's auspices to capture "what some Technology women have done." A few alumnae outlined their accomplishments as research assistants, research fellows, high-school teachers, college professors, and published authors. Others reported going the way of most women: marrying, managing a household, nurturing a family. "What knowledge of Chemistry I obtained" at Tech, wrote one respondent, "has been put to the most practical use possible—that of housekeeping." "My time has been occupied in rearing our children and keeping house," wrote another. Alice Tyler ('84), wife of mathematics professor Harry Tyler, wondered if those, like her, whose lives revolved around homemaking could be encouraged to share photographs of their sons and daughters, as "illustrating our work. . . . To those of us who are not brilliant stars . . . it will be a comfort to look at our children's pictures—for they may become distinguished if we have not." The tone of these responses ranged from upbeat to regretful.

At Tech, women straddled the social and professional margins. The Technology Club allowed female students and alumnae in, along with wives and girlfriends, only on ladies' night, once a month or so. Some courses were essentially off-limits to women, who gravitated toward architecture, chemistry, geology, general studies, physics, and biology. Naval architecture acquired one female student in 1900 and, in 1904, there was one in mechanical engineering, but women steered clear of engineering subjects. The adventurous two—Lydia Weld (naval architecture) and Lahvesia Packwood (mechanical engineering)—raised a few eyebrows among male classmates, who in 1902 suggested that if women, "having taken up the technical courses," were to go in for engineering they ought to do so in "domestic engineering or household science." Required subjects might include "sweepology," "principles of horseless perambulators, or mind the baby," "hash analysis," "washineering," "kindergartnering," and "theory of periodic dusting."

This was all in good fun, with just a hint of irritation about women edging too close for comfort to male–female boundary lines. But it did not help that the Tech woman's primary role model, Ellen Richards, was herself marginalized as an instructor without faculty status, that her field of expertise—home economics—reinforced a common female stereotype, and that she opposed women's suffrage ("You know my sentiments," someone overheard her say in 1904, "when women have pockets, then it's time for them to vote"). And when Pritchett, in a 1907 commencement speech at Simmons College, suggested that only unreasonable women complained about unequal pay for equal work, then who knew where to turn. "Do not be made unhappy by arbitrary distinctions in the financial rewards that come to you, but accept the world frankly and graciously. Do not quarrel with the order of things you find in the business world or get into a self-conscious spirit that is a minute form of selfishness. Take the hard knocks as they come and lead a womanly life in whatever work you do." How seriously such comments affected Tech's own women students is anyone's guess, but the fact remains that their number declined by more than 70 percent during Pritchett's administration.

His attitudes on race trended slightly more progressive. Pritchett never mentioned black students as a group, in the way he referred to Chinese students, but Tech admitted at least a half-dozen between 1900 and 1907. Among them were Daniel Smith Jr., Wendell Terrell, Eldridge Baker, Dallas Brown Jr., and siblings Henry Turner Jr. and Marie Turner. Four chose engineering disciplines, either Course II (mechanical) or Course VI (electrical), while Henry Turner opted for Course XIII (naval architecture) and Marie Turner, Tech's first known black woman student, selected Course IV (architecture). Smith, Terrell, and Brown earned bachelor's degrees, Baker and both Turners left without graduating.

While female students outnumbered blacks at Tech, in some respects black male students were better integrated. Their choice of courses was wider, freer, more flexible. Terrell and Brown joined the Mechanical Engineering Society while Lahvesia Packwood did (could?) not. When Smith graduated in 1903, Tech helped him with job placement. "Wants to get a position in connection with power transmission or electric light company. Has had some practice in putting up telephones and wiring houses." But racial issues cropped up just the same. Baker was expelled after faculty complained about him as "obstinate . . . bound to have the last word . . . uncomfortable boy to get along with. He seems to have a chip on his shoulder" (this last phrase common code for "uppity negro"). In September 1902, tempers flared when Southern members of a national engineers association meeting at Tech demanded a whites-only policy for membership. William Ripley, professor of sociology and economics, brought his racial views into the classroom. Under Ripley's influence, J. Bradford Laws ('01) wrote his senior thesis on a group of blacks in Louisiana, drawing conclusions that mirrored age-old racial stereotypes: "As a race . . . they are not bold, nor yet cunning enough to be successful thieves. . . . Very few of them appear capable of deep emotion; sorrow over the dead dies with the sun; resentment passes with the night; gratitude and local attachment they know nothing of. Yet they are often faithful servants, and in advancing years seemingly much attached to those whom they have served. . . . The young children seem bright, but progress ceases at an early age. They appear to have little intellectual and moral capacity. . . . They are grossly animal in their sex relations, both in and out of their families."

Pritchett's perspective on race and ethnicity proved broader than that championed by Ripley, his mentor Francis Walker, and their disciples, all of whom assumed Anglo-Saxon biology, culture, and values as inherently superior. Pritchett grew close to several Jewish intellectuals, the brothers Simon and Abraham Flexner among them, and later, in the early 1930s, excoriated Hitler for his treatment of Jews, and for fomenting anti-Semitic hatred in Germany, long before talk of holocaust or final solution filtered into the public consciousness. In 1904, Pritchett urged President Roosevelt to hold fast to "the full spirit of Lincoln" and to halt the Republican Party's slide into racial hostility and indifference toward the plight of American blacks. The eminent black leader Booker T. Washington spoke at Tech more than once, on Pritchett's invitation, to inform the community about efforts underway to educate Southern black youth in practical trades at Washington's school, Tuskegee Institute in Alabama. The first speech that he gave at Tech, in the spring of 1901, inspired students to contribute to a Tuskegee scholarship fund. Pritchett called a holiday in December

1905 so that students and faculty could hear another plea by Washington: "The deeper down an individual or a race may be, the more forsaken or unpopular it happens to be, or the more difficult the task of elevating it, the more it should challenge our deepest interest and highest courage."

Pritchett's interest in educating blacks, like Washington's, was confined mostly within the racially segregated guidelines laid down by the U.S. Supreme Court's separate-but-equal decision (Plessy *v.* Ferguson) in 1896. Both men saw Tech as a feeder school that could supply black graduates, albeit in small numbers, to help build programs in practical trades at Southern black schools. Tech's first known black graduate, Robert Taylor ('92), had become one of Washington's close associates at Tuskegee. Another black student, William Smith ('00), planned to go there—"the great incentive of his life arose from his meeting Booker T. Washington"—but died before graduating. Daniel Smith built Tuskegee's first electrical plant, while Wendell Terrell joined the faculty in mechanic arts at another black school, Prairie View State Normal and Industrial College in Texas.

Opportunities for postgraduate work grew under Pritchett. The total number of advanced students—those with first degrees under their belts—more than doubled between 1901 and 1906 (98 and 200, respectively). Many were Tech graduates who stayed on for further study, but a substantial group also came from Harvard, the U.S. Naval Academy (these drawn almost exclusively to the naval architecture program), Princeton, Yale, and elsewhere. While only five graduate students were candidates for advanced degrees in 1900, that number soared to 32 in 1906. Over 70 master's degrees were awarded between 1901 and 1907, about three times as many as the total awarded between 1886 and 1901. Most were in architecture (32), followed by naval architecture (24), mechanical engineering (6), chemistry (5), electrical engineering (2), sanitary engineering (2), and physics (1). Donald Belcher earned the first master's in sanitary engineering in 1903. The first three master's in naval architecture (construction option) went to William Ferguson Jr., William McEntee, and John Spilman in 1904, and the first in naval architecture (regular) to Donald Battles in 1906. In 1904, eight candidates—all in chemistry—were accepted for doctoral studies, the first Ph.D.'s going to Raymond Haskell, Robert Sosman, and Morris Stewart in 1907.

Three programs—the Research Laboratory of Physical Chemistry, the School for Engineering Research, and the Sanitary Research Laboratory and Sewage Experiment Station—set the stage for an increasing emphasis on research. In 1901, Pritchett, Arthur Noyes ('86), and George Ellery Hale ('90) discussed ways to boost Tech's profile in this respect. Noyes, already on Tech's chemistry faculty, had suggested that Pritchett help him create a research division—"an audacious proposition," he conceded, but one that he was certain would be worth the risk. Hale, an astronomer, soon joined in. He was then at the University of Chicago's Yerkes Observatory, which he had founded in 1897, but he said he would move to Tech if conditions were right. The plan called for a loose coalition of efforts, as the fields that Noyes and Hale worked in were so different. But talks went on hold when Hale, who doubted the Institute would sustain its commitment, learned of a new agency, the Carnegie Institution of Washington, founded by steel magnate and philanthropist Andrew Carnegie in 1902 to support such programs. Hale turned to Carnegie on Yerkes's behalf—it "seemed almost too good to be true," he later recalled—and left the Tech plan dangling.

Pritchett and Noyes went ahead anyway. On February 10, 1903, Tech's executive committee approved establishment of the Research Laboratory of Physical Chemistry, which opened in the fall with a staff of eight research associates and assistants under the direction of Noyes, Willis Whitney, and Harry Goodwin. The lab was funded by grants from Hale's father William (a wealthy elevator manufacturer whose contracts had included lifts for the Eiffel Tower), from the Carnegie Institution, and from Noyes himself. Whitney left after a year to found General Electric's research laboratory in Schenectady, New York, replaced by William Coolidge who came on board as assistant professor of physicochemical research. Coolidge soon followed Whitney to GE. But Noyes continued to build a remarkable group that included William Bray, Gilbert Lewis, Richard Tolman, Miles Sherrill, Edward Washburn, and Charles Kraus, along with Yogoro Kato of the

University of Kyoto and Wilhelm Böttger of the University of Leipzig. Wilhelm Ostwald, soon to receive the Nobel prize in chemistry, came from Leipzig as a guest lecturer in 1905. The lab had a strong interdisciplinary bent, drawing on staff in both physics and chemistry.

The School for Engineering Research opened, like Noyes's outfit, in the fall of 1903. Billed as the first program of its kind in America, it was intended to offer advanced students a place to work, experiment, publish, share ideas, and build a knowledge bank for engineering-related disciplines. The doctorate in engineering (Eng.D.) would be offered, but the focus was to be on research not credentials. Pritchett modeled the school after the German polytechnics, especially the Royal Technical College at Charlottenburg. In October 1903, he returned from a tour of Germany inspired by what he found there. "The time has come," he wrote, "when the Institute must be not only a teaching body, but it must well lay the foundations for a school of investigation in the physical sciences. . . . How important is the development of the research spirit as a part of national progress we are only just beginning to realize." But Tech found few qualified candidates. Only one, Harold Smith, head of electrical engineering at Worcester Polytechnic, made the cut that first year, and the school was soon phased out.

Tech's Sanitary Research Laboratory and Sewage Experiment Station opened on April 1, 1903, with startup funds from Sarah Hughes, who preferred to remain anonymous. Hughes had begged Pritchett to mobilize efforts to address the sewage problem—"The Neponset River" which ran, she said, not far from her home in south-suburban Boston, "smells so horribly you can't sit near it." William Sedgwick, as lab director, surrounded himself with a capable group of biologists, chemists, and bacteriologists, among them Dwight Porter (consulting engineer), Henry Talbot (consulting chemist), Charles-Edward Winslow (biologist-in-charge), and Earle Phelps (research chemist and bacteriologist). The group carried out an exhaustive study of the chemical and bacteriological composition of city sewage, along with treatment experiments using sand-filtration techniques and septic-tank purifiers. The first volume of *Contributions from the Sanitary Research Laboratory and Sewage Experiment Station of the Massachusetts Institute of Technology* appeared in June 1905 and sold so fast that it went out of print within a few weeks. When Hughes asked for simple leaflets to supplement the scientific tomes full of words that she "had to fly to the dictionary to discover the meaning of," Sedgwick dutifully churned out pamphlets with titles like *Why Dirt Is Dangerous, Why Dirty Milk Is Dangerous*, and *Why Dirty Water Including Sewage Is Dangerous.*

As he pushed Tech toward high-end research, Pritchett also reinvigorated the paraprofessional framework that John Runkle had introduced in the 1870s through auxiliary programs in industrial design and practical mechanics. The Lowell Institute School for Industrial Foremen opened in 1903 as a cooperative venture between Tech and Harvard professor A. Lawrence Lowell, a member of the Tech Corporation since 1896, soon to succeed Charles Eliot as president of Harvard. Lowell, who had taken over from his late father, Augustus Lowell, as trustee of the Lowell Institute, gave funds through the family trust while Tech provided space and teaching manpower. Classes met at night to accommodate working men who, after finishing up at their day jobs, would grab a quick dinner and head over to Boylston Street for a long evening of lectures, recitations, and lab experiments under the guidance of Tech faculty members. The students, all men, came from service jobs in industry. From motormen to machinists, electrical workers to draftsmen, they ranged in age from teenagers to fifty-somethings. The teaching staff, also coming off a full day's work, got extra pay for staying on at night. Many were less interested in the income than in the notion that this was a benefit to the community and that it would help keep them abreast of changing conditions in local industry. Pritchett viewed the program as an affirmation of Tech's hands-on tradition—a sidelight, not to draw attention or energy from his research goals, but important nonetheless. Tech's chief duty was to "work upward" toward research excellence, even as it "reached down" to serve the needs of "men who stand between the unskilled worker and the engineer."

Pritchett also put his mind to the core needs of the undergraduate curriculum. The 13 courses, in place since Francis Walker's time, remained constant. Between 1901 and 1907, mechanical engineering was the most popular (342 graduates) followed by civil (251), electrical (233), mining and metallurgy (177), architecture (133), chemistry (113), naval architecture (112), chemical engineering (77), physics

(36), sanitary engineering (31), general course (18), biology (15), and geology (6). One new option in physics—electrochemistry—emerged in 1900. Also that year, municipal sanitation joined the subject list in biology. Courses in Spanish were added to the modern-language offerings, because of increased interest following the war with Spain. Electrical engineering broke ranks with physics to form a separate department in 1902, led briefly by Louis Duncan, then Harry Clifford (acting) followed by the forceful, irascible Dugald Jackson, whom Pritchett recruited from the University of Wisconsin in 1907 and whose tenure as department head would last nearly three decades. Clifford would defect to Harvard in 1909. William Hovgaard, Danish naval officer and one of the world's foremost authorities on ship design, joined the faculty in 1901 to strengthen Tech's program in naval architecture. Among the new recruits from outside were Edwin Bidwell (E. B.) Wilson for mathematics and Gilbert Lewis for physicochemical research. A new category—nonresident professor—was introduced in 1902 for the cream of the special part-time lecturers. Nonresident professors taught, at their pleasure, without stated engagements. The first two were Elihu Thomson in applied electricity and Percival Lowell in astronomy. Both men were Corporation members as well. Lowell, son of Augustus Lowell and brother of A. Lawrence Lowell, shared academic and scholarly interests with Pritchett, but Pritchett was drawn as much by the Lowell-Harvard connection as by mutual ties to astronomy. Willis Whitney, the third to hold nonresident professor status, taught electrical theory.

While attending to shifts in structure and personnel, Pritchett challenged his teaching staff to think deeply about the status quo, reform, and change. One faculty member worried that because the "spirit most generally prevalent in our present student body is for the study of something they think they are going to practice with pecuniary results," efforts ought to be renewed "to cultivate in our students a higher and broader estimate of the worth of education than the monetary value." Two potentially conflicting visions for Tech recurred: on the one hand, that "the Institute should remain an undergraduate institution offering to students fixed and rather rigid courses of undergraduate study leading to the professional work of the engineer, the chemist, the architect," and on the other that it should "preserve the undergraduate work as its great heart and center, but . . . make this the foundation for a great school of professional, graduate and research work."

Pritchett, meanwhile, pushed the Institute toward "a somewhat greater elasticity," a culture more like a college or university than a narrow technical school. He tried to persuade Phi Beta Kappa, a leading symbol of academic distinction, to establish a Tech chapter. PBK had been chartered in 1776 at the College of William and Mary to recognize excellence in liberal scholarship. But could Tech convince PBK that its students merited the label "liberally educated"? Pritchett's first application, submitted in 1902, was rejected. There was no evidence, PBK officers said, that the curriculum demanded rigorous enough work in nontechnical areas. Furthermore, Tech awarded only the bachelor of science degree while PBK member institutions all offered the bachelor of arts. In his second application, a year later, Pritchett proposed that while Tech did not—and likely would never—offer the B.A., perhaps the honor society would consider the Institute's new programs leading to the Ph.D. as liberal enough in name and scope. This was positive, came the reply, but while PBK did not consider the B.A. a hard-and-fast requirement, if any institution not offering that degree applied for membership, its path to acceptance would not be easy. Sure enough, PBK denied Tech's application a second time.

Pritchett was of two minds, anyway. His philosophy, echoing William Rogers's, was that Tech should "provide a full course of scientific studies and practical exercises for students seeking to qualify themselves for the professions" while furnishing "such a general education . . . as should form a fitting preparation for any of the departments of active life." Every scientist or engineer should be generally educated, in other words, but at Tech general education could never substitute for scientific focus and rigor. Pritchett did not subscribe to Walker's view that the Institute might reasonably offer studies leading to doctorates in modern languages and other humanistic disciplines, just as Harvard did. Tech, Pritchett held, was not like Harvard in this respect; its mission was, and should remain, science-oriented. So strong was his commitment to this premise that he yanked Walker's favorite program—Course IX, general studies—away from liberal, elective subjects toward science with a liberal component. In a clear reversal of Walker's emphasis, Course IX was reworked as general science in

1904—not economics, history, and English with a dash of science, but science supplemented by select humanistic disciplines. It was a move guaranteed to increase PBK officers' skepticism had Pritchett submitted a third application, which he did not.

Under Pritchett, Tech moved along in this hybrid, fluid way as a technical institute with goals beyond what that term often implied. Some of his changes were structural, some content-driven, others cultural. In 1902, he added three administrative posts—dean, registrar, and recorder—to handle duties thus far carried out in an ad hoc way by faculty secretary Harry Tyler, whom Pritchett asked to refocus his energies on building the mathematics department. The first dean was Alfred Burton, with Walter Humphreys as registrar and O. F. Wells as recorder. Pritchett expected Burton, one of Tech's most popular teachers (topographical engineering), to help create a more nurturing environment, one that resembled the Tech of earlier days, when the community was small and the faculty learned as it went along, experimenting with ideas and techniques and staying one step ahead of students. Over the years, as the curriculum had grown more prescriptive and standardized, particularly in engineering subjects, and as the number of students rose without a parallel increase in faculty, work in the classroom had assumed a more efficient, hardnosed template. Professors taught, students listened and followed orders. Anecdotes about aggressive teachers and cowed students were common.

First-year students took the brunt. "My! but those Profs are grumpy," a typical freshman's reaction on opening day. One recalled his ordeal—a fatal struggle, he called it—at his first algebra class, October 1901. He sat in a corner at the back. The professor barked an order that the student did not hear, but when his classmates headed toward the front, he joined the line. The professor demanded registration cards. The student had left his at home and, when he mumbled something in apology, the professor lashed out—"Home? Well, what are you up here for? Didn't you hear me say if you haven't got it with you, bring it next time?" The student slunk back to his seat, humiliated. The class proceeded. "Brown," the professor growled, "define a logarithm." When Brown hemmed and hawed, the professor waved him off and moved around the room. The new student awaited his turn in terror: "I couldn't have told my name if I had been asked."

When the class ended, he breathed a sigh of relief. "Meanwhile," he vowed, "I am keeping a wide tract of land between myself and that professor in the hopes that when he finds me . . . at the next recitation he won't recognize me."

Students rarely complained and some saw this style as another proud, distinctive contrast with Harvard, "that casino for the young across the Charles." But Pritchett began to build through his new dean a gentler, more interactive model. Burton greeted all freshmen on opening day, calming them and their parents with cheery words. He reinforced the adviser system. Those first few weeks were critical, Pritchett and Burton both believed, a time when "many [students] are meeting unaccustomed responsibilities . . . when it is of great importance—difficult though the task is—for us to secure adequate acquaintance with them." Burton boosted Tech's efforts to help students find suitable room and board, and to provide a rudimentary employer placement service. According to Pritchett, these changes helped humanize the Institute and transform it into "a live center of intellectual and moral influence."

Pritchett's vision for Tech brought renewed pressures with respect to finances and space. The Institute ran operating deficits despite substantial donations that included $50,000 from the estate of Robert Billings and $10,000 from George Gardner in 1901, $10,000 each from brothers A. Lawrence Lowell and Percival Lowell in 1902, and $10,000 each from Mrs. William Putnam, Mrs. T. J. Bowlker, and Amy Lowell (sister of A. Lawrence and Percival) in 1903. These amounts, however, were earmarked—Billings's for scholarships, the rest to erect a new building for electrical engineering—so the budget for general expenses, and for special projects that Pritchett kept adding, limped along with treasurer George Wigglesworth working hard to keep it not too far from break-even. The annual state grant of $25,000, first awarded under Walker, was renewed in 1901, but made barely a dent. Pritchett was no fan of public financial support anyway—the risks of excessive state control, political interference, and lowering standards in his view outweighed the benefits—so one of his goals was to push Tech toward reliance on dependable, private sources.

Discussions that had begun under Crafts, about Tech possibly moving out of Boston, resumed in earnest as the need for space grew critical. In spring 1902, the Institute purchased a 12-acre tract in Brookline adjacent to land donated by Samuel Cabot for the athletics program. But what purpose this property would serve was unclear. The Cabot land hosted Field Day and became known as Technology Field in 1904, while the rest awaited some decision on development. The property was too small to accommodate a self-contained campus, which Pritchett believed would require no less than thirty acres, far larger than the 70,000 square feet of undeveloped, Tech-owned land at Trinity Place. To make such a purchase the Institute would have to sell off its Boylston Street property, which could happen only if the state legislature removed restrictions on the original deed-of-gift so that Tech owned clear title. The legislature granted Tech's petition, filed in spring 1903, but this got tangled up in the courts when abutters, including Tech's former ally the Natural History Society, sued to prevent the Institute from gaining title, selling, or erecting new buildings on Boylston Street.

That summer Pritchett, convinced the courts would rule in Tech's favor, led a group of Corporation members in search of a new site. His role as chairman of the Committee on Charles River Dam, soon to recommend creation of a river basin bordered by parklands, familiarized him with historical, legal, and environmental issues surrounding land policy and development in Massachusetts. He and his group looked at properties in Jamaica Plain, Watertown, Cambridge, and possibly elsewhere. The Cambridge site, ironically, forty acres along the Charles River Parkway near the Harvard Bridge, was the one that Tech would eventually move to. But Pritchett preferred the Watertown location for its fishing pond, orchards, and distance from the noise and roar of traffic yet within easy reach of the city. Then in the fall, as the court case dragged on, he suggested going nowhere and erecting instead a five-story building between the Rogers and Walker Buildings, all three connected by passageways. This further inflamed abutters, who filed suit to stop that plan. The Tech community, meanwhile, endorsed Pritchett's efforts to resolve the space issue whether on or near the present site or at some other location, as long as Tech's unique culture remained intact.

> There's a whisper in the air, you can hear it everywhere:
> The Institute is soon to move from Copley Square.
> In the country it will rise to an unexampled size,
> With a beauty and completeness that will win a world's surprise.
>
> Now it all seems pretty nice—it seems cheap at any price—
> Yet there is a point on which we'd like a little more advice.
> The thing we have in mind is the students plug and grind—
> 'Tis that which makes Technology—it must *not* be left behind.
>
> If expansion means decay, we must turn the other way—
> 'Tis quality, not quantity, that makes this business pay.
> If it's possible to plan to preserve the real Tech man,
> We'll move faster than the earth does—but we *must* be in the van.

Tech continued to erect new buildings on Trinity Place, although with the future uncertain these came to be viewed as temporary, short-term fixes. The Lowell Laboratory of Electrical Engineering, a sprawling, one-story structure, went up in the summer of 1902 and housed not only the new electrical engineering department but part of chemistry and all of modern languages as well. Without this 40,000 square feet of additional space the Institute would have been unable to move forward on several of Pritchett's key programs. Another building went up in 1903 to house naval architecture, the physical chemistry research laboratory, and mineralogy. This structure, sometimes referred to as Engineering C, was made of brick but constructed in the simplest, cheapest form possible. A new gymnasium on Garrison Street, next to the machine shops, replaced the old one on Exeter Street in 1905. This was as temporary as buildings come, erected for just $2,000, a stop-gap as fundraising for a Walker Memorial Building moved forward. The Walker Memorial, in tribute to Francis Walker, was conceived not simply as a gymnasium but as *the* social center of student life. Originally planned for the corner of Trinity Place and Stanhope Street, its construction was put on hold pending a decision as to whether or not Tech would move out of Copley Square.

Pritchett's initiatives set some people's nerves on edge. Emma Rogers asked William Sedgwick in October 1903 for his take on all this talk of a move, expansion, legal jeopardy, of new programs outside Tech's mainstream: a recipe for confusion, she thought, likely to undermine the hard-won stability achieved under Francis Walker. Sedgwick, her closest confidant on the faculty (he had taken time from sanitary research to help compile and edit her husband's *Life and Letters*), tried to calm her even as he voiced concerns of his own. "No one knows better than I," he wrote, in reference to Pritchett, "that our eminent friend needs watching."

What troubled them most, in light of Harvard's ongoing designs on Tech, was how Pritchett and Charles Eliot had struck up a cozier-than-advisable relationship. Eliot was best kept as Walker had kept him—at a safe distance. But Eliot edged his way in early, telling Pritchett that his appointment as Tech president was "one of the pleasantest" bits of news in a long time. He warned him to ignore malcontents, mostly on Tech's side, who liked to paint the Harvard-Tech dynamic as destructive, negative, and friction-filled. Eliot, confident that the two schools could "prosper abundantly side by side in hearty cooperation," sensed an opening that might work in his (and Harvard's) favor. With Alfred Burton's shift to the deanship and with the retirement of William Niles, Tech's versatile geologist-cum-geographer, the geology program was left with William Crosby in charge and few qualified personnel. To fill the gaps, Pritchett and Eliot arranged for Tech's department to be carried on in cooperation with Harvard's. In 1902, Crosby was joined by five Harvard faculty members: Thomas Jaggar (experimental and field geology), Nathaniel Shaler (general geology), Robert Ward and Frederick Wilder (climatology), and Jay Woodworth (glaciology).

Whether Pritchett had suggested or been drawn into it, the arrangement raised a red flag for Tech loyalists. They were also bothered by the way he liked to compare Tech to Harvard, or to some other college or university, calling attention to ideas that the Institute might adopt. In 1902, he dug up Harvard's and Yale's admissions requirements from fifty years earlier to see how these institutions had changed with the times. The point was not, he said, that Tech should embrace Yale and Harvard's content wholesale, merely that it should open up to new ideas within its own context. But some did not read his thrust this way. Let Tech's unique curriculum be, they urged, leave it alone. Some saw his attempts to tinker with terminology—his request, for example, that workshops be referred to as laboratories—as out of sync with Tech's traditional culture. Workshops were what they were, students insisted, places where to learn parts of one's profession one went to get one's hands dirty, where for decades students had proudly carried out grubby functions in grubby settings. As one student wrote sarcastically in 1902: "A fellow used to walk through the mud of the alley into the shops, then jump into his overalls, and start to work like a man. Now he must promenade through the avenue into the 'laboratories,' clothe himself in protective raiment, and practice the mechanic arts. The proper dress for such an occasion is a black frock of either thibet or worsted, patent or enamel shoes, white or black bow tie, opera hat, and a John Drew hair-part. It is not considered in good taste to wear a turn-down collar. On leaving the 'laboratories' cards should be left with the janitor, one for each member of the instructing staff. Cards sent by mail, however, will not take the place of a call." None of Pritchett's sons attended Tech—Ted went to West Point in 1903, Leonard to Harvard in 1904—leaving some to wonder what this implied about their president's confidence in the value of an Institute education (after Harvard, Leonard studied electrical engineering at Tech, 1907–9, but by this time his father had moved on). In contrast, four of Francis Walker's sons were Institute graduates.

When talk of amalgamation with Harvard resurfaced, then, in January 1904, it came as no great shock. Eliot floated the idea and Pritchett grabbed at it. What surprised some, however, was the interest aroused outside the institutions. Ranged behind Eliot this time were powerful public and private concerns. Philanthropists Andrew Carnegie and Henry Lee Higginson both weighed in, urging union and promising support to bring about a greater, grander, combined institution that would benefit the larger cause of technical education. Henry Marcy, part owner of forty-plus acres on the Charles River Parkway, also backed the idea and hinted that he would help facilitate a purchase on reasonable terms. On top of this, a multimillion-dollar legacy to Harvard from shoe-machinery mogul Gordon McKay, targeted specifically for engineering education, promised substantial benefits. Eliot and Pritchett, meanwhile, used Boston media contacts to sway public sentiment. The *Advertiser* and *Globe* became ardent proalliance advocates, with the *Globe* editorializing that merger would "give to students the advantages of two great institutions at the price of one." Pritchett hired a publicity firm to spread the word on Tech's behalf. The firm prepared and distributed syndicated articles to newspapers nationwide, not only on "the Tech–Harvard merger," as the writers baldly referred to the Pritchett-Eliot plan, but also on what the Institute was up to in other respects: technical training for new professions, chemical research, ship design, Panama Canal engineering, locomotive fuels, "germ hunting as a business," and other topics. As compared with previous merger talks, there was more at stake now on both sides with respect to resources, expectations, and outcomes. Whatever decision was hashed out could reshape, possibly redefine, both institutions along with the entire landscape of twentieth-century research and education in the applied sciences.

As the process unfolded, each side thought in narrower terms, too, about what amalgamation might mean for its own interests. For Eliot, Tech held the key to invigorating Harvard's applied science programs, which had slowly improved since 1891, when Nathaniel Shaler took over as dean of the Lawrence Scientific School, but still lagged behind Tech's. Pritchett saw the plan as a one-fell-swoop solution to the chronic, twin problems—money and space—that had plagued Tech for many years and that promised to spiral out of control if left unaddressed. But the process was messier than before. Tech could move only if the courts decided the real-estate title question in its favor, money would flow only if the McKay bequest survived an expected legal challenge from heirs and others about Tech reaping benefits.

Another wild card was community support. In previous negotiations, Harvard and Tech faculty, students, and alumni had sat on the sidelines as the presidents and board members worked through the issues, holding each other to account. But this time, with so much agreement at the top, they were inclined to take more notice. The Harvard faculty, never satisfied with the state of the Lawrence School, supported the plan while Shaler and his Lawrence colleagues fought it, livid over the prospect that Tech might invade their territory and possibly take over. "Here," Shaler observed, "are two stout, trained, and seasoned wheel-horses pulling this eternal load of education at a good pace. Shall we swap them for a merger mule? 'A critter with no pride of ancestry or hope of posterity?'" Harvard students and alumni, meanwhile, did not involve themselves much one way or the other. Eliot had difficulties to contend with for sure, but Pritchett faced steeper odds: a Corporation deeply divided and a faculty, student body, and alumni association united in opposition, convinced that this latest Harvard-hatched plot required a vigorous counteroffensive with signs pointing to their own president's defection to the enemy camp. Word went out that Pritchett had maneuvered to take over as president of the combined institutions, once Eliot retired. Eliot asked William Ripley, a defector from Tech's faculty to Harvard's in 1902, "what the Institute Faculty really thinks of him." Ripley's response left Eliot wondering if Pritchett might face a popular uprising.

The arguments for alliance were familiar: mutual assistance, larger enterprise, competition eliminated, duplication done away with, resources pooled, Harvard to gain from Tech's relative strength in the applied sciences and Tech to benefit from Harvard's liberalizing influence. The counterarguments were also familiar: erosion of Tech's special culture, creeping absorption by Harvard, sacrifice of control, loss of alumni support and interest. Negotiations got under way in May 1904 with Pritchett and A. Lawrence Lowell, Harvard loyalist and a Tech Corporation member, representing Tech and for the Harvard side, Henry Walcott and Charles Francis Adams.

The first of Tech's groups to go on the warpath was the alumni association. The process, billed as a fight to the finish even by mild-mannered Harry Tyler, involved much verbal saber-rattling. "Dr. P. shows signs of pushing the Harvard plan," Tyler wrote to his wife, "and there may be bloodshed before you get this"; then later, "The civil war smolders with increasing intensity," along with other sentiments couched in equally colorful, bellicose terms. A delegation of alumni met with Pritchett as early as January 1904 to warn him that while they might support a move out of Boston, never would they "consent to a merger that would sink Tech's individuality." If money was a problem, they would work hard to match the McKay millions. To underscore this point, they staged a symbolic event at a dinner meeting of western alumni in Chicago in February. The keynote speaker, railroad magnate James Hill, arrived after dessert trailed by baggage carriers each lugging a brass-bound chest. He urged the by now well-fed alums to give generously to their alma mater, then ordered his men to empty the chests. Wads of hundred-dollar bills floated to the floor until a million dollars lay in a heap, four feet high and twenty feet long. Then off came the disguise to reveal Hill as Edward Huxley ('95) and the bills as stage money borrowed from a local theater. Alumni filed out wondering what each could do for Tech in its hour of need, an object lesson that paved the way for the creation of the Technology Fund later that year.

In April, the association canvassed alumni by mail to sign a petition demanding that the Corporation "entertain no proposition to unite, ally or associate itself in any way, financially or otherwise, with any other educational body." By mid-May, 1,637 responses were in hand. The vast majority, 1,157, signed, while 47 expressed support (without signing) and the rest either were not sure or liked the idea of some form of cooperation with Harvard, short of merger. Only 8 backed Pritchett's initiative. The Corporation, meanwhile, softened merger talk by resolving, on May 4, only "to ascertain whether any arrangement can be made with Harvard University, for a combination of effort in technical education, such as will substantially preserve the organization, control, traditions, and the name of the Massachusetts Institute of Technology." But this did not appease the alumni. At their June reunion, in Boston's Symphony Hall, they worked themselves into an anti-Harvard frenzy with loud, repeated choruses of a ditty penned by Gelett Burgess ('87), referencing the rival schools' colors and sung to the old Civil-War tune, "John Brown's body":

> You can't make crimson out of cardinal and gray etc.
> As Tech goes marching on.
> We don't give a d— for Ha-a-arvud etc.
> As Tech goes marching on.

The night before, Arthur D. Little ('85) had been shouted down when he suggested that one great technical school made more sense than two competing for scarce resources.

Current students wanted to give Pritchett leeway—the benefit of the doubt, even—in light of how much he had done to improve Tech's social life. But the passions vented by alumni quickly pulled them in. Students, overall, supported the idea of Tech moving to Cambridge,

Watertown, Allston, Brookline, wherever, to ease the space crunch. A 1904 survey showed 62 percent favoring a move, 14 percent opposed, and 24 percent on the fence (not unless absolutely necessary). All agreed that geographic proximity to Harvard was no obstacle—"We must . . . seek our competitors and not avoid them; we must carry the war into enemy's country"—so long as Tech preserved its identity, dodging Harvard's infectious "spirit of easy indolence." They satirized Pritchett's fondness for beer, Germany, and Harvard. Pritchett, they joked, might well map out and teach a subject called analytical prestidigitation with content along these lines: "The combining of two antagonistic forces into one united body; the ability to drink beer while having a good time and not getting drunk; the changes of the American language into German; and finally, the different results gained by treating cardinal and gray with other compounds." Prestidigitation implied sleight of hand, manipulation, contortion, a new side of the president they had always thought of as open, transparent, and straightforward.

The faculty watched and waited, postponing adoption of an official position. Some who were concerned about the path laid by Pritchett and Eliot shared their frustrations privately. "The main and sufficient restraint now," as Harry Tyler outlined the faculty strategy, "is the necessity of great caution in order to serve our cause best." But after November 14, 1904, when Pritchett's negotiations produced a so-called Tentative Plan of Co-operation between Harvard University and the Massachusetts Institute of Technology, the faculty no longer felt compelled to remain quiet. They dismissed Pritchett's preferred terms—"treaty" or "concert of effort" or "cooperation"—as euphemisms for union, merger, or amalgamation. "We are more and more convinced," lamented Tyler, "that Dr. P. will carry his scheme by hook or crook if he can & I'm afraid he has a fair chance."

The alumni shared Tyler's fears. By November the Technology Fund raised $120,000 to maintain independence and promote development, with Pritchett canceling his own contribution when fund organizers refused to drop "maintain independence" as a key goal. Also that month a group took options on 120 acres as a possible new home for Tech, near the Neponset River and the Blue Hills reservation, about as far from Cambridge as they could get without moving out of convenient reach of Boston. A hundredth-birthday commemoration of founder William Rogers on December 4 turned into a rally for independence. When Pritchett spoke at an alumni gathering one week later, a motion of thanks and confidence was amended by striking out the "confidence" part. According to Harry Tyler, the president showed "wear and tear considerably." By this time Burton, Pritchett's erstwhile ally as dean, had defected to the other side. He and Tyler talked about Pritchett "a good deal as a unique psychological phenomenon," someone with "a great command of the vocabulary of sincerity" who played fast and furious with the truth and who "evidently deceives himself as well as many others." Loyal donors like Sarah Hughes threatened to withdraw support. "I regret," she advised Pritchett, "to see the idea of any affiliation between Harvard & the Tech still pending. I have always told you I have no use for Harvard in any form as far as the Tech is concerned & don't believe any sort of mutual pull would be of mutual advantage. . . . I stand ready to help on any entirely independent organization that the Tech may plant itself on permanently." And, on another occasion: "I hate to think of Harvard & the Tech, one, or driving rival teams to cut one another's throats. I still maintain that they fill two different needs to different sorts of men, & I feel rather more in sympathy with the Tech sort."

Confident, however, that his plan would move forward, and with continued backing from Andrew Carnegie and other interested philanthropists, Pritchett sought faculty consensus and proceeded by March 1905 to explore options on vacant land in Brighton, adjacent to Soldiers Field (present-day site of the Harvard Business School), as Tech's new home. But the faculty went on record as overwhelmingly opposed to the whole scheme—56 to 7—at a volatile meeting on May 5. Tyler and Sedgwick, the only two faculty members absent (both were in Europe at the time), did not vote but also registered their opposition. Joining Pritchett in support were Henry Fay, Thomas Jaggar, Frank McKibben, F. Jewett Moore, George Swain, and William Walker. Jaggar was already Harvard-connected—part of Pritchett's arrangement with Eliot on geology—and Walker was beholden to Pritchett for having recruited him to reenergize Tech's industrial chemistry program. Both McKibben and Swain would soon leave, McKibben to join the faculty at Lehigh and Swain at Harvard. Pritchett also referred the plan to the alumni, with little expectation of a more favorable outcome. The question was put (by mail) to 3,200 graduates and 1,858 responses came back: 1,374 against, 462 for.

Pritchett pushed the measure through in the face of opposition from these two key groups. On June 9, 1905, a fractious Corporation voted 23 in favor to 15 opposed, the executive committee voting 5 for and 2 against. Among the Corporation members backing Pritchett's plan were six alumni: Robert Peabody ('68), John Ripley Freeman ('76), Charles Hubbard ('76), Eben Draper ('78), A. Lawrence Rotch ('84), and Charles Stone ('88). On the nay side, four alumni—James Tolman ('68), Howard Carson ('69), Francis Williams ('73), and James Munroe ('82)—denounced the outcome. Munroe, a former protégé of Francis Walker, called it an "act of defiance" against the faculty and alumni, with Pritchett quick to respond that support for the measure outweighed opposition among alumni members of the Corporation. "I did not realize," Munroe told Emma Rogers, "how far the canker of greed and 'brute money-power' had eaten into men of the high standing of our Corporation." The vote, according to Sedgwick, was a "heavy blow," "a most unwise and injudicious step," as clear a signal as any that "faculty and alumni opinion alike go for naught"—"wrong as respects the Institute, wrong as respects the community, wrong as respects technical education, wrong through and through."

Pritchett's conduct was called autocratic, deceitful, selfish, and, worst of all, a sell-out of Tech's rich culture, tradition, and legacy to the high bidder across the river. Whatever trust he had built, through efforts to develop Tech's research programs and to improve its social environment, all but vanished. The opinion of one former faculty member, who had warned Emma Rogers about Pritchett the year before, now looked prophetic: "I felt all along that he is one of the kind in for number one, first, last and all the time. . . . If the interest of the Institute being advanced tends to secure his own prestige and advancement, very well and good. Otherwise he is likely to see that trend as being best, which would advance himself." A harsh portrayal, to be sure, signaling just how far Tech's confidence in Pritchett had plummeted.

Emma Rogers, venerable symbol of all-things Tech, became a lightning-rod for both sides. Even after she threw her support behind the antimerger side, the promerger camp continued to seek her out.

Pritchett wrote to her right after the June 9 vote acknowledging how much pain the news would bring, yet promising that "this step will not lead to any loss of independence on the part of the Institute and . . . will result . . . in its development along those lines Mr. Rogers would have chosen." Munroe and others said not to worry, the battle was far from over. History professor John Sumner assured her that the Corporation had "a long road to travel before the 'merger' can become an actual fact—and a road with a good many difficult places." Others were confident that the courts and legislature would shut Pritchett's plan down. Sedgwick found it difficult to imagine that state officials would allow Tech to be "merged or confused, or amalgamated, or entanglingly allied, with any other institution." Ellen Richards issued a call to arms. The Corporation vote, she told Mrs. Rogers, "has freed our lips—we can speak out as we could not before & we shall. Diplomacy does not always succeed—& when insult is added to injury the musket may turn and rend the attacking party." Munroe announced the formation of a Tech defense league and took the pulse of the McKay trust chairman, who, he was happy to find, opposed the plan and promised to help prevent the Institute from being "humiliated and dragged into bondage." The defense league, run by alumni, would fight to the bitter end, vowed Munroe, as the faculty continued to conduct business in a spirit of "dignified belligerency." Ernest Bowditch ('69) urged alumni to ensure that "the Harvard element and incidentally the Pritchett element is gradually superseded by a Tech element that in the hereafter will not even squint in the direction of an alliance." Current students, meanwhile, turned much of their anger on Eliot:

> You'd never expect the Tech to wreck
> Her forty years of fame,
> By wagging the tail of the Harvard's dog
> Or selling her good old name.
> If Eliot thinks he's got us cinched,
> We'll give him a chance to see,
> For he'll dwell in hell before we'll sell
> The yell of M.I.T.

Tech dispersed that summer with everyone on edge, uncertain about what the future held. But by the time school reopened in the fall, the matter had been settled quietly, bluster-free, without a dramatic climax. The state supreme court handed down a decision on September 6, 1905, denying Tech's right to title on the Boylston Street property, barring it both from selling the property and from expanding or erecting new buildings there. This represented a death knell for Pritchett's plan, dependent as it was on property-sale to help finance property-purchase in Cambridge. He learned the news while traveling in Europe. Despondent, he advised Corporation members at a meeting on October 11 that Tech had no choice but to call the whole thing off. The Institute, he told Eliot in a letter written that same day, found it "impossible to proceed with the plan of cooperation." Jerome Greene, Eliot's secretary, replied later that month with regrets that the plan had been "brought to naught." Tech's antimerger camp, relieved that the supreme court had preempted a drawn-out battle, returned to business as usual savoring the victory, but with little overt celebration and no apparent wish to taunt their defeated opponents. The following year, Harvard dissolved the Lawrence Scientific School, absorbed its undergraduate students into the regular college, and created the Graduate School of Applied Science targeting advanced students in engineering.

Many had assumed that Pritchett would resign if the Harvard plan failed. Instead, he promised to continue as president but with modified goals: "It seems clear that under the conditions which now hold the Institute must remain, for a number of years at least, in its present site. . . . I trust we may all turn heartily and enthusiastically to the future development of the Institute in its old home." Pritchett predicted that Tech would stay in Copley Square at least until 1925 or 1930, perhaps later. While not promising to serve as president that long, he pronounced himself ready to move forward with a new strategy: acquire more land in and around Trinity Place; raise money to finance construction of new buildings, and to start an endowment; erect dormitories on Tech's vacant Brookline property next to Technology Field; and "develop with sound judgment and true perspective in education the intellectual work of the Institute," by which he meant both undergraduate teaching—Tech's "heart and center"—and programs essential to "a great school of professional, graduate, and research work."

The suggestion that their president might settle in for the long haul did not sit well with either alumni or faculty. Biologist and librarian Robert Bigelow guessed that Pritchett's friends on the faculty had dwindled to no more than five. The faculty's spirits were buoyed, however, by persistent rumors that Andrew Carnegie wanted Pritchett to head one of his agencies. So when the announcement came, in December 1905, that he would become president of the Carnegie Foundation for the Advancement of Teaching, they breathed a collective sigh of relief.

Pritchett resigned as Tech president on December 12. At month's end he grew emotional in a farewell address to alumni, describing how difficult it was "to speak of parting from you, and most of all from the student body, without these choking sensations." But he got little sympathy in return. Some alumni and faculty went about proclaiming that his resignation stemmed directly from the failure of the Tech–Harvard merger. Pritchett's position, someone told the *Globe*, was "analogous to that of a prime minister whose pet political plan has been defeated [and] he has no choice but to resign." Someone else called him "not the ideal administrative head of the Institute, mainly because he has been too much controlled by the German idea of technical education." His frequent absences in Europe were called damaging: "There has been considerable feeling that Tech was not getting the best service which it might expect of its president, by reason of his great interest in foreign institutions, which has led him to devote a considerable share of his time to lectures before German and Scotch and other foreign technical schools." But through all this, Pritchett never felt forced out. His decision to leave was motivated more by the pull of fresh opportunities elsewhere. Carnegie promised him a broader sphere of influence, ways to come "into touch with the world at large," an environment where parochial self-interest—either personal or institutional—mattered less than grand-scale causes.

Carnegie had had his eye on Pritchett for a while. In 1903, he caused a stir by publicizing his desire to recruit Pritchett as head of the Carnegie Technical Schools (later Carnegie Tech) in Pittsburgh. "Carnegie knows mighty well he never could get him," said one Tech insider, confidently. The two actually did not meet until the spring of 1904, when Pritchett joined Carnegie and others at the White House to go over administrative issues surrounding the Carnegie Institution of Washington, founded as an independent, nongovernmental agency yet with President Theodore Roosevelt and all members of his cabinet on its first board of trustees. The next day Pritchett and Carnegie sat together on the Washington-to-New York train and grew better acquainted. That summer, the Pritchetts spent time with the Carnegies at their summer residence in Skibo, Scotland. Pritchett and Carnegie passed much of the three- or four-day visit deep in discussion, as they wandered the heath. Carnegie was eager to hear Pritchett's views on what American college teachers needed most. A good pension system, said Pritchett, citing Augustus Lowell's innovative but distinctly local scheme already in place at Tech.

Carnegie followed up that next spring. For financial advice, Pritchett placed him in touch with his friend Frank Vanderlip, well-known banker and former assistant secretary of the treasury under President McKinley. On April 18, 1905, Carnegie announced a gift of $10 million to create the Carnegie Foundation for the Advancement of Teaching with the primary goal of providing pensions for teachers in universities, colleges, and technical schools throughout North America—the United States, Canada, and Newfoundland. The first board of trustees consisted of twenty-two college presidents (Pritchett among them), two bankers (Vanderlip joined by Robert Franks), and Carnegie's nephew, T. Morris Carnegie. Their inaugural meeting took place at Andrew Carnegie's home, 2 East 91st Street in New York City, on November 15. Pritchett and Vanderlip presented a report recommending a simple administrative structure: president, treasurer, chairman, vice chairman, board secretary, and a six-member executive committee. The trustees voted their approval and Pritchett was elected president, with Charles Eliot as chairman. Rounding out the team were William Harper (president, University of Chicago) as vice chairman, T. Morris Carnegie as treasurer, and an executive committee consisting of Vanderlip, Franks, Nicholas Murray Butler, Charles Harrison, Alexander Humphreys, and Princeton University president Woodrow Wilson.

Carnegie's disappointment over the collapse of the Tech–Harvard plan, which he had strongly supported, was offset by the knowledge that Pritchett could now abandon Tech's internal squabbles for a career of broader service to education. But Pritchett did not leave right away. While many wished him gone, cooler heads prevailed and persuaded him to stay until a successor could be found. For him, this meant juggling two full-time jobs in different cities with a grueling commute by train from New York to Boston and back three times a week. He agreed partly out of duty, but also to repair if possible some damaged bridges. In 1905, with Tech alumni flexing more muscle in the wake of Pritchett's failed plan, Corporation bylaws were modified to add members elected for term appointments by alumni from within their own ranks. As a result of the first election, nine new alumni joined the Corporation in 1906. Charles Main ('76), Frederick Wood ('77), and T. Coleman du Pont ('84) came aboard for five-year terms; Frederick Copeland ('76), Joseph Gray ('77), and Frank Locke ('86) for four-year terms; Eben Stevens ('68), Richard Soule ('72), and Frederick Newell ('85) for three-year terms. In March 1906 alumni association president Everett Morss ('85) declared that the "recent warm contest" over merger had "left no sores" and that everyone was now working "harmoniously to make a bigger, better, and busier Tech." He and others were confident, with the alumni now showing critical mass on the Corporation, that Harvard would never again come this close to swallowing the Institute whole. A generation of Tech students to follow would salute the alumni as heroes-to-the-rescue. "Having awakened within them the spirit of chivalry," went the legend, "they came forth as one body to do battle with a common foe, stripped him of his disguise, and saved for Tech its name and honor for all time."

Pritchett hoped to leave in June 1906, with Morss's conciliatory words signaling smoother sailing ahead. But the arrangement stretched another year and would have gone on longer had he not set June 29, 1907, as a firm deadline. The Corporation accepted the news with regret in April 1907. In May, looking drawn and tired, Pritchett blamed overwork for undermining his health. A mild attack of typhoid fever had laid him low for weeks. On his last day in office, he rushed off a relieved note to his old friend and ally George Wigglesworth, himself retiring as Tech's treasurer to make way for Francis Hart ('89). "I write to say that, so far as I can do it, matters have been cleared up and the few outstanding things which could not be settled—on account of the absences of certain people from the city—I have turned over to Professor Noyes." Arthur Noyes, who shared Pritchett's vision for building Tech into a great research institution but not his commitment to a coalition with Harvard, took over as acting president. It was a razor-thin transition, Noyes agreeing to step in less than three weeks before Pritchett left for good.

Tech still had not found a new leader after a year-and-a-half-long search, and with the net cast wider than usual. This time alumni joined the parade of serious candidates. The first offer went to George Ellery Hale in the spring of 1906. Arthur Noyes and Harry Goodwin, his old friends from student days, egged him on with visions of the three working together for the larger good of Tech, science, and scientific education. But Hale declined, "absolutely contented" with his work at the Mount Wilson Observatory in California. The Institute presidency impressed him nonetheless as an opportunity to craft an innovative vision for scientific and technical education. "I am still regretting," he told Noyes in the fall, "that I could not see my way clear to telegraph you favorably. New possibilities for introducing important work at the Institute strike me daily, especially about five-thirty A.M. when I am likely to be thinking over matters of this kind." He put some of his ideas on record in a 1907 piece for *Technology Review* entitled "A plea for the imaginative element in technical education." Hale warned Tech and similar institutions—Throop Polytechnic Institute, for example, which before long he would help develop as the California Institute of Technology—about the risks of narrowness. A high-quality technical education must, he argued, aim for scope, breadth, and creativity, "the largest possible proportion of men capable of conceiving great projects and the smallest possible proportion of men whose ambition can be completely satisfied by the work of executing them."

After Hale, fifty more candidates were vetted by October 1906. Four finalists emerged—Francis Walker ('92), James Garfield, Harry Garfield, and Andrew Fleming West. Walker, an economist like his father, Tech's adored president, became a sentimental favorite but lost out even though his work as an examiner for the U.S. Department of Commerce and Labor was highly regarded. The Garfield brothers, sons of former U.S. president James Garfield, a key supporter of Walker's efforts to modernize the census bureau in the 1870s and 1880s, were Williams College graduates like their father. James, a close adviser to President Roosevelt, was commissioner of corporations at Commerce and Labor, while his brother Harry, a lawyer and political scientist, served on the faculty at Princeton. Neither was offered the Tech post, but both assumed increasingly prominent roles in government and academe—James as interior secretary under Roosevelt (1907–9), Harry as president of Williams College (1908–34).

The Corporation's unanimous vote went to Andrew Fleming West, the only candidate with no ties to Tech or to President Walker and, on paper anyway, the most unlikely choice. West, a professor of Latin at Princeton, was a militantly old-fashioned classics scholar whose editions of works by the Roman comic playwright Terence were standard fare in college classics courses. His dismal view of academic trends was well known. The purpose of education, he argued in a series of acerbic, hard-hitting essays, was to promote mental rigor, moral character, and civilizing influences, not to serve up easily digestible pabulum or to prepare students for careers. None of this elective, liberal arts nonsense for him—an educational lunch counter, he called it, in a direct slap at reforms implemented by Charles Eliot at Harvard—and, at the other end of the spectrum, he frowned on specialized, professional programs in applied science, engineering, anything that smacked of the practical, the useful, or the utilitarian. German academic innovations in chemistry, chemical engineering, and other fields that Pritchett had been so impressed by struck West as crude, too tied to industry or government, outside the scope of a real university. He had low regard for technical schools. "You're nice boys, but it's too bad you never went to college," he told more than one Tech graduate over the years.

Some wondered what the Corporation could possibly have been thinking. Francis Walker and the Garfields, at least, had strong scientific credentials—social science, to be sure, but science nonetheless, following on President Walker's campaign to reinforce the position of economics, political science, business, and law in Tech's curriculum.

But West's specialty, Latin, had never crossed Tech's threshold. Besides, his acrimonious public disputes with Princeton president Woodrow Wilson over control of graduate-school policy suggested that he was difficult to get along with, authoritarian, hardly a team player. He was intolerant, too, from a religious point of view—he once opposed the appointment of a Unitarian to the faculty—and his passion for ceremonial flamboyance in the Harvard-Oxbridge style seemed anachronistic, even for Princeton's relatively conservative culture, much less for Tech's more modern spirit.

The Corporation felt drawn by West's magnetic personality, forceful style, proven record in fundraising ("three-million-dollar West" was his Princeton nickname), and, after what Tech had just gone through, by his combative relationship with Charles Eliot. But some scientists outside Tech voiced concern. "It is a long way," J. McKeen Cattell wrote in October 1906, "from the chair of Latin in a classical and monastic college to the presidency of the Massachusetts Institute of Technology. . . . It would seem to an outsider that in the present emergency the [Institute] needs for its president one of its own men, imbued with its methods and traditions, a man bred to science, believing in science as the chief factor in culture and in life, knowing that pure and applied science must go forward hand in hand." The offer surprised West as much as it worried Cattell, and while he did not dismiss it outright—its very incongruity tempted him—he declined on October 30 when Wilson begged him to stay at Princeton.

Other candidates rejected offers early in 1907. Henry Fine, professor of mathematics at Princeton, was among these, along with Tech's own John Ripley Freeman ('76), engineer and business executive, and Corporation member Frederick Fish, AT&T president, whom Pritchett called exceptionally desirable. Also in the mix were Victor Alderson, president of the Colorado School of Mines, and economist Jeremiah Jenks, of Cornell's faculty. The final candidate in this round, Benjamin Ide Wheeler, president of the University of California and, like Andrew West, a classics scholar, agreed to be interviewed in April. Alumni George Hale and George Baldwin ('77), on the Corporation's behalf, paid Wheeler a visit at his San Francisco office. Wheeler asked why a technical school would want someone with his background, to which Hale and Baldwin replied that Tech preferred "a man of broader academic training and experience" over "a purely technical man." Wheeler, still skeptical, stopped at Tech on an otherwise scheduled east-coast swing in mid-May, and quickly said no. "It certainly is, however, a very important and a very honorable post, and the man ought to emerge who can take it." In June 1907, a last-ditch effort to recruit John Bates, former governor of Massachusetts, fell short when Bates vowed undying loyalty to Boston University, his alma mater.

Pritchett lived comfortably with his wife, daughter Edith, and five live-in maidservants at 22 East 91st Street in Manhattan, a few doors down from Andrew Carnegie. But he missed Tech. Once, when Robert Bigelow stopped by his office at 522 Park Avenue, home of the Carnegie Foundation, Pritchett felt quite touched after an hour or two reminiscing. He grew wistful about mistakes made, opportunities missed, strategies gone awry, wondering how his two great predecessors might have handled this or deflected that: Rogers, who inspired "all that has followed and all that will ever follow," and Walker, who had nourished Tech "into the full vigor of institutional life." What he did not miss was the incessant worry over money. "Bigelow, you don't know what a comfort it is to have more money than you need to spend."

In 1908, Carnegie added $5 million to his original gift of $10 million. But the foundation's pension plan outgrew these resources, too, and in 1918 Pritchett helped to incorporate the Teachers Insurance and Annuity Association (TIAA) as a life-insurance company, with additional Carnegie support. He also developed a series of in-depth policy studies, mostly relating to higher education. These were the famous Carnegie Foundation bulletins, 25 in all published between 1907 and 1930, dealing with medical, dental, legal, technical, engineering, and vocational education. The best-known bulletin, Abraham Flexner's *Medical Education in the United States and Canada* (1910), inspired radical changes in medical-school curricula and requirements. Another published in 1910, Morris Cooke's *Academic and Industrial Efficiency*, criticized Tech and six other academic institutions for wasteful management practices, while praising Columbia University's excellence in this regard. Two bulletins in 1929 and 1930 warned that the sinking of large sums of money into college sports—the King Midas of athletics, as Pritchett labeled the problem—would undermine standards, distort academic values, and change the face of American education.

Pritchett retired in 1930. He spent his remaining years in Santa Barbara, California, where he and his family had vacationed since 1911. The tall palms and blue ocean attracted him. He built a house at 2417 Garden Street in the mid-1920s. "We have been reveling," he told a friend, "in the beauties of our garden and the luxury of a daily ocean dip. . . . Truly California is a wonderful country." The children—all but Ted, who was killed in a car accident in 1916—came and went regularly. It was a placid retirement in some ways, but Pritchett kept active, alert, engaged with public affairs. Trends in his beloved Germany grieved and angered him. "It is not easy to estimate Hitler's character," he observed in 1933, "but he has certainly governed Germany with an iron rod and his treatment of many of the Jews has been unnaturally cruel and unjustified." Then, in 1936: "The abuse of civil liberty by the German government is, in my judgment, one of the most atrocious governmental proceedings that has been enacted in modern times, and the great danger in it lies in the fact that this action may cause to spread to other countries the tendency to restrict the personal liberty of the citizen." Pritchett did not live to see the full tragedy of his predictions played out. He died of heart failure on August 28, 1939, just three days before the German invasion of Poland plunged the world into war.

"Thoroughly sure of himself"

Richard Cockburn Maclaurin, 1870–1920

Arthur Noyes agreed in June 1907 to serve as acting president on condition that a replacement be hired soon, preferably within a year. Besides wanting to get back to his research, he considered himself less than ideal for the job. "I have no hesitation in saying that I would rather be president of the Institute than to hold any other position in the country, provided I felt myself well fitted to fulfill the duties of the place. I have, however, clearly recognized that this would not be for the true interests of the Institute; for it needs at its head a man with a far larger working capacity, with a greater aptitude for the public and social sides of the work, and with certain other important qualities more highly developed." By working capacity Noyes did not mean physical stamina—he was one of Tech's most notorious toilers, in his lab from 4 a.m. to 8 p.m. most days—but, rather, the demands on presidents to politic, to circulate, to plead for money, and to negotiate with difficult, recalcitrant characters on a daily basis. He disliked small talk, large groups, and public functions and gatherings. Typically, his nonwork time was passed either with his mother, Anna, who resided with him permanently after the death of his father, Amos Noyes, in 1896, or in the company of a small, close-knit circle of friends: likeminded scientists, favorite students, fellow yachting enthusiasts, all men. The presidency was no job for a confirmed bachelor like himself, when expectations of the first lady—from Emma Rogers through Eva Pritchett—had played so pivotal a role in Tech's social life. "The qualifications for the office," he wrote, "are so numerous and varied that their combination in a single person is extremely rare."

But Noyes threw himself into the position with such aplomb that the urgency to replace him quickly faded. With Gilbert Lewis, his trusty second filling in as acting director of the Research Laboratory of Physical Chemistry, Noyes felt free to focus on Institute-wide affairs. Initiatives that he shaped and promoted at the start suggested that he would be no mere caretaker president, no ceremonial stand-in.

Noyes reorganized the faculty to include a chairman, elected annually, along with new committees on business, rules, educational policy, and courses of instruction—all intended to shape a more democratic, deliberative body, and to increase its proactive role. General education requirements were modified to lay more emphasis on English and history, less on modern languages, while instruction in mathematics underwent a radical revision favoring the integrated approach, obliterating old divisions between the branches. In 1908 Tech introduced the universal language Esperanto to its subject list, a move hailed by the *Boston Globe* as a sign of the Institute's "progressive policy . . . toward a judicious combination of the liberal studies with the professional." An option in steam-turbine engineering was added to Course II (mechanical engineering) under Cecil Peabody. A new five-year undergraduate program allowed more electives in natural sciences, social sciences, and the humanities, leading to double-bachelor's degrees in a specialty and general science. The adviser system was revamped to promote closer instructor-student relationships. A summer work requirement was added, part of Tech's commitment to providing not only "a liberal education, a thorough training in fundamental scientific subjects," but also "sufficient technical knowledge to enable students to enter at once upon the practice of their profession." Master's degree programs were expanded and systematized. Opportunities for research and advanced work grew through doctoral programs (Ph.D. and D.Eng.) in the physical and applied sciences, reinforcing Tech's role as "a scientific institution of university scope." Noyes encouraged faculty to publish more and suggested that promotions be pegged to productivity in this regard. He pushed to create two new research laboratories, one for chemical engineering and the other for physical geology. The Research Laboratory of Applied Chemistry opened in 1908 under the direction of William Walker, staffed initially by Walker, Henry Talbot, Willis Whitney, Augustus Gill, and Frank Thorp, with the assistance of two young research associates: Warren Lewis ('05), just back after two years of graduate work at the University of Breslau, and William Guertler of the University of Göttingen. Noyes convinced the Corporation to appoint a committee for the promotion of student welfare, to tackle problems ranging from housing to social events, athletics, and overall school spirit. In September 1908 the Technology Union moved to new quarters on Trinity Place, between the Pierce Building and Engineering C, with a spiffy club house, dining rooms, offices, conference rooms, post office, coat room, lavatory, and "a large social or living room, where students may gather in their spare time for reading and conversation and where evening entertainments may be held." Noyes traveled about speaking to alumni and other groups in an effort to forge closer community ties.

His was as packed a schedule as any permanent president might follow. Noyes steered clear, however, of the two largest, most difficult challenges facing Tech—endowment and site—while acknowledging them as the next president's top priorities. When he sensed the Corporation dragging its heels, he reminded them that he was a research man and must get back to his lab. To keep him happy, '08 class president Harry Rapelye, the type of popular, outgoing athlete that Noyes liked to draw into his private circle, was hired as a personal assistant to relieve him of burdensome routine and to foster friendly, constructive relations between the administration, faculty, and student body. The Corporation, meanwhile, kept its collective eyes peeled for a replacement.

On February 2, 1908, Ernest Fox Nichols, professor of physics at Columbia University, hosted a dinner at his home on Manhattan's upper west side in honor of Richard Maclaurin, just arrived from New Zealand to succeed Robert Woodward as professor of mathematical physics. Woodward had left Columbia in 1904 to serve as president of the Carnegie Institution of Washington. After a long, fruitless search, Nichols had persuaded president Nicholas Murray Butler to hire Maclaurin, whom he had met while on sabbatical at the University of Cambridge, despite Butler's skepticism about chasing after a candidate halfway around the world. Among Nichols's guests that evening were George Wendell, a Tech alumnus ('92) and physics professor recently transplanted to the Stevens Institute of Technology, and ex-president Henry Pritchett, now president of the Carnegie Foundation for the Advancement of Teaching. Wendell, who shared with Pritchett, Noyes, and others a desire to see Tech transformed from a fine engineering school into a great scientific university, left Nichols's apartment convinced that Maclaurin was the man for the job. He met with him several times in the next few weeks. They talked Tech, and nothing but Tech. Maclaurin wondered why the obsession. Wendell popped the question early in March. Would Maclaurin allow himself to be vetted for president? Maclaurin, barely settled at Columbia, wondered if things in America always worked this fast. But he agreed to go with Wendell to Boston, meet Tech folks, look the place over. Wendell got Pritchett to write ahead with an introduction to Noyes.

They went up by train on March 18. Tech physics professor Harry Goodwin hosted a dinner. Maclaurin liked the people he met—"all delightful fellows," he said, "much traveled and full of fun and good fellowship." Goodwin stacked the group with cosmopolitan types, those who like himself and math professor Harry Tyler had studied abroad, in an effort to make Maclaurin feel at home. But aside from Wendell, only Noyes knew the real reason behind the visit. As cover, word went out that Maclaurin was there to get a sense of how American technical education compared with systems in other countries. This way Maclaurin could get a feel for the Institute in relative peace. Wallace Sabine, who had succeeded Nathaniel Shaler as Harvard's dean of applied science, hosted a lunch for Maclaurin at Tech's sometimes not-so-friendly competitor across the river.

The visit coincided with a talk given by Noyes to the instructing staff on March 20. Maclaurin sat incognito in Huntington Hall, listening as Noyes outlined Tech's guiding philosophy. Our aim, Noyes told the assembly, "is to produce men who have the power to solve the industrial, engineering, and scientific problems of the day—men who shall originate and not merely execute." He also underscored Tech's responsibility to build character, to nurture ideas, to shape values, and to promote cultural depth. "Our task is a much larger one than that of imparting a knowledge of our particular subject, and . . . it is a broader one even than that of developing the power of dealing with its problems. . . . The most important and difficult part of our undertaking consists in cultivating sound habits of thought and work, in developing breadth of interest and good judgment, in molding character, and in creating a high moral purpose." He reminded colleagues that their duty to promote scientific method and scientific spirit outweighed their allegiance to mere subject matter.

The next day, when the two met in private, Noyes turned from ideals to problems. Tech's tiny endowment, inadequate to equip labs properly, or to pay faculty a decent wage, or to maintain buildings. Cramped quarters, with options for expansion in downtown Boston all but exhausted. A new campus essential, but the millions of dollars needed to acquire and build it apparently out of reach. A hand-to-mouth existence, the future uncertain. Noyes welcomed Maclaurin warmly, sensing across the desk someone like himself—scholar, teacher, scientist, a man of action who thought broadly and deeply—but he did not want him, or whoever the next president might be, to come in with any illusions.

To Maclaurin, the problems looked hard but solvable. He imagined himself going at them, untangling strands, sorting them out. But he heard nothing from Tech for months after George Hale ('90) spent an evening with him in late April and raved to Noyes about his "pleasing personality" and "sound commonsense." In June Maclaurin, his wife Alice, and their infant son William Rupert headed to Britain for the summer. George Wendell kept singing his praises: "He believes in a training that will stimulate the mental and human faculties in a way to develop the power to meet and solve the new problems as they arise in the world of affairs." Wendell also testified to Maclaurin's personal qualities, his "simple and unassuming manner [that] carries with it an assurance of sincerity," his "keen sense of humor," and the "richness and variety of his experience and information [that] lend an unusual charm to his conversation." A number of eminent men, including three (Lord Kelvin, Sir George Stokes, and Joseph Larmor) whom Noyes referred to as the greatest English scientists of their generation, sent glowing appraisals. Kelvin remarked on Maclaurin's "businesslike power of seizing on the essentials of a problem," while Stokes called his interests "extraordinarily wide" and Larmor judged him "a skillful and profound mathematician, quite of the first rank."

Another round of interviews took place that fall. Tech went to Maclaurin this time, a steady stream of visitors at Columbia, one by one, then in twos and threes. Columbia president Butler, who did not want to lose his new mathematical physicist, tried to keep Tech at bay, recommending two alternative candidates, both Columbia faculty members whom he called highly qualified but in reality considered more expendable than Maclaurin. Charles Stone ('88) reached New York ahead of the others. He and Maclaurin hit it off so well that Stone called fellow executive committee members Francis Hart and Frederick Fish to say, "I think we've found our man." Hart and Fish took Stone at his word. Fish, instead of asking Maclaurin tough questions on academic or administrative affairs, spent an evening with him chatting about Robert Browning's poetry. Maclaurin impressed them all with his self-confident, calm demeanor, the way his "quiet, steady gaze . . . seemed to fix and hold a visitor."

Tech's executive committee sent his name up on November 11, 1908, and the full Corporation voted unanimously to confirm on November 23. They placed confidence in Maclaurin's ability to apply his swath of talents—"unusual knowledge of many different phases of our civilization . . . a broad foundation of scholarship, comprehensive in the important fields of law and physics and extensive in all branches of human effort"—at this critical juncture in Tech's history. The choice surprised some outsiders. Who *was* this foreigner? A Victorian throwback for sure, with that quirky straw hat, bushy mustache (which he shaved off soon after he got to Tech), his odd English accent and mannerisms. Who did he know here, and what administrative experience did he bring? His academic credentials were impressive, but had he ever run anything? He was young, just 39. A slender, frail-looking man with dark hair and calm bluish-gray eyes, he carried himself in a proper-English way. His voice was soft-toned yet authoritative, wasting no words, covering the essentials, and his firm handshake suggested "a man thoroughly sure of himself." But he was an unknown quantity, some said, out of his element. Insiders countered that Maclaurin's very foreignness was more asset than liability, with the Institute's reputation trending international, its community increasingly cosmopolitan, its sense of global self emerging. Besides scholarship, Noyes pointed to Maclaurin's knowledge of systems of higher education worldwide as proof that Tech was "about to enter upon a new epoch in its history . . . characterized by an unexampled development in all directions."

The Maclaurins came from Lindean, Scotland, a tiny village south of Edinburgh, near the English border. Born there on June 5, 1870, Richard was the son of Robert Campbell Maclaurin, a free-thinking Presbyterian minister, and Martha Joan (Spence) Maclaurin, daughter of a Shetland Islands surgeon. Robert Maclaurin had dipped into science, literature, religion, and other fields as a student at the University of Glasgow, where one of his classmates was soon-to-be-famous engineer, inventor, and mathematical physicist William Thomson (Lord Kelvin). An amateur scientist in his own right, Robert claimed to be a direct descendant of eighteenth-century mathematician and physicist Colin Maclaurin, whose works on oblate spheroids and power series are classics in the history of science. In 1874, after a blowup with local church authorities over doctrine, Robert packed up and moved his large family—wife and ten children—to New Zealand, where he did missionary work, taught, and studied botany and geology in his spare time.

Richard Maclaurin was a slight, delicate child, but what he lacked in physical strength he made up for in mental prowess. He attended a rural country school in Hautapu, about a hundred miles from Auckland, and then, on scholarship, the prestigious Auckland College and Grammar School. In 1887 he entered University College, Auckland, where he graduated B.A. in 1890 and, a year later, earned the M.A. with first-class honors in mathematics and mathematical physics. When his math professor W. Steadman Aldis urged him to continue his studies at the University of Cambridge, he sailed for England in 1892 and was admitted on scholarship at St. John's College, Cambridge, that fall. He could have gone to Emmanuel College, which also offered him a scholarship, but he chose St. John's for its renown in mathematics.

Maclaurin gained a reputation in the larger Cambridge community as a serious student who gave little if any time to athletics or social life. But those who got to know him well told a different story. These friends—mostly men, like himself, from the colonies—included the South Africans Jan Smuts and Etienne de Villiers, later eminent statesmen and jurists, and fellow New Zealander and physicist Ernest Rutherford. According to de Villiers, Maclaurin's "talk ranged over all the ambit of literature, history, current politics and social affairs . . . [with] much general information, and a natural wisdom." The colonials formed a close-knit group and sometimes took joint walking tours of London, Paris, Antwerp, Strasbourg, and the Swiss Alps.

Maclaurin graduated B.A. (honors) in 1895 and M.A. in 1896, at which point he began work on a research problem that he planned to submit for the coveted Smith's Prize, whose past recipients had included John Herschel, Lord Kelvin, Lord Rayleigh, James Clerk Maxwell, and Maclaurin's own former professor at Auckland, Steadman Aldis. Mathematics, Maclaurin liked to say, was a science of inspiration and imagination. The problem he set out to solve involved physical applications of elliptic coordinates, a classic overlap between math and physics. While working on it, he crossed the Atlantic and spent several months in Montreal as tutor to the son of wealthy railway magnate, John Ross, and a few other engineering students at McGill University. Back in England by the spring of 1897, following a brief stopover on the U.S. east coast, he submitted his essay and was awarded the Smith's Prize that June. Then, when his friend Smuts persuaded him that mathematics was too dry a subject to commit one's life to, Maclaurin read philosophy for several months at Strasbourg and returned to England intent on studying law. He joined Lincoln's Inn and worked up yet another brilliant essay, this one on the history of realty law from the Norman conquest to the present, for which he was awarded Cambridge's Yorke Prize in December 1898. His law studies gave him a break, he said, from the abstractions of mathematics and a chance to probe themes rich with human interest.

Maclaurin found himself pulled in several directions. His election as a fellow of St. John's College in November 1898 meant that he could stay on indefinitely as a Cambridge don. Then, with the Yorke Prize followed in January 1899 by the master of laws (LL.M.)

The Boylston Street building, ca. 1870—named the Rogers Building, after MIT's founder and first president, in 1883.

"Founder and father is his title perpetual, by a patent indefeasible"—
ca. 1850 (left) and ca. 1870 (below).

With the Savage family—Emma and, seated, James Sr. and James Jr.— in the drawing room at 1 Temple Place, Boston, ca. 1860.

"A kinder gentleman treads not the earth," 1875.

With his youngest children, Gordon and Eleanor, ca. 1885.

At his favorite occupation, 1901—"Now, gentlemen, I am going to show you one of the most beautiful and interesting things you ever came across."

Francis Amasa Walker

Chasing after Southern rebels, 1862—"never winced whether he was facing men with guns in their hands or with pens, or with words of bitter and sharp controversy in their mouths."

"soberly and seriously in dead earnest," ca. 1896—sometimes whipped off his spectacles and shook them at students to stress a point.

James Mason Crafts

"A choice must be made between administrative and scientific occupations, and it is the latter which I choose"—
ca. 1872 (left) and ca. 1898 (below).

"The Institute must be not only a teaching body, but it must well lay the foundations for a school of investigation in the physical sciences," 1902.

Eva Pritchett, with daughter Edith, ca. 1902.

Arthur Noyes, acting president, 1907–9 —as packed a schedule as any permanent president might follow.

Elihu Thomson, acting president, 1920–23 —signed documents and presided at official functions.

Ernest Fox Nichols with Albert Einstein, May 1921—"I should like to be called the 'step-president' of Technology."

Richard Cockburn Maclaurin

"a businesslike power of seizing on the essentials of a problem"—
1909 (left) and 1919 (below).

Cornerstone for the Cambridge campus, 1914—(left to right) Maclaurin, Everett Morss, Harold Kebbon, Allyne Merrill, Alfred Burton, J. R. Lotz, Welles Bosworth, Walter Humpheys, Albert Smith, and Horace Ford.

Alice Maclaurin, with sons Rupert (left) and Colin, in front of the president's house, ca. 1917.

The military, 1884—"a beautifully organized, coordinated piece of living human machinery."

Some felt intimidated by his steely gaze, 1923.

Samuel Stratton with (left to right) Edward and Lois Bowles, Morris Parris, and an unidentified friend in rural New York, summer 1930.

Fellow students James Killian (right) and David Shepard, 1926—Killian would become MIT's tenth president, 1948–59.

Julius Stratton, ca. 1932—MIT's eleventh president, 1959–66.

A junior at the College of Wooster, 1906.

The Compton clan, ca. 1937—(clockwise from top left) Wilson, Karl, Arthur, Mary, Elias, and Otelia.

With Samuel Prescott (center) and Vannevar Bush (right) at freshman camp near Lake Massapoag, 1932.

With wife Margaret (right), Gertrude Hinckley ('10), and Redfield Proctor ('02), MIT Alumni Day, 1936.

Informal time with students, ca. 1954.

The Great Court—at the heart of what Richard Maclaurin visualized as "a great white city" facing the Charles River.

degree, a career as a London barrister looked plausible. But when he was offered the professorship in mathematics at newly established Victoria University College in Wellington, New Zealand's capital city, he chose to head home. He had been gone seven years. "There is so much danger of us growing apart," he wrote to one of his sisters. "With all my English friends—I still have fits of loneliness, and then especially I long for something warmer and more personal from those with whom I spent my youth."

Maclaurin was elected to chair Victoria's first faculty, which consisted of four members (including himself). They doubled or tripled up, taking on extra subjects as needed. The initial student group numbered about a hundred. Besides mathematics, Maclaurin taught jurisprudence, constitutional history, and astronomy. In 1907, when a law school was added, he became its first dean. Outside the classroom he pursued interests in politics, legislation, and other affairs of state—New Zealand's status within the British Empire became increasingly independent after the turn of the twentieth century—and sought outlets for intelligent, probing dialog, which in his view was "the simplest and shortest way of avoiding the narrow mind." He joined efforts to reform an educational system mired in a retro-colonial time warp favored by cultural conservatives. His reading tastes grew broader, more eclectic: *The Golden Treasury*, Alfred Lord Tennyson, Robert Browning, reams of parliamentary blue papers. In 1903 he met Margaret Alice Young, an art student and daughter of an Auckland merchant (of Scottish origin, like the Maclaurins), whom he married in 1904. Also in 1904 he returned to England to receive the Cambridge LL.D.—"a few more odd letters tacked to my name," he irreverently observed—and crossed the American continent en route, with stops among other places at Stanford University and the Massachusetts Institute of Technology, partly in pursuit of data for his work on educational reform. The only scientific paper he published during this period was one in *Philosophical Magazine* (July 1903) on "the influence of stiffness on the form of a suspended wire or tape." But that year, in a recommitment to science, he began a massive project on the theory of light aiming to trace all the known connections between light, mechanics, and electricity. This work, published initially as seven papers in *Proceedings of the Royal Society* (1905–8), was highly regarded among physicists, including those who saw Maclaurin as a worthy successor to Robert Woodward at Columbia.

But by the time he reached New York, Maclaurin had just about lost interest in the work, at least in that part of it that appealed to a small, super-sophisticated audience of physicists and mathematicians. The first volume of *Theory of Light* was published by Cambridge University Press in 1908, but progress on the second went slowly and Maclaurin eventually dropped it. "This is . . . a somewhat demoralizing city," he confided to his wife Alice. "There are so many interests and so much life that the attractions of pure science have an unreal ring about them and one is occasionally depressed with the feeling that after all what is this struggling in matters scientific really worth. . . . Just now I am oppressed with the idea that pure science has its limitations, that life has so many other and more human problems to be attacked and solved that it seems foolish to be beating one's head against the hard wall of scientific difficulty."

Maclaurin circulated widely, well beyond Columbia's campus. At one dinner he sat next to community settlement organizer Jane Addams, who pulled him into talk about poverty, social problems, and economic justice in Australia and New Zealand. He visited local courtrooms to observe America's legal system in action. And, as his interests shifted outward, he went through the motions at Columbia. "He was patient," recalled one physics student, "although I know that many times he was irked by thoughtless questions." The intellectual challenge that he enjoyed most was not his work on light theory, or the thesis topics undertaken by his several graduate students, but the course of ten lectures that the Jesup Foundation invited him to give at the American Museum of Natural History. This was a public-outreach effort much like the programs run by the Lowell Institute in Boston. Maclaurin wanted to unravel the mysteries of science for a lay audience, to showcase science as "an elaborate work of art," to communicate the spirit of science to "the man of intelligence who lays no claim to scientific knowledge but who wishes to know what all the talk of science is about, and, in particular, why the physicists make such strange postulates as ether and electrons, and why they have so much confidence in the methods that they employ and the results that they obtain." When the Tech offer came, then, in the midst of all this, he saw it as a way to further expand the scope of his work.

A peeved Nicholas Butler refused to let him out of his contract, so Maclaurin was obliged to finish out the academic year at Columbia. But he was well into Tech mode by January 1909. Some of his spare time that winter was spent studying Emma Rogers's *Life and Letters of William Barton Rogers*, which he borrowed from the library at Columbia and then got as a gift from Mrs. Rogers herself, now 84 years old. "I had often heard people speak of your husband," he wrote thanking her, "but it was not until I read [your volumes] with avidity, that I learned to know him personally & realized his breadth & charm. The knowledge that I have thus acquired makes me more than ever proud to preside over the Institute, the foundations of which he laid so solidly & well." Maclaurin was struck by how closely his own perspective matched Rogers's: the importance of uniting the special and the general, the professional and the liberal, while at the same time learning-by-doing and problem-solving. That winter he spoke to several Tech alumni groups, whose meetings often began with the Institute's traditional yell, blood-curdling in its volume and intensity. "*We are happy; Tech is hell . . .*" climaxed by a cacophonous, ad-lib crescendo of whoops and hollers. The sound reminded Maclaurin of bagpipes, Scotland's "national weapon."

That spring brought more frequent trips to Boston. Often Maclaurin stayed at the St. Botolph Club on Commonwealth Avenue, just around the corner from the Rogers Building, but one weekend in April Charles Stone and his wife put him up at their townhouse on the corner of Beacon and Dartmouth Streets. He loved the view out back, over the Charles River, yet his eye fixed more on what lay on the other side: a stretch of empty, barren land with a stone wall at the shore, behind that a single row of stunted trees, an open field parallel to the river, and behind that a railroad, factories, and smokestacks puffing soot around the clock. This area had once been part of a grand vision for the riverfront laid out in the 1870s by railroad magnate Charles Davenport, but while the Boston side—the Back Bay—had thrived, the Cambridge side fell victim to the panic of 1893, lying fallow since, a swampy, industrial wasteland. "Why isn't *that* a good site for Technology?" Maclaurin asked his stunned host. Many reasons, said Stone. Waterlogged soil, excavation and construction problems, health hazards (factories, sewage, mosquitoes), and certain opposition by Harvard and others to burdening Cambridge with yet another tax-exempt institution. The whole idea was wildly impractical as far as Stone was concerned, but Maclaurin tucked it away for future reference. Meanwhile, he leased a riverfront residence at 187 Bay State Road for himself and family, Alice and little Rupert.

Tech put him to work on June 1, 1909. He signed diplomas—all 250-plus—for the class of '09, while Noyes disappeared to make preparations for a year-long leave in Europe with companions William Bray, research associate in the physical chemistry research lab, and Richard Ranger ('11), physics student and *Tech* editor, along for the ride. What followed was a blur of lunches, dinners, meetings, commencement activities, alumni reunion events, a special Boston Pops concert. Ernest Nichols, who had just accepted the Dartmouth College presidency, stopped by for a chat. Governor Ebenezer Draper and his wife hosted a reception at the State House. Maclaurin dined with the Lowells, Eliots, and others at Harvard to mark another changing of the guard. Charles Eliot had just stepped down as president, succeeded by A. Lawrence Lowell. "They are really most delightful people," Maclaurin wrote to his wife about the Lowells, adding—"I had a good deal of talk with Eliot and liked him very much indeed . . . certainly a much more pleasing person than I had expected" (Tech gossip had predisposed him to find in Eliot a cold, calculating, ruthless character). The day before his inaugural he squeezed in a tromp over the Harvard Bridge, to get a feel for that intriguing stretch of land first seen from a distance—swampy, trash-littered, junky, as Stone had warned, but in Maclaurin's view a fine prospect just the same.

"I for one am perfectly definite that the thing to do first is to secure a new site and on that site raise a new Technology with all the characteristics of the old Technology." Otherwise, his inaugural address stuck to sweeping brush-strokes: the interdependence between science, technology, culture, philosophy, and social values, the importance of teaching science and technology "in such a way as to make for that broad and liberal outlook on the world that is the mark of the really cultured man." In conclusion, he cited the recent Tech–Harvard struggle. No more talk of merger, he said, but "we should be . . . false to every precept of common sense if we failed to do our utmost to co-operate with Harvard wherever such co-operation is possible. I believe that in the domain of applied science there is much that we can do for our mutual help, but, to make co-operation real and practical, we must be strong enough for independence." This last phrase brought the house down.

As Noyes predicted, Maclaurin's first order of business—relocation and financial stability—turned into a long-term preoccupation. During his presidency, the value of Tech's property rose from $1.7 million to $10.7 million and the endowment skyrocketed from about $2 million to just under $15 million. In May 1909, a month before officially taking up duties in Boston, he approached the General Education Board, part of the Rockefellers' philanthropic empire, which declined to help. Next, he turned toward state funding. With the $25,000 annual state grant set to expire, Maclaurin went after renewal on a larger scale. He considered the current grant "absurdly inadequate . . . paltry compared with what we need," a travesty considering Tech's contribution to the state's industrial and commercial vitality and compared to levels at which California and other state governments routinely supported their colleges and universities. To prove his point, Maclaurin mobilized Tech personnel to compile a record of the Institute's role in national, state, and municipal development. The list included nearly a hundred items: power transmission, telephone technology, locomotive efficiency, control of steel and iron corrosion, recovery and disposal of industrial waste, reinforced concrete, fire-resistance, synthetic rubber, sewage management, traffic control, and much more. No one, Maclaurin believed, could fail to recognize that current funding levels lay orders of magnitude below the value of services supplied. Woodrow Wilson, governor of New Jersey and soon to run for U.S. president, supported Maclaurin's arguments—"I am very much interested in the effort you are making to put the Institute of Technology upon a firm footing. Technical education of the highest sort is of course absolutely essential to the successful development of our industries. It is not a luxury, but a necessity . . . everywhere recognized in the world of education and of business, and it would be a misfortune which the whole country would recognize if anything should happen to impair the usefulness of the distinguished Institute over which you preside." Legislators evidently agreed, approving on May 20, 1911 a million-dollar grant to be paid in annual installments over ten years.

Meanwhile, Maclaurin had begun work on relocation. The present Boston site was noisy and dirty, and because the buildings were so scattered, much time was lost moving students and faculty across crowded, dangerous, increasingly commercialized thoroughfares. The campus had an incoherent feel. The older buildings were in a state of disrepair, some of them crumbling, while the newer ones, erected on-the-cheap, had not been meant to last. Maclaurin appointed a Corporation committee to identify a site large enough (at least 25 acres), affordable, and within reach of Boston—not inside city limits, necessarily, but close enough and able to accommodate "a dignified group [of buildings] worthy of the Institute's importance."

The committee evaluated many options. Among the sites discounted early because of limited size, high cost, or inconvenience were tracts in Hyde Park, Newton, Winchester, Waltham, Brookline, South Boston, Chestnut Hill, Forest Hills, and Cambridge (Fresh Pond and Norton's Woods, the latter eventually home to the American Academy of Arts and Sciences). Serious candidates included sites in Brighton (the Allston Golf Club and the Soldiers Field property originally considered by Pritchett); Cambridge, along the riverbank east of Massachusetts Avenue; Boston's Fenway neighborhood, bounded by Brookline Avenue, the Riverway, Avenue Louis Pasteur, and Longwood Avenue; Jamaica Plain, overlooking Jamaica Pond and bounded by Perkins Street, South Huntington Avenue, Centre Street, and Jamaica Way; West Roxbury, at Allendale and Centre Streets; and Dorchester, between Morton and West Selden Streets. All but three of these—the Allston Golf Club, Fenway, and Cambridge riverbank sites—were eliminated for various reasons, the Dorchester land, for example, because it was swampy and unattractive. Maclaurin liked both the Allston and Jamaica Plain sites for their physical beauty, but he agreed with committee members that one drawback to Jamaica Plain was its "total lack of the 'advertising' opportunities of the Allston site, as well as the convenience in relation to the best residential part of the city with its great institutions." The Allston site was too tied, perhaps, to the abortive Tech–Harvard merger plan, still rankling in recent memory. But Corporation member John Ripley Freeman ('76) backed the Soldiers Field site, directly across the river from Harvard, as a way to consolidate cooperation between the institutions short of merger. "I hope that none of the old prejudice against the alliance will be brought to bear. . . . I firmly believe that tho [sic] the absolute independence of Harvard and Technology in matters of government continue to the end of time, there are great advantages in the association with Harvard College and that this distance of three-quarters of a mile is just about the correct degree of separation." Maclaurin vetoed Freeman's proposal as politically unfeasible, pointing out that Tech–Harvard cooperation, while an important goal, did not require geographic closeness.

What appealed to Maclaurin most about the Cambridge riverbank site was ease of transportation to and from Boston, plus its "excellent outlook over the river and very conspicuous location." But there were concerns, too, about high cost, undesirable proximity to factories and workers' tenements, and likely objections from Harvard and city officials. Sure enough, Lowell quickly informed Maclaurin that Harvard opposed Tech's relocation to Cambridge as "a very serious peril to both institutions," a financial risk that was "so great as to be prohibitive." All in all, then, the Fenway site—prime location, near downtown Boston, an area poised for dignified development, relatively noncontroversial—looked like the best bet if cost and certain structural difficulties could be overcome.

But other sites kept popping up. Late in the fall of 1910, Maclaurin and his group gave up on the Fenway and opted for a 35-acre tract elsewhere in Boston, between Commonwealth Avenue and the Boston & Albany Railroad tracks and ending at the Cottage Farm Bridge to Cambridge over the Charles River. To fund the purchase, Maclaurin made the unusual (for him) impolitic blunder of approaching Andrew Carnegie. Carnegie, still seething over the failure of the Tech–Harvard merger, brushed him off—Scotchman to Scotchman—in spicy brogue: "Ye're no blate. . . . I enjoy the joke! . . . If I mistake not, I am part owner of that ground that my friend Lee Higginson and some of us purchast [sic] to unite the two institutions, *which should be done.*" Maclaurin turned instead to Coleman du Pont, who did not like the look of the proposed site—too small, he thought, and encroached on by a row of auto showrooms creeping down Commonwealth Avenue—but he pledged half a million dollars to purchase it anyway, provided Tech acquired ten additional acres adjacent. Later, in a thinly veiled swipe at former president Pritchett, Coleman's cousin Pierre would write about how pleased he was to work with an administrator as "careful and competent" as Maclaurin.

Other cities, meanwhile, urged Tech to come to them. In January 1911, Springfield, Massachusetts, offered a 30-acre tract on the east bank of the Connecticut River between Springfield and Chicopee. Worcester joined the list of suitors, the president of its local board of trade urging a union between Tech and Worcester Polytechnic "under either name or a new one." The Metropolitan District Commission proposed placing Tech on a manmade island in the middle of the Charles River. Even Chicago piped in. "We could support a 'Boston Tech,'" claimed the *Chicago Evening Post*, "with our loose change, and we wouldn't, like some cities we know of, have to search all the hinterland roundabout to find the money."

Cambridge underwent a change of heart. Concern about another educational institution further reducing tax revenues was offset by the certainty that Tech, ensconced along the riverfront, would slow if not halt the spread of unsightly, smoke-spewing factories and tenements in that direction. The only holdouts were a few politicians "who are without influence in any way, and who accomplish nothing more than attracting attention by their noise." Cambridge mayor William Brooks stated on the record that Tech would be "a blessing and not a burden upon the community." By October 1911 a deal for 46 riverfront acres owned by Henry Marcy, in partnership with thirty-four other owners, was set in motion. The final price of $750,000, paid over in March 1912, was $225,000 less than the original asking price. Coleman du Pont agreed to commit his half-million toward this purchase, with the rest made up by $100,000 from Maria Evans and the estate of her husband, Robert Dawson Evans, and cash contributions from several Corporation members. One member, Hiram Mills, worried about the site sinking long-term and about its structural soundness for building, but Maclaurin persuaded him that technical solutions could be found for whatever problems might arise and that Mills's concerns, if he went public with them, would play into the hands of "men of the baser sort in this community who would welcome any opportunity to lower the Institute's prestige" and to throw its plans into disarray. "I should have supposed," Maclaurin soothed Mills, "that the ingenuity of the engineers of a few centuries might be expected to cope." Boston University eventually snatched up the Commonwealth Avenue property across the river, and the Cottage Farm Bridge was renamed the Boston University Bridge. "Boston is sorry to lose the institute," the *Globe* remarked, "but since it will be in greater Boston it will still be in the family."

Maclaurin's site-and-fund drive drew energy, to some extent, from an accident of history: Tech's upcoming fiftieth anniversary. Here was an opportunity to celebrate, but also to highlight achievements and potential. Maclaurin borrowed the idea from Arthur D. Little ('85), who a year earlier had proposed an event to raise awareness of Tech's value to the community. A two-day function, called Congress of Technology, was planned for April 10–11. "Will bring a bunch of men to the Congress; can't tell how many now. A great many Western Alumni will meet . . . and come with us," wrote Van Rensselaer Lansingh ('98) from Tech's New York club on April 3. Folks from out west would connect by train from Rochester, Cleveland, and Chicago, with laggards catching up on their own. The next day Lansingh's classmate Charles Wing telephoned to say that a sizeable group would come up from New Bedford, riding together in a specially rented train car draped with Tech banners.

Monday, April 10, 1911, fifty years to the day after Tech's charter was granted. A convocation in Huntington Hall, free and open to the public. Following Maclaurin's welcome address, two faculty members—William Walker and Charles-Edward Winslow—spoke on problems and prospects in their respective fields, chemical engineering and public health. Groups then drifted off to attend separate class dinners at hotels, cafés, clubs, and restaurants around the city, from Copley Square to Arch Street to Columbus Avenue and points in between. A parade of alumni converged on Symphony Hall at 7:30 p.m. Here, at 8, the start of a traditional smoker—"a jolly evening," in the words of organizers, "full of music, mirth and mystery." Inside the hall, stereopticon views of Tech past and present flashed onto a vast stage screen, a band struck up popular tunes stage rear, light refreshments made the rounds, pipes and tobacco sold fast at booths in the lobby. The hall was decked out informally, Pops-concert style, with tables and chairs pushed together. Wives, children, and underclassmen were exiled to the balconies, at half price, while alumni and current senior-year men mixed on the floor, puffing away and sharing Tech lore. The air lay heavy with smoke, mixed with the scent of Moxie, muffins, and warm scones.

The next day, April 11, it was down to serious business at Huntington Hall—morning and afternoon sessions, papers by dozens of experts on the technical, commercial, and industrial applications of science, with special reference to the Institute's contributions. The theme of the Congress, a half century of applied science, highlighted Tech's focus on the practical side of science. Speakers included alumni in the railroad industry, architecture, building, electrical engineering, banking and finance, chemical engineering, sanitation, and technical education. Then a banquet back at Symphony Hall, open to the public with seating for a thousand and tickets at five dollars a head. Among the guests were governor Eugene Foss and Boston mayor John ("Honey Fitz") Fitzgerald. Fitzgerald, grandfather of U.S. president John F. Kennedy, praised Tech's populist style, its democratic openness, its scorn for "the nonsense of caste" and "the foppery that goes with idle wealth." The chef of the exclusive Eastern Yacht Club and the Country Club, specially hired for the occasion, churned out caviar, turtle soup, roast jumbo squab, and sweetbreads glace, capped off by fancy frozen pudding ices, coffee, pecans, and olives. Alumni went away with renewed pride in their alma mater, but more—a sense of urgency that here, in its hour of need, was an opportunity to repay an old debt, to fulfill an obligation postponed or till now taken for granted.

Maclaurin sometimes referred to 1911 as Tech's annus mirabilis. Whether or not he actually believed in miracles, it was a year when hard work and sheer good fortune joined forces. Besides Coleman du Pont's half-million and the million-dollar grant from the state, bequests from the estates of Emma Rogers and Francis Greene totaled around a half-million dollars each along with $100,000 left by Mrs. Frances Perkins and $50,000 by Nathaniel Thayer. These funds effectively plugged Tech's deficit, which would otherwise have reached a new high in 1911. "Everything seems to be coming the way of the Massachusetts Institute of Technology just now," *Globe* editors remarked, "well, it's a deserving institution."

On January 3, 1912, Cambridge's mayor inked final approval for Tech's move across the river. Two weeks later Maclaurin set out on a promotional tour through New England, New York, and the Midwest. The goal was straightforward enough: spread news of Tech's good fortune and raise money to build on the new site. Maclaurin's first stop was New York City, where Thomas Alva Edison told alumni at a special banquet that "more nearly than any other school or college in this country, the Massachusetts Institute of Technology meets the demands of modern American life." Edison's son Charles was a Tech student ('13), and the wizard of Menlo Park sang the Institute's praises as Maclaurin's campaign pressed forward. In March the *Yale Daily News* quoted him lashing out at traditional colleges and lobbying hard for the Tech brand. "I would employ almost any graduate of that institution who came to me," he told a stunned interviewer, but "in my business, if a Yale or a Harvard man should come to me for employment I should probably say that there was no place vacant." Edison counted thirty or more Tech graduates in his enterprise at any one time, with a mere handful of hires from liberal-arts backgrounds.

The longest stop on Maclaurin's tour was Rochester, New York. Frank Lovejoy ('94), top assistant to the Eastman Kodak Company's founder, George Eastman, persuaded Maclaurin to stay over for two days, February 2–3. A number of Tech grads had worked at Kodak over the years and Lovejoy believed that Maclaurin's visit would inspire them. Eastman, like Edison, thought of Tech as a hunting-ground for technical talent. His first hire from there had been Darragh de Lancey ('90), who designed and supervised the construction of Kodak's original plant and managed its operations for nearly a decade. In 1891, Eastman had asked Tech professor of chemistry Thomas Drown to recommend a young chemical engineer or industrial chemist for the firm's nascent research and development program. "I do not," he wrote Drown, "want anyone who is not painstaking, thorough and thoroughly reliable. Harum-scarum youths are not of any account in this business." Tech graduates streamed to Kodak in rapid succession by the mid-1890s. After de Lancey came Lovejoy, Harriet Gallup ('94), James Haste ('96), Henry Tozier ('96), Edward Woodworth ('97), Charles Flint ('01), Albert Sulzer ('01), Martin Eisenhart ('07), and others.

Eastman happened to be out of town during Maclaurin's visit, but Lovejoy wrote later that month to say that his boss was inclined to lend a helping hand in light of "recent developments that had made a new Technology possible." Maclaurin introduced himself by letter on February 29. Two days later Eastman asked Maclaurin to meet him in New York City. Maclaurin telegraphed ahead on March 4, grabbed the mid-morning train, pulled into Grand Central Station around mid-afternoon, and headed for the Belmont Hotel, where Eastman was staying. The two met briefly and hit it off right away.

Next morning, Eastman pledged two and a half million dollars for building construction. Maclaurin wondered if he had misheard. Surely the Kodak king meant two and a half thousand, maybe two hundred and fifty thousand? Check the number of zeroes, how many and in what order. But no mistake, $2.5 million. Flabbergasted, Maclaurin made his way back to Grand Central. On the ride home he played out the scene over and over. Almost too easy—no probing, little discussion, few questions, no pre-conditions. Both men agreed the money should be used to design and construct a group of functional, simple buildings, with Maclaurin confirming "the inappropriateness of the Institute indulging in any extravagant architectural features and the desirability of getting breadth of effect, more by the proper grouping and general design of the buildings than by elaborate details." The new campus must project Tech's unique culture, its differences from the traditional college or university, its distinctness from gothic, baroque, ivy-covered Harvard a couple of miles down the road.

Eastman's insistence on anonymity—he did not want his gift to elicit requests from other institutions—sparked a public guessing game that continued for several years. "May I tell my wife?" Maclaurin asked. "Well, yes, but no one else," replied Eastman, a lifelong bachelor. Their secretaries, clearly, also had to be let in on the secret. Maclaurin coined the alias Mr. Smith, the most ordinary, nondescript moniker he could come up with, and both men found no end of amusement in all the frenzied speculation revolving around the benefactor's identity. Two wealthy New Yorkers, suspecting each other, tried to pry the truth out over dinner. One woman convinced friends that her husband was Mr. Smith. Maclaurin floated clues, enough to discount certain suspects but never enough to point toward Eastman. Mr. Smith was not a Tech graduate, did not reside in Massachusetts, had never supported education at this level before. The latter hint pulled attention from Andrew Carnegie and John D. Rockefeller Jr. toward men like Thomas Edison, coal mogul and art patron Henry Frick, banker J. P. Morgan, and diplomat John Hays Hammond. The chatter so annoyed Edison that in June 1912 he issued an emphatic denial: "I can use my money to a thousand times better advantage than any college in the country." Press coverage kept Tech in the public eye, and Maclaurin played it to the hilt. "Mr. Smith," said Maclaurin, hoping that others would follow his example, "showed the modern business man at his best, ready to do a great thing quietly and unostentatiously."

Eastman gave Maclaurin carte blanche to proceed with building plans. His suggestion for architect—Charles McKim, of McKim, Meade & White, creators of the Boston Public Library, Boston's Symphony Hall, Columbia University's Morningside Heights campus, New York's Penn Station, and other notable landmarks—did not appeal to Maclaurin, not because Maclaurin had anything against McKim's work but because he preferred a Tech alumnus for the job, someone steeped in the Institute's culture. Maclaurin's first choice was Cass Gilbert ('80), designer of the much-talked-about Woolworth Building in Manhattan. But when negotiations broke off early in 1913, William Welles Bosworth ('89) was hired based on glowing testimonials from banker Frank Vanderlip, AT&T president Theodore Vail, and especially John D. Rockefeller Jr. Bosworth had just completed a residence for Rockefeller on West 54th Street in Manhattan, which, as Rockefeller told Maclaurin, captured Bosworth's "strong leaning towards simplicity and dignity in architecture, rather than complication and ornateness." This was exactly what Maclaurin wanted to hear, intent as he was on hiring someone whose "trained taste . . . can be relied upon to produce fine effects without ornamentation or frills of any kind." Maclaurin made a special trip to New York to inspect Rockefeller's new mansion, and came away impressed. Two other alumni—James Knox Taylor ('79), a member of Tech's architecture faculty and a former partner of Cass Gilbert's; and Harry Gardner ('94), also on the architecture faculty—were brought in as consultants. When some complained about the project's snail-like pace—Bosworth did not sign on until nearly a year after Eastman's money was in hand—Maclaurin blamed engineering and other complications. "The dirt is not flying," he declared in October 1912, "because a careful and elaborate study is absolutely necessary." Symbolic first ground was broken on September 15, 1913, and workers poured the first concrete on April 9, 1914.

Bosworth's plans met with widespread approval, particularly his concept for a simple, continuous, efficient building—or series of interconnected buildings—organized on a grid with passageways intersecting at right angles, and adaptable for future expansion. He worked from Maclaurin's brainchild, a great white city around a central courtyard facing the Charles River. Bosworth's volatile temperament sometimes got in the way, but Maclaurin reined him in with help from Corporation members Charles Stone and Edwin Webster, whose company Stone & Webster was contracted for the engineering tasks, while Bosworth focused on the esthetic. Some worried about Bosworth's carefree attitude toward cost overruns, others that his beaux-arts background would send him off on artsy tangents. Because of these concerns, the Great Dome, one of Tech's best recognized landmarks, narrowly escaped being axed from the plans. The expense, first of all, was phenomenal—$700,000, almost a third of Mr. Smith's total gift and just slightly short of the purchase price of the land. Why spend this kind of money for mere architectural effect? How about something functional, maybe one more wing or another lab? Even granted that sprawling corridors might need some unifying feature, was a dome the best choice? To some, it seemed wasteful but also like an asynchronous add-on, a Roman pantheon plunked overhead, a sign of the architect's obsessive taste for the neoclassical. Why not forgo all echoes of the past—classic, baroque, gothic—and showcase a modern, progressive esthetic? Maclaurin supported Bosworth on this, however, arguing that the dome made for a stable nucleus. Opposition eased when the dome's most expensive feature, a grand auditorium under the rotunda, was scrapped in favor of a working library. The dome's impressive scale—smaller than St. Peter's in Rome, but larger than St. Paul's in London and the Washington (D.C.) Capitol—sparked much public comment.

The new buildings went for efficiency over ornament. Constructed of reinforced concrete faced with Indiana limestone, they radiated a geometric elegance, the spare esthetic—knowledge stripped to its barest essentials—that appealed to engineers, scientists, and mathematicians just as baroque fussiness and gothic mysticism resonated with some humanists. Squares, rectangles, circles, cylinders, concentric shapes—all pulled together by the sphere, the Great Dome, that Maclaurin had salvaged. The contours were designed to fit the needs of the 2,500 or so people who would study and work here. The inner corridors, uniform in width and height, connected at right angles without accent or decoration at the bends, all very frugal. Dreary, grayish-brown wall paint was chosen not for its stylishness or appeal but to resist scuffing. Exposed pipes stretched overhead in clumped ranges, veering off at sharp angles into labs and offices. Doors along the corridors were hard to distinguish one from the other, the Tech insider's scheme of numerical coding for buildings and rooms—no names—the only clue to what went on inside. Each office had a blackboard, with rugs and paintings in scarce supply. Life revolved around apparatus of varying shapes and sizes tucked between rows of long wooden tables laid out in neat, space-saving rectangles.

Outsiders got the impression that there was little human texture to the place, that it resembled a factory more than a school. Yet its plain, austere image was a source of self-pride, symbolizing the value of results over frills. The Great Court that stretched between the dome and the river would remain a bleak patch of gravel for some time—treeless, shadeless, flowerless, no comfortable place for students to sit around. Bosworth's plans had called for grass plots, greenery, shrubs, and fountains, but these never materialized. In 1917, a tree-planting campaign promoted by Corporation member Frederick Fish did not get very far.

Maclaurin recognized, however, that certain quality-of-life issues had to be addressed. Whether or not Tech would evolve as a fully residential campus—its culture had always been otherwise, commuter-based—he laid plans for a student quarter that would snake either along the riverfront or down Massachusetts Avenue. A 1913 survey showed 40 percent of students wanting to live on campus and 60 percent preferring to stay in Boston and commuting either over the Harvard (Massachusetts Avenue) Bridge or via subway into Kendall Square over the Longfellow Bridge. There was little money to spare, but Maclaurin persuaded Coleman du Pont and an anonymous donor to underwrite the construction of Tech's first dormitory—what became known as Senior House—as a pilot project to accommodate about two hundred students. Charles Stone and Edwin Webster, meanwhile, offered a president's house. While the Maclaurins had always enjoyed entertaining students in their rented townhouse at 187 Bay State Road, there were benefits to an official residence in the heart of campus. The president, Maclaurin said, "should take an active part in the social life of the students and a suitable house upon the grounds of the Institute should prove a most valuable aid." Maclaurin managed to keep Bosworth from indulging in excesses of size, height, vaulted hallways, and four-thousand-dollar bronze entryways. The president's house, situated next to the dormitory, was completed in 1917, reflecting the unfussy elegance that Maclaurin preferred. Its design, according to one observer, was "typically Technology—utilitarian . . . quiet without being stately . . . homey in spite of its large spaces [with] richness and coziness well combined." One friend teased Maclaurin about moving into "your new palace (or do you call it a shanty)?" The cornerstone for the much-delayed Walker Memorial was laid by Francis Walker's widow, Exene, on June 12, 1916, with four of their children in attendance. The Walker building, placed between the main group and the president's house, was completed in 1917 as a center for student social activities.

While he opposed ornaments on principle, Maclaurin made one exception. Clarence Ward ('72) had asked if an inscription could be placed over the main entryway, something like "Let no one ignorant of Geometry presume to enter here," a variant of the phrase believed to have greeted students at Plato's Academy. The question caught Maclaurin off guard. While he had allowed Bosworth to place a few lion-head gargoyles and acorn finials along the outer facade, widely dispersed and high enough so as not to draw attention from the overall plain lines, he worried about adding yet another frilly, decorative feature. But while inscriptions were ornamental, they could also project a sense of mission. Plato's phrase, of course, would never do. Too many people thought of Tech as unapproachable, and a warning directed at those weak in mathematics would compound this problem. But Ward's general suggestion, some message to capture what the place was all about, made good sense.

For the main entrance, Maclaurin went for the stark minimalism of name and date. MASSACHVSETTS INSTITVTE OF TECHNOLOGY on the horizontal slab atop the ten Ionic columns, MCMXVI halfway up the dome's base. When Bosworth, in April 1915, urged him to consider inscriptions for the attics of the Great Court's four pavilions, Maclaurin suggested using names of men famous in science, engineering, and allied fields. But which men, and which allied fields? Maclaurin knew how controversial this question could prove when he whipped off a letter to department heads asking them to discuss the matter with colleagues and get back to him. As the carvers were already at work, he needed a quick response. The letter went out on a Thursday and Maclaurin wanted to hear back by Monday or Tuesday. He imposed just two limits: no living candidates, and only those whose reputations would survive the scrutiny of generations. He also asked that candidates be placed in pecking order, as the attics had two tiers: the lower tier would carry ten names in large script, the upper about a hundred names in small(er) script.

Some departments pulled together lists by Monday afternoon, others came in not far behind. The lists shared little in common except for acknowledged giants such as Newton, Darwin, and Lavoisier. The result, an unwieldy pile of names far in excess of what could be used. How to winnow it down? Maclaurin sought advice from executive committee members, but their views were in conflict too. Maclaurin himself lobbied hard for Shakespeare, arguing that "most of the students are prospective engineers and we should suggest to them not only that they should be interested in literary form, but especially that they should be interested in human nature and I should place Shakespeare before them as showing pre-eminently an insight into humanity in its limitless phases." George Kittredge, a literary scholar at Harvard, was asked for an authoritative opinion on which spelling was preferable—"Shakespeare" or "Shakspere"—but the great bard went by the boards in the end, attracting no supporters other than Maclaurin. If Shakespeare, one faculty member wondered, then why not Homer, Virgil, Dante, or Goethe? Some urged Maclaurin to include the living, for without men like Thomas Edison and Alexander Graham Bell, few if any Americans would make the grade. But Maclaurin stood firm against the idea as a Pandora's box likely to embroil Tech in unwanted controversy. His final choice for ten names in large lettering were Aristotle, Newton, Franklin, Pasteur, Lavoisier, Faraday, Archimedes, Leonardo, Darwin, and Copernicus. Maclaurin chose Benjamin Franklin, the lone American in this cohort, because in his opinion the faculty's most popular American choice, physicist Joseph Henry, did not measure up next to the other nine. Leonardo's versatility as scientist, architect, engineer, mathematician, and astronomer vaulted him over another popular artist-cum-scientist, Michelangelo. Maclaurin included Joseph Henry, along with several other Americans, in his final cut for the hundred-or-so (actually, 105) names engraved in small lettering. Meanwhile, he had the inspirational phrase "Alia initia et fine"—roughly translated as "a fresh start from every finish," culled from Pliny the Elder's *Natural History*—carved above the fireplace in his new office.

Maclaurin had hoped to sell off most of Tech's Boston real estate to help finance building, upkeep, and future expansion in Cambridge. But in 1915 the state supreme court, which awarded the Institute its long-sought title to the Boylston Street property, placed so many restrictions on use, sale, and disposition that Maclaurin had to choose between holding on or selling for far less than the $3 million he expected to get. He held on, for now. The Rogers Building became home to Tech's architecture department while the Walker Building was leased to Boston University to house its business administration school. This setback was offset, for Maclaurin, by the sense that Tech would now retain more than symbolic ties to Boston and that older alumni, distressed over their alma mater throwing off so central a part of its heritage, would feel relieved. Some alumni found it hard to let go, to reconcile themselves to this "uprooting of traditions, of sentiments, of loyalties that have been digging themselves deep for half a century." Tech's engineering buildings on Trinity Place went up for sale in January 1916 and were sold in part, along with the property on Garrison Street, by the end of the year.

Maclaurin wanted to hold a simple, dignified dedication ceremony that June. His appointment, however, of Tech's architecture professor Ralph Adams Cram to manage the affair guaranteed that it would be anything *but*. Cram, known for his ultra-gothic tastes, had designed churches, cathedrals, and part of Princeton University's campus in the rich, ornate styles that Maclaurin disliked. Charles Stone warned that Cram would need watching "to prevent him from doing the foolish things that people with artistic temperaments are apt to do." In April, when Maclaurin criticized Cram's ideas as bloated, Cram threatened to walk away but stayed on when Maclaurin let him retain control over the pageant scheduled for June 13. Maclaurin took charge of plans for the dedication ceremony next day. Cram's pageant concept so alarmed Maclaurin that he washed his hands of it (the alumni association took over), withdrew himself and the Corporation from any official role, and advised all that while he did not endorse the event, they were free to take part as individuals.

The pageant fully lived up (down?) to Maclaurin's expectations. In the early evening the *Bucentaur* set sail from the Boston side of the Charles River with dignitaries and Tech's 1861 charter on board. The 100-foot-long vessel, modeled after an ancient barge built for Venetian nobility, held in its prow a massive carving—"Technology enlightening the world," by Cram's mythology—with a T-square in her left hand and a torch in her right. Surrounding her were carved Neptunes, Venuses, nereids, fish, dolphins, an eagle. On either side a frieze of seahorses, dolphins, mermaids, cupids at play in the waves. Four Tech seniors carried the charter in a gold casket down the gangway on the Cambridge side, trailed by four others bearing a bronze beaver, nature's engineer, adopted as Tech's official totem-mascot in January 1914. The state governor, Boston and Cambridge mayors, other dignitaries proceeded to thrones in the Great Court. A grand masque followed—cast of over a thousand—allegorizing age-old conflicts between man and nature. Song, dance, chorus, orchestra, poetry, searchlights, fireworks, horses, triumphal marches. Actors portraying the elements Earth, Air, Electricity, Fire, Water, Steam; the vices Greed, Vainglory, Selfishness, War; the virtues Will, Wisdom, and Righteousness. Cram as master of ceremonies, holding sway (and forth) as Merlin the magician. Boston had rarely if ever witnessed a spectacle like this, and the alumni loved every minute of it. Some, vying with Cram and each other, had devised inventive ways to reach Boston. The New York delegation came via chartered steamship, rechristened *M.I.T. Ropolitan*, escorted by tugboat whose pilot was an alumnus costumed as Neptune. Another group planned to fly in, another to bring with them a trained whale. On the eve of the pageant, Godfrey Cabot ('81) spelled out his class numerals in smoke and vapor from a massive hydroplane buzzing overhead. As enthusiasm rose, some such ideas—impractical, complicated, or outlandish—fell by the wayside.

Maclaurin felt relieved, anyway, that Cram's extravaganza did not turn Tech into a laughing stock. The muted, two-hour-long ceremony next day was more to his taste, an opportunity to showcase Tech as "an institution without nonsense and without frills devoting itself always with seriousness of purpose to the business on hand." Four trenchant speeches followed a brief invocation by Maclaurin family friend and minister, George Gordon, of Boston's Old South Congregational Church, whom Maclaurin admired for his spirituality combined with commonsense. First up Maclaurin, then Massachusetts governor Samuel McCall, Harvard's president Lowell, and U.S. senator Henry Cabot Lodge. Ample time before and after for visitors to stroll around the Wright brothers' original flying machine, the one that had flown for twelve seconds at Kitty Hawk in December 1903, on loan from special guest Orville Wright. That evening, a banquet at Symphony Hall. Telephone lines linked the hall to venues in thirty-two other cities. "Hello, New York." "This is New York." "How many have you there?" "We have 130 members and guests." And so on down the roll. Maclaurin spoke, then Alexander Graham Bell and Orville Wright followed by Coleman du Pont and Henry Pritchett. Warm applause for Pritchett, a not-so-long-ago butt of alumni anger, signaled resolute bygones for a new era.

While some found it difficult to break with the past, others were ready to move forward with few regrets. Over the next several years, Tech or Boston Tech would become known, finally, as MIT, a switch that captured something about the Institute's new geographic, intellectual, and cultural contours. A number of old-timers stuck with Tech, anachronistic though the term was, while newcomers latched onto MIT. A song from 1907, "Dear Old M.I.T.," had proven popular but not popular enough to budge the vernacular. It was up to one old-timer, inspired by the move to Cambridge, to set the change in motion a decade later. Arthur Hatch ('91) penned this tribute in 1917:

Come, raise a song of M.I.T.,
And let it ring right loyally,
With mighty cheers, like ocean's roar,
For our dear home on the Cambridge shore.

Unfurl our colors here today,
Cardinal red and silver gray;
Our flag to all good courage brings,
So join with us while old "Tech" sings.

M is for Mass., for her we root;
I for the dear old Institute;
T for Tech, Technology (you see).
Now all together:—M.I.T.

After that, older school songs—"Tech forever" (1903), "Tech men" (1914), "Technology rag" (1915), "O Institute, Technology!" (1916)—made way for "Alma mater" (1923, with references to both Technology and M.I.T.), "The courts of M.I.T." (1925), and "Hail! M.I.T." (1927). By the late 1920s, the shift was all but complete.

Maclaurin worked hard to heal lingering sores opened under Pritchett. In November 1909, he began to talk with Lowell about how Tech and Harvard might cooperate. He did so on the sly, out of view, with residual tensions on both sides still palpable four years after the Pritchett–Eliot fiasco. In public he couched the matter in broader terms, as cooperation between neighboring institutions—no mention of Harvard—aimed at streamlining efforts, reducing duplication and waste, "a problem of some difficulty and delicacy . . . that cannot be solved satisfactorily without care and patience." A year later Harvard crept into his announced program, alongside several other local institutions. By 1910 Tech's staff taught part-time not only at the Lowell Institute School for Industrial Foremen, an arrangement in effect since Pritchett's time, but also at the Young Men's Christian Association, Franklin Union, Wells Memorial Institute, and People's Institute. William Sedgwick and Samuel Prescott lectured on biology and public health at Simmons College. Maclaurin encouraged a loose framework for faculty exchange and student cross-registration with Harvard. In September 1910, when Harvard asked if Tech's F. Jewett Moore would teach an organic chemistry course there, Maclaurin told Moore that this was "entirely proper," provided his other work "did not suffer," and that it would serve a larger end: to improve relations between the institutions to the "lasting benefit" of both. Other Tech faculty who taught courses at Harvard in 1910–11 included Désiré Despradelle, on architectural design; Davis Dewey, on monetary policy; E. B. Wilson, on applications of probability in theoretical physics; and Gilbert Lewis, on Einstein's new, controversial theory of relativity. Harvard students went to Tech for Hervey Shimer's class in paleontology, while Tech students went to Harvard for physiography.

"The cooperative spirit is evidently at work," Maclaurin observed in January 1911, "how far it should lead us is one of the larger problems that confront us." For him, it was mostly a question of trajectory control: press far enough to generate results, brake soon enough to avoid identity compromise—find that tenuous spot where "anything much less would be ineffective in practice, and anything much more would mean amalgamation." But when Harvard's dean of applied science, Wallace Sabine, implied to Maclaurin in 1912 that Harvard, "with its greater financial resources, can put Technology out of business," prospects looked grim. Then, with Tech's position strengthened by the purchase of land for a new home, by George Eastman's enormous cash gift to build, and by the lifting of fear among Institute loyalists that they were "in danger of being swamped," Maclaurin was able to brush Sabine aside and take a tougher stand with Lowell.

Their negotiations grew prickly. Lowell parried, Maclaurin counter-parried. "You're holding a pistol to our head." "I'm not in a position to hold a pistol up to Harvard, even if I had any inclination to do so." "It is clearly impossible to agree upon distribution of a cake if each of the claimants feels entitled to the whole of it." "It would seem to me that we are discussing the *making* of a cake rather than its distribution, and I agree that the cake will not be very palatable unless the cooks restrain their individual preferences for the placing of the ingredients." After this heated exchange in March 1912, discussions broke off until Maclaurin approached Lowell once again the following January. By this time Lowell had had a change of heart, for he wrote back in February inviting Tech to join with Harvard to form a joint school of applied science bearing the name of both institutions, supported in part by funds from the Gordon McKay trust. The agreement he drew up, however, was so skewed toward Harvard that in April Maclaurin called it a nonstarter. Lowell revised it in May, and Maclaurin sent it back with additional revisions on June 18. After yet more back and forth, officers for both institutions signed off—simultaneously—on January 9, 1914.

The agreement called for cooperative teaching and research efforts in four engineering programs: civil and sanitary, mechanical, electrical, and mining and metallurgy. Harvard would contribute its fifteen engineering faculty and about $100,000 annually in income from the McKay trust. Tech would supply the infrastructure—buildings, equipment, other facilities—and take charge of all instruction. Faculty would be listed as members of both institutions, while retaining their primary affiliations. Students would be awarded both Tech and Harvard degrees. Each institution would consult with the other on new faculty hires, but appointments, salaries, and other personnel issues would remain separate. Tech's president would function as executive head of the joint school, reporting annually to Harvard's president. Harvard's president would have the right to consult on the appointment of any new Tech president. The agreement could be terminated by either side with at least five years' notice to the other or, if mutually acceptable, within a shorter time frame.

Maclaurin felt confident that the plan would "put an end to all misunderstandings between Harvard men and Technology men and should end, once and for all, the difference between merger and anti-merger men." At last—"an alliance between independents, and no merger," an agreement "spelling cooperation and nothing more." Many shared Maclaurin's optimism. One widely held view among Tech students was that Harvard had come to its senses in acknowledging the Institute as "its equal, if not its superior, in certain lines, co-operating with us and not absorbing us, as was its former policy." But there were skeptics, too. Charles-Edward Winslow ('98), in a speech to fellow alumni in January 1914, warned that the institutional cultures were so far apart that mixing them in this way risked a destructive clash of values. When Maclaurin urged him to soften his tone in publishing the speech, Winslow agreed out of deference to the president and willingness to give him the benefit of the doubt. Lowell faced deeper opposition at Harvard, not only from Wallace Sabine but also from the McKay trustees and from alumni. What did Harvard get out of it? This question was hurled regularly at Lowell, who found himself always on the defensive and sometimes responding testily. "She gets plenty out of it. . . . But that is not a question to ask; that is not important." A delegation of Harvard men descended on Henry Pritchett's office in New York, complaining that Tech had "put over a deal which robs Harvard of its scientific school without giving anything adequate in return." In April 1914, the McKay trustees demanded that Harvard seek an opinion from the state supreme court as to the legality of (ab)using the trust in this way.

Meantime, the venture went ahead on an interim, experimental basis. Opponents did not insist that it shut down pending a judicial decision in part because another collaborative Tech–Harvard program, a school for health officers, had opened to much acclaim and without the controversy that swirled around the joint engineering school. Harvard and Tech agreed to create the health-officers program in April 1913, and the first class entered that fall. Five students graduated in 1914, each with a certificate in public health (C.P.H.) signed jointly by the presidents of Tech and Harvard. The administrative board consisted of William Sedgwick (chair), Milton Rosenau (director), and George Whipple (secretary). Rosenau and Whipple were both on Harvard's faculty, but Tech deemed the balance satisfactory with Sedgwick as chief executive and Whipple, a Tech alumnus ('89), a close ally of Sedgwick's. "The school appears to have met a real and pressing need," Sedgwick wrote at the close of the first year, "to have found a ready response with the public, and to have exhibited a quick and natural growth." Maclaurin, confident that the school would emerge as "a real power in the land," hoped that it would change the minds of some skeptics on the larger issue of Tech–Harvard cooperation.

But as Winslow and others on the Institute side soft-pedaled their concerns, Harvard-based critics grew more shrill. Their case, some of it grounded in erroneous assumptions—that Tech, for example, was "a school of technical education for the training of pupils of either sex in practical arts"—aimed to convince the court that the agreement flew in the face of Gordon McKay's objective to bring together engineering and culture at *one* school, Harvard, that embodied "unique and transcendent prestige." A single justice took testimony in February 1917 from Lowell, Maclaurin, and others, even 83-year-old Charles Eliot. Four of seven justices heard arguments on October 15, 1917. The court ruled in November that McKay's original intent prohibited the use of trust funds for any institution other than Harvard, a decision that voided the Tech–Harvard agreement.

For Maclaurin, the ruling was a crushing disappointment. He had invested so much time and effort, with positive results. In four years around three hundred joint bachelors' degrees and fifty joint higher degrees in engineering were awarded. There had been rough spots, to be sure, some of them echoing Winslow's predictions about culture clash. Institute faculty grumbled that Harvard's faculty got higher salaries and better perks (Harvard subsidized secretarial assistance, Tech did not). Harvard faculty thought Tech's mathematics and chemistry prerequisites stringent to the point where Lowell, in trying to defuse this issue, began to "feel like an oil can for purposes of lubrication." Sabbatical policies were difficult to reconcile. Lowell insisted that leaves be granted for recreation, study, or book writing only, while Maclaurin argued that Tech's policy—allowing a faculty member to consult for industry, teach somewhere else, work in a laboratory overseas, conduct experiments, get involved in whatever would help keep him abreast of developments in his field—made more sense from the engineering educator's perspective. Maclaurin felt pestered by Lowell's insistence on volunteers to lead Harvard's morning prayers, a tradition unknown at Tech. When Lowell fussed about the physical dangers faced by students in field work, Maclaurin told him that all would be well "provided our men do not play the fool." Even the issue of what letterhead to use provoked debate. When Harvard suggested "Harvard University and the Massachusetts Institute of Technology Department of— Engineering," Tech countered with "Massachusetts Institute of Technology Department of— Engineering maintained by Harvard University and the Massachusetts Institute of Technology." But despite these and other remnants of an age-old competitive struggle, the program moved well from an academic point of view.

Maclaurin hoped to salvage the agreement on some other basis, but signs looked ominous. For one thing, both he and Lowell were preoccupied by the war, and by the growing role of their respective institutions in national defense. Harvard's attitude hardened, too, drawing energy from Wallace Sabine's not-so-long-ago mantra about putting the Institute out of business. Much tension arose over redeployment of faculty, who would be allowed to stay at MIT (part-time) and who would not. None of Lowell's ideas on how to pick up the pieces and move forward appealed to Maclaurin, or to the Institute's executive committee. According to Maclaurin, Lowell's suggestions left "real problems practically untouched." Someone wondered about bringing in an arbitration team consisting of three or four sensible leaders in education led, perhaps, by Henry Pritchett. But that suggestion went nowhere. In February 1919, Maclaurin heard gossip to the effect that MIT had churlishly rebuffed generous proposals by Harvard. Later that month he halted negotiations, convinced that the atmosphere had turned too poisonous for good-faith efforts to proceed. The Institute, he said, would not serve "as a temporary convenience" for Harvard, nor sign onto any one-sided proposal where Harvard "paid for and controlled the instruction and research" while MIT "provided the buildings" and "paid the coal bills." The demise of Tech–Harvard cooperation—evidence, Maclaurin sadly observed, of a retreat into provincialism on Harvard's part—did not affect the school for health officers, which kept going until Harvard, with support from the Rockefeller Foundation, formed a separate school of public health in the early 1920s. A proposed collaboration between the architecture departments was scrapped in 1918.

Maclaurin shared his predecessors' preference for broad-gauge approaches to scientific and technological education—overlap between liberal and professional studies, importance of research (pure and applied, in appropriate balance), and science as both a cultural force and a framework for career development, industrial service, and other practical lines. In his first year as president he referred often to the ties between science and humanism, to his notion of science as the "crown of culture . . . the ability, or even merely the effort, to frame a reasoned synthesis of all that reality is." He defined research as "the very breath of life to a scientific school," and by this he meant research both pure and applied. For Maclaurin, Arthur Noyes's work in the Research Laboratory of Physical Chemistry and William Walker's in the Research Laboratory of Applied Chemistry exemplified the best of both traditions, which he saw as complementary rather than in conflict.

But as financial pressures mounted, so too did Maclaurin's emphasis on the product-oriented, the income-friendly, the practical. Most lines of research introduced on his watch were engineering-based, not science-oriented. In 1911, the electrical engineering department studied ways to improve electric vehicles while naval architecture worked on ship propulsion. A research division in electrical engineering, analogous to the laboratories already in existence for physical chemistry, applied chemistry, and sanitary engineering, was established in 1913 with support from AT&T and other companies. One of its inaugural projects looked at problems of speech clarity in telephone transmission. Also in 1913 Maclaurin arranged with U.S. Navy secretary Josephus Daniels for Jerome Hunsaker, of the naval constructors' corps, to come to Tech to train aeronautical engineers and to promote research on "flying machines" and manned flight. A research laboratory of aerodynamics was founded in 1914, with a wind tunnel—the first of its kind in America—to evaluate the impact of air currents at then-breathtaking velocities up to 40 m.p.h. Results laden with military applications flowed to the Navy in a series of confidential reports. One study in 1915 assessed the influence of gyroscopic stabilizers on the motion of aircraft exposed to high wind gusts. The problems tackled by these various laboratories arose, Maclaurin observed, out of "the actual difficulties of our industrial life" and presented Tech with key opportunities in light of how rapidly such problems continued to crop up. Engineering research was vital to "the whole field of profitable enterprises," and with German supplies cut off because of the war, chemical engineering was uniquely positioned to stimulate America's next industrial boom.

Maclaurin's emphasis on applied fields spilled over into the teaching program, where what he called the Institute's "especial aim . . . to bring our men into direct touch with the actual conditions of life" drew renewed interest. An option in industrial physics, added to Course VIII (physics) in 1913, was intended to meet growing demand for graduates trained in the application of physical knowledge to industrial problems. The joint Tech–Harvard school for health officers was created. Also in 1913, talks that had gone on ever since Francis Walker's time with respect to a possible course in business management moved into high gear. Course XV (engineering administration), more narrowly construed than earlier concepts, accepted its first class in the fall of 1914. Students chose one of three options—civil engineering, mechanical and electrical engineering, or chemical engineering—and spent about three-fourths time on technical subjects, one-fourth on business methods, economics, and law. In 1916, the Institute opened a school for chemical engineering practice intended to link the curriculum directly with industry. The school, conceived by Corporation member Arthur D. Little and made possible by a seed grant from George Eastman (again anonymously), assigned students to different stations, ordinarily a company plant or factory, to acquire hands-on experience in various processes, operations, and production methods. "In this profession," declared Tech's visiting committee on chemistry and chemical engineering, "one needs to get into the water to learn to swim." William Walker handled overall coordination. Each station was managed by a faculty member, designated as director. Station A, in Bangor, Maine, dealt with paper manufacture (Eastern Manufacturing Co.); Station B, in Everett, Massachusetts, with high-temperature processes (New England Gas & Coke Co.); Station C, in Niagara Falls, New York, with electrochemistry (Carborundum Co.), Station D, in Stamford, Connecticut, with dyes (American Synthetic Color Co.); and Station E, in Allentown, Pennsylvania, with grinding and crushing processes (Atlas Portland Cement Co.). An analogous venture, a cooperative course between MIT's electrical engineering department and the shop-factory facilities of General Electric in Lynn, Massachusetts, was established in June 1917. By 1919, an option in industrial biology was added to Course VII (biology).

On the scientific side, prospects looked bleaker. Some science faculty had the foresight to sense this near the start of Maclaurin's administration. Reginald Daly, one of Tech's prime assets in geology, defected to Harvard in 1912. Also that year two of Arthur Noyes's protégés in physical chemistry, Gilbert Lewis and William Bray, accepted offers at the University of California, Berkeley, taking with them prized graduate student Merle Randall ('12). In 1913, research professor of chemical biology Earle Phelps ('99) left to take charge of a federal laboratory in Washington, D.C. Charles Kraus ('08) resigned to head the chemistry department at Clark University in 1914. Noyes had thought about going to California himself—Berkeley offered him the chemistry department chairmanship in 1908—but on advice from George Hale, and out of loyalty to Tech, stayed with a view to building what Hale called "the most successful laboratory of chemical research in the United States if not in the world." Both men later regretted the decision, as Maclaurin's program veered away from their core academic values. They worried, too, about the part they had played in Maclaurin's appointment as president.

Noyes brooded in silence until January 1916, then dropped a bombshell. "During the last few years," he advised Maclaurin, "and more definitely during the last few months, I have come to feel that a serious situation confronts the Institute in the respect that its scientific work is being more and more subordinated to its engineering work—that instead of the earnest effort being made which is essential to the development of the scientific courses and research work there are stronger tendencies than formerly operating in the opposite direction.

As the logical outcome of these tendencies the Institute is becoming a technical institute in the narrow sense instead of a broad scientific school in which the principle is recognized and acted upon that science and engineering must be coordinately developed, if the institution is to be even a technological school of the first rank." Noyes accused Maclaurin of dragging the Institute down a path radically different from that established as far back as William Rogers's time, and he threatened to resign—or, at least, "cease to feel a vital personal interest"—unless Maclaurin took action to restore the old balance. Noyes had learned of Corporation visiting committees advancing prejudicial views about the role of pure science at Tech. "It is not, of course, the object of the Institute to train mathematical teachers or pure mathematicians," wrote the mathematics committee in 1912, "but to give to its students such a grasp of the subject and such a general training in attacking practical problems from the abstract or mathematical side as will be most useful to them in the special careers which they may have chosen"—this, a throwback to when math had been viewed as a mere service subject.

Maclaurin took Noyes's criticism to heart, but did not act right away. He hinted in his address at the dedication of the new campus, in June 1916, that he grasped the dilemma and would take steps to address it. The Institute's primary mission, he said, was to "strengthen American industry at the base by fixing it firmly on the rock of science. . . . Hence we must have industry linked with science, not merely for the benefit of industry but for the sake of science. Of course, our American science will never grow as it should if it is cramped by a short-sighted policy as to what is useful. But if the value of science to industry be generally appreciated, science will be free to expand in any direction." Maclaurin recruited George Hale and Willis Whitney to weigh in on the issue, and, like Noyes, neither bit his tongue. In October 1916, opening day on the new Cambridge campus, Hale pleaded with an eager audience of students, faculty, and Corporation members. "May we not hope . . . that research may at last be raised to the high place which Rogers, with characteristic insight, felt that it should occupy? And may we not expect that pure science . . . will be recognized at its true value as the foundation stone of the Institute, and be permitted to assume equal importance?" Whitney's remarks to an alumni group at dinner in January 1917 were even more pointed. "I want to talk about pure research because we Americans seem to know so little about it. Nothing in the world is so important to engineers. Although ours is the greatest engineering school, it is the home of few research men." Such a position, voiced by General Electric's research director, surprised some who assumed that commercial and industrial concerns would harbor a far narrower outlook. When AT&T's Frank Jewett ('03) urged Maclaurin to find a "broad gauge physicist" rather than an industrial-applications type to replace retiring Charles Cross as physics department head, Maclaurin agreed not to consider anyone who failed to demonstrate high regard for the scientific endeavor at all levels. The president continued to reiterate, however, that "this Institute exists . . . for service, and particularly service to industry."

Noyes, Hale, and Whitney joined a group formed by Maclaurin to review the Institute's research goals, but made little headway. The problem was not so much lack of motivation, even though Noyes suspected that Maclaurin considered the effort as window-dressing, but that everyone was preoccupied by the unfolding European crisis. The war, a source of deep anxiety for Maclaurin ever since the opening salvoes of August 1914, compounded his worries over fundraising, the Harvard alliance, and Tech's move to Cambridge. It pulled him also toward short-term, pragmatic, product-oriented objectives and away from the broad-vision goals enunciated by Noyes, Hale, and Whitney.

Maclaurin had applied for naturalization as a U.S. citizen sometime between 1910 and 1914, but never followed through. The war in Europe revived ancient loyalties and he remained a British subject—partly out of solidarity, national pride, a desire to serve the cause, but also because he resented his adopted homeland's apathy, its self-centered isolationism. A 1916 survey suggested that many in the Institute community viewed the war as little more than a topic for casual conversation. Tech students, in a 1916 poll, slightly favored Germany over the allies for fear that an allied victory might disrupt the flow of imported beer. Pacifists struck Maclaurin as foolish, uninformed, and ostrich-like. The world, he cautioned, was "so bound together that to a certain extent one part must suffer with the rest."

His warnings fell, mostly, on deaf ears. The European conflict failed to resonate much even when it touched close(r) to home, as reports of alumni casualties trickled in. Paul Vignal ('15) died at Ypres in January 1915, Kenneth Weeks ('12) at Cabaret Rouge in June 1915, and Henry Lamy ('13) at Champagne in September 1915. Arthur Adams ('90) was among those who went down with the *Lusitania*, a civilian cruise-ship torpedoed by German forces on May 7, 1915, and in March 1916 Edward Huxley ('95) narrowly escaped the sinking of the *Sussex*, another cruise-ship. A few alumni contributed to the war preparedness movement. In 1915, Frederick Woods ('95) wrote about war as inevitable in his ironically titled book, *Is War Diminishing?* Lester Gardner ('98) founded a magazine *National Defense*, first published in January 1916, while in the wake of attacks on the *Lusitania* and the *Sussex*, George Hale helped persuade U.S. president Woodrow Wilson to create the National Research Council (NRC) as an arm of the National Academy of Sciences to coordinate defense-related scientific programs and policies. On April 26, 1916, Wilson appointed Hale to chair the NRC's organizing committee. Noyes became a member, too, having already, with Willis Whitney and Tech Corporation member Elihu Thomson, served on a U.S. war department committee to determine the adequacy of the nation's nitrate supplies. Once the United States joined the conflict, Whitney's voice of reason helped check some of Washington's hawkish excesses. "We are playing in a dangerous fire," he warned Maclaurin in October 1917, "and are forcing our young men to certain death before we ask our most intelligent and representative men to get together. I'd rather see more buried pride than buried patriots."

But Maclaurin waged an uphill battle to persuade government officials that the Institute had much to offer on the military side. His main argument, that America, like it or not, would be drawn in eventually, took a while to register. "Personally," he told George Eastman in March 1917, "I think that this country must get into the war in order to retain its self-respect. It seems to me indecent that we should have to rely on the protection of the allied fleet, and it seems to me altogether unworthy of a great and generous people to rest content with making money when the free peoples of Europe are giving their very lives with the utmost lavishness for a cause that is just as much ours as theirs." While Maclaurin's was a minority position—Wilson won reelection as U.S. president in 1916 by vowing to retain neutrality and to keep the country out of war's reach—his technical arguments were difficult to ignore. He laid groundwork that convinced the government of MIT's strategic military value. The problems of modern warfare, he suggested, were 90 percent a matter of straight engineering.

Maclaurin's theory intrigued Wilson's secretary of war Newton Baker, who appointed a board to survey the Institute's course offerings and to assess what the military branches might gain from them. The board's report, issued in March 1917, called MIT's resources in several areas—civil, sanitary, mechanical, electrical, chemical, and engineering administration—remarkably impressive, far better than those in the military academies. In addition to recommending that members of the Reserve Officers Corps be required to take MIT courses whenever possible, the report also urged the Institute to devise additional courses of a shorter, more specialized type—the chemistry of explosives, for example—which officers could be assigned to register for on a limited basis. None of this implied a push toward international engagement, but it did confirm a need to build a nucleus of engineers to serve in times of stress. Maclaurin, meanwhile, drove himself hoarse in speech after saber-rattling speech. As he told a church gathering on March 25: "When war comes no time will be lost. We shall be struck quickly by a powerful enemy! We must see to it that there is no more vacillation or hesitancy in expressing our national will. The honor of this country is at stake. Let everyone know that we mean to act. . . . The whole history of this country has been a history of shameless unpreparedness." And at another venue three days later: "If you don't raise your boy to be a soldier, somebody else's boy will have to fight for him."

When Congress declared war on Germany in April 1917, Maclaurin telegraphed Washington to commit the Institute's buildings, personnel, equipment, whatever resources he could muster. The campus shifted from civilian to military purposes almost overnight. "At last we are in the war," Maclaurin sighed to fellow Brit and University of Toronto president, Robert Falconer. MIT's labs and offices turned into makeshift barracks, khaki uniforms replaced wing collars, Norfolk jackets, and Chesterfield overcoats. A 25-member joint student-faculty committee on national service, proposed by students Leon McGrady and Edward Brooks (both '17), was appointed by Maclaurin as a clearing house for war-related information and to advise students on military duties. Maclaurin created a number of special schools in collaboration with the war department. The Army School of Aeronautics, maintained by the Signal Corps, opened on May 21, 1917, followed by the Naval Aviation Department on July 23. That fall the aeronautics schools each serviced 200 students at a time, with Washington eager to boost the number to 300. The course lasted two months. Students arrived on a staggered basis, 25 new entrants weekly. MIT also ran programs for engineers and navigational officers in the merchant marine, in naval architecture, and in field operations, the latter conducted at Tech's summer camp in East Machias, Maine. There were schools, too, for cadet ensigns, radio engineers, and sanitation workers. Campus labs, particularly those in chemical, mechanical, and electrical engineering, turned to ordnance testing as the war department placed ever-higher priority on ways to offset German advances in military technology. As the war pressed on, Maclaurin proudly brought Corporation members and other dignitaries to campus for a look at "what we are doing by way of training men for warlike ends."

Maclaurin tried to retain a civilian presence, but the odds fell against him. In October 1917 MIT reported a 13 percent decline in student enrollment, while Harvard and Yale each lost 40 percent. As draft and enlistment rates rose, the Institute's losses mounted and quickly matched those of other colleges and universities. A list compiled early in 1918 showed 182 students from MIT's class of 1918 on active duty, 84 from the class of 1919, 42 from the class of 1920, and 3 from the class of 1921. Two students, Norwood Johnston and Kimberly Stuart (both '19), were awarded the French *croix de guerre* for heroism in 1918. By midway through the year, 22 Institute students had died on Europe's battlefields. That number soared to 106 by the time the armistice was signed on November 11, later revised to 126. Scores of Institute students and alumni received citations for bravery, gallantry, and other acts of heroism.

The civilian exodus accelerated as government recruiters drew away staff for war-industry projects. MIT's roster of 277 faculty and other instructors for 1917–18 dwindled by almost a half. Some of these remained on campus attached to one or more of the reserve officers' schools, but most went away on assignment. By far the largest number, 34, were from chemistry and chemical engineering. William Walker, commissioned as commanding officer of the chemical warfare service, based at the Edgewood Arsenal in Maryland, took with him 18 members of the department, including James Norris, Samuel Mulliken, Frederick Keyes, and Miles Sherrill, to work on poison-gas manufacture and assay, while the rest, led by Warren Lewis, were involved in gas defense projects (dispersants, antidotes, masks, and protective gear). Civil and sanitary engineering sent 21, focusing on construction design and supervision, land reclamation, public health, housing, and aviation parts. The 18 from mechanical engineering, applied mechanics, and mechanic arts worked primarily on gas engines, fabrics, and airplane trusses. Mechanical engineering department head Edward Miller helped design airplane bomb sights for the Navy and took charge of secret research on flame projectors and other weapons. Of the 19 from electrical engineering, most worked in communications, radio, and welding. But duties assigned to electrical engineering department head Dugald Jackson, in consultation with secretary of war Baker, included a plan to generate emergency power supplies to industrial cities. Smaller numbers of faculty on war service came from mining engineering and metallurgy, architecture and architectural engineering, biology and public health, physics and electrochemistry, naval architecture and marine engineering, drawing and descriptive geometry, English, and economics.

A worn-out Maclaurin planned to take a long break in the summer of 1918 with Alice and their sons Rupert and Colin—hiking, swimming, resting in rural New Hampshire. He hoped, too, to refine some postwar plans. Just as he had been prescient about the war, he was already thinking ahead to its aftermath. But the moment he reached his vacation cottage, the telephone rang. On the other end was Baker. Would Maclaurin come to Washington immediately to coordinate educational programs for the newly established Student Army Training Corps (SATC)? Maclaurin could not very well say no. He had been out front at the start, and the government considered him foremost among academic leaders on military affairs. SATC, forerunner of what would become ROTC, the Reserve Officers' Training Corps, ran from April through December and was demobilized following the armistice. At its peak, SATC covered 524 educational institutions registering around 150,000 students.

Maclaurin found the job harassing, not to mention a distraction from MIT business. He had to balance the interests of colleges and universities nationwide, while fending off incessant complaints about lack of supplies, personnel, funding, and tirades about military intrusion in academic life. "Inwardly," one of his assistants observed, "he must often have been boiling over—as . . . when a college president . . . took an hour of his time on one of his busiest days protesting against the 'militarizing' of the colleges." But Maclaurin stood his ground. When it was all over, he reiterated his view that academic institutions, particularly those with strong science and engineering programs, must play a central role in national defense. "It should never be permitted again," he told an audience in Boston at war's end, "that the country be called upon suddenly to provide a large army of technical experts for the needs of war. . . . If war is to remain a possibility we must face the fact that it will tend more and more to be a war of applied science, and the nation that does not take that lesson to heart by training men to apply science to war-like ends . . . will surely go down in the next conflict."

Under Maclaurin, the number of students saw a gradual rise in the years leading up to the war. In 1909 there were 1,481; in 1916, 1,957. The number dropped to 1,689 (around 1913 levels) in 1917 as students were drawn into war service, then exploded to 3,078 in 1919. While most students still came from Massachusetts and other states in the Northeast, raw numbers from other regions—the South, Midwest, and West—more than doubled between 1910 and 1919, and in the case of the South, more than tripled. The number from foreign countries also more than doubled during this period and sustained a growth rate undisturbed, remarkably, by volatile world events. The drop-off among students from western Europe after 1914 was offset by arrivals from other regions: Asia, Central America, South America, the Middle East, and eastern Europe (Russia sent ten students in 1918). Maclaurin took special pride in the large proportion of foreign students at Tech compared to other American colleges and universities. Figures for 1912 showed the Institute's student body about 7 percent foreign, with the University of Michigan at 2 percent, Yale and Harvard 3 percent, Cornell and the University of Pennsylvania 4 percent.

While there were more than twice as many students in 1919 as in 1909, the proportion of engineering to science majors remained constant. Engineering students comprised more than 80 percent of the student body and science students less than 10 percent during this period. Architecture students slipped from around 10 percent in 1909 to 5 percent in 1919. Of the engineering fields, mechanical was the most popular followed by electrical and chemical (a close 1,975 and 1,900, respectively, between 1914 and 1920), then civil, administration, mining, naval architecture, electrochemical, and sanitary, in that order. Most science students majored in chemistry, followed by biology, physics, and geology.

The number of advanced students soared, from 181 in 1910 to 354 in 1920. Most of these were graduates of other colleges registered for specialized, professional studies at the undergraduate level, but the number working toward higher degrees tripled (32 in 1910, 91 in 1920). A total of 291 master of science degrees were awarded between 1910 and 1920, more than double the number awarded between 1886 and 1909. Hou-Kun Chow, from China, earned the first S.M. in aeronautical engineering in 1914, while the first in electrochemical engineering went to Eastman Weaver in 1915. The first S.M.'s in naval construction, five in all, were also awarded in 1915. The S.M. in mathematics was awarded for the first time in 1919, to Israel Maizlish. Harold Osborne earned the first doctor of engineering degree (D.Eng.), in electrical engineering, in 1910. In all, eight D.Eng. degrees were awarded. After the last two, in 1917, doctoral candidates in engineering earned either Ph.D.'s or Sc.D.'s. Jerome Hunsaker earned Tech's first (and only) D.Eng. in aeronautical engineering, in 1916. The first doctorate in electrochemical engineering went to Edward Cyrus Walker III in 1917. The Sc.D. was awarded for the first time in 1920, when Frederick Brooks, William Davy, and Fredrik Hurum graduated in aeronautical engineering, geology, and metallurgy, respectively. Besides the eight D.Eng.'s, 31 Ph.D.'s were awarded between 1910 and 1920—20 in chemistry, 8 in geology, 2 in physics, and 1 in biology. The first Ph.D. in geology went to Charles Clapp, in 1910, in biology to Eugene Howe in 1911, and in physics to James Ellis in 1916.

Maclaurin focused as much attention on social aspects of student life as on the academic side. Students gravitated toward him and his family, just as an earlier generation had felt drawn to the Pritchetts. Alice Maclaurin was an eager surrogate mother, visiting sick beds and hosting many a Saturday-evening get-together at the president's home on Bay State Road, then, as of 1917, at the official residence in Cambridge. These events usually ended with folks gathered around the piano belting out "Take me back to Tech" and other patriotic songs. Students adored Mrs. Maclaurin's style, the way her "glow warmed the hearts of many a weary, homesick freshman, and cheered to better endeavor some lagging upper-class-man." At Christmas the Maclaurins regaled students who could not get home for the holidays with food, drink, gifts, and a special surprise treat—in 1910, for example, a quasi-staged reading of Dickens's *A Christmas Carol* by professor of English Arlo Bates. Rupert and Colin were every Tech student's little brothers. Two weeks after Colin was born in December 1914, an alumni group presented a silver porringer and spoon underscoring how deeply affection for the family ran.

Maclaurin spurred a broad range of extracurricular activities. The Technology Orchestra, which had performed in fits and starts since the 1880s, was reenergized, alongside some new musical groups—the so-called Chocolate Soldiers, for example, a blackface revue that flourished briefly around 1911. An honorary society, the Baton, was founded in 1914 to recognize musical achievement among undergraduates, while Frieze and Cornice—for final-year students in architecture—appeared by 1918. With Maclaurin's support, a movement to establish crew (rowing) was spearheaded in March 1910 by Nathanael Herreshoff Jr. ('12), son of "Captain Nat," himself a former Tech student ('70), famous for having revolutionized yacht design. A swim team appeared for the first time around 1911. Maclaurin's cautious yet supportive approach to sports, revealed the year before his arrival at Tech—"I don't approve of the extreme vogue of athletics in certain institutions, when it goes so far that a man thinks and dreams and talks of nothing but athletics"—prevailed throughout his presidency.

Student publications grew increasingly ambitious. *The Tech* turned from a triweekly into a daily in September 1909 and stuck to this strenuous schedule for four years, after which it went back to triweekly status and then, for one year during the war, became a twice-weekly before resuming as a triweekly. Students banded together in the fall of 1909 to produce a guide, *Concerning the Massachusetts Institute of Technology*, intended to smooth the transition for incoming freshmen but, more broadly, to dispel myths about the Institute and to make it better known to the public. The guide proved popular enough that a second edition appeared in 1912 and a third in 1919. The first number of *Technology Monthly*, a magazine meant to crystallize school spirit, was published in April 1914. When Maclaurin and Lowell signed the Tech–Harvard affiliation agreement in 1914, the magazine joined with the *Harvard Engineering Journal* to form the *Technology Monthly and Harvard Engineering Journal*, but after the agreement foundered in 1917, its name was changed to *Technology Monthly Engineering Journal*, then back to *Technology Monthly*. The *Woop-Garoo*, a comedic magazine launched in February 1918, bonded loosely with the *Monthly* to form *Voo Doo*, whose first issue appeared in March 1919.

Student professional societies continued to fill both career and social needs. Under Maclaurin, at least two such groups ventured into the innovative technologies of radio and manned flight. Students founded the Wireless Society on April 30, 1909, a month before Maclaurin took office, and devised plans for a powerful transmitter atop the Walker Building by the following fall. The Aero Club appeared in October 1910, part of Maclaurin's push for aeronautics coursework, and was succeeded later by the Aeronautical Engineering Society. A chapter of the honorary journalistic fraternity, Pi Delta Epsilon, was formed in 1910. The Architectural Engineering Society was founded in November 1911, and Vectors—a group that brought together "men whose interests are not limited to the technical aspects of electrical engineering"—appeared by 1913. Chapters of engineering fraternity Theta Tau and chemical fraternity Alpha Chi Sigma were formed in 1912 and 1919, respectively. Engineering administration students created the Course XV Corporation, registered under Massachusetts statutes on May 27, 1916, "for the purpose of holding literary, engineering, scientific, economic or social meetings, and the discussion of economic and scientific problems." The Political Engineering Society, whose philosophy differed starkly from that of the Course XV Corporation, was founded in March 1913 by "men of socialistic tendencies."

New fraternities emerged. A chapter of Kappa Theta was formed in 1908, shortly before Maclaurin arrived, followed by chapters of Zeta Beta Tau in 1911, Delta Kappa Phi in 1912, Lambda Chi Alpha and Beta Theta Pi in 1913, Kappa Sigma in 1914, Sigma Alpha Mu in 1917, Phi Kappa and Tau Delta Phi in 1918, and Tau Epsilon Phi in 1919. Newbury Street, an easy walk from the Rogers Building, remained a popular locale for housing, with each fraternity residence accommodating about twenty men. A couple of the younger fraternities, Tau Delta Phi and Sigma Alpha Mu, set up residence across the river, not far from the Cambridge campus. Other new social groups included the Cosmopolitan Club, founded on March 1, 1910; the Latin-American Club, in 1913; the Chinese Club, around 1913; the Menorah Society, for the study and advancement of Jewish ideals and culture, by 1915; and in 1919 the California Club, to help recruit students from the west coast. The Association of Undergraduates was organized in 1913, partly to unify these diverse constituencies.

Maclaurin favored the Cosmopolitan Club for its efforts to promote internationalism and tolerance regardless of race, ethnicity, nationality, or creed. The club, proposed by an American civil-engineering student, Gorton James ('10), served as a foil to the exclusive—and exclusionary—spirit that permeated a number of the other social groups. While Maclaurin valued fraternities for the residential options that they offered, and for their tightly knit, independent, self-supporting style, he fretted about the "danger of . . . clannishness, which can easily develop into snobbishness." Local chapters were bound by restrictive covenants imposed by the parent group. The most common qualifiers concerned race (white) and religion (Protestant). Zeta Beta Tau, which accepted only Jews, came about partly because few fraternities admitted Jews. Sigma Alpha Mu, Tau Delta Phi, and Tau Epsilon Phi were also all-Jewish. Phi Kappa admitted only Roman Catholics. Issues of this sort arose around nonresidential social clubs, too. The Southern Club served regional interests and, as such, accepted only

Southerners, but when "only Southerners" turned into "only Southerners when approved by members of the club," it was clear that other criteria—race, for one—were part of the equation. The Cosmopolitan Club, conversely, strived for a multicultural membership and hosted programs ranging from Arabic recitations by Syrian students to plays written and performed by Chinese students (in Chinese) with American and Chinese flags flying overhead.

For similar reasons, Maclaurin admired the work of the Technology Christian Association (TCA). He liked its nonsectarian, inclusive style which stood in contrast, for example, with the Catholic Club, known to look little beyond Boston's cardinal William O'Connell and other church prelates for guidance. TCA's roster was broadly eclectic. When its guest-speaker choices leaned controversial—as in 1915, when one speaker announced his plan to describe a program for large-scale conversion of Confucian atheists to Christianity—Maclaurin gently suggested not that he be barred but that in everyone's interests he might care to soften the aggressively evangelistic side of his message. "The broader the basis that he can give to it, the larger will be the number of students likely to be interested." TCA agreed, and the speaker obliged. Other helpful efforts conceived and carried out by TCA included a housing placement service (begun in 1912), a student-peer advisory program (begun in 1914), an English-as-a-second-language program for foreign students, faith-based forums, community outreach strategies, a book exchange, an employment bureau, and a handbook nicknamed the "Tech bible" that oriented new arrivals to schoolwork and to life in the community.

Maclaurin's efforts to build an inclusive Tech largely missed two groups, blacks and women. The only blacks known to have graduated during his presidency were Anselmo Krigger ('17) in civil engineering, Bertram Jones ('18) in chemistry, and Hosea Smith ('20) in chemical engineering. Three others entered postwar—Lee Purnell ('21, electrical engineering), Emmett Jay Scott Jr. ('21, civil engineering), and Harry McGee ('22, architecture). Women also remained on the margins. The number admitted during Maclaurin's early years as president, 1910–14, trended downward and remained well below numbers that had prevailed in prior administrations. Alfred Burton's address to the entering class in October 1913 was perhaps telling in this regard—"We do not want mollycoddles," he said, "we want strong men." In 1911–12, there were only seven women in a student body totaling 1,559. The number grew slowly after 1914, then more than doubled in 1918 (to 45 from 19 in 1917) as male students were drawn off for war service.

The number of women earning degrees also rose after 1917. Ten graduated between 1910 and 1917, then four in 1918, six in 1919, and seven in 1920. Women continued to lean toward architecture, chemistry, general science, biology, and public health. Before the war, engineering remained off-limits to all but a venturesome few. Marion Rice graduated in chemical engineering in 1913, while three other women—two in chemical engineering, one in mining engineering—either dropped out or transferred to other courses. A marked shift occurred postwar. Edith Clarke earned a master's degree in electrical engineering in 1919. In 1919–20, four women were regular undergraduate students in engineering disciplines (mining, chemical, electrochemical, and engineering administration). But in a survey carried out that year some male students offered grudging support for coeducation only so long as "the damsels be good to look upon."

In his first year as president, Maclaurin relied on Ellen Richards, Tech's informal dean of women, to attend to women's interests, occasionally sending her circulars on fellowships for women and other relevant items that came across his desk. Women, he observed in 1912, were unlikely to enter the Institute in substantial numbers partly because all-women's colleges had improved radically, but also because women "hesitate to enter the professions to which the Institute's courses are designed to lead." He maintained, however, that Tech erected no barriers based on gender. The Institute, as one informational brochure pointed out in 1916, "has a graduate school, offers some summer courses, and enrolls a few young women."

What Maclaurin failed to acknowledge, in referencing barriers, were pressures exerted by male students whose caustic remarks about women's suffrage (Congress finally passed the 19th Amendment, granting federal voting rights to women, in 1919), about women usurping men's roles, the dreadful prospect of women smoking pipes, and so on, sometimes showed up in campus publications like the *Tech* and *Technique*. Women students who supported suffrage either kept their opinions to themselves or furtively posted leaflets in Rogers and other buildings. Still, Sybil Walker ('18) became class secretary, likely the first woman elected to Tech's student government, followed by Celeste Brennan ('19). The lone female electrochemistry student, Florence Fogler ('20), played a prominent role in several campus groups. Besides serving as vice president of the sorority Cleofan, she was secretary of the Electrical Engineering Society, secretary of the Cosmopolitan Club, and secretary-treasurer of *Technique*, while also active on the staff of *Tech* and as a member of the Chemical Society. Fogler, said to be the only female electrochemical engineer in the country, recalled being warned by faculty that "woman's place was anywhere but in that profession," then struggling to prove that she could shovel coal, stock a muffle furnace, work a milling machine, and operate a lathe on par with her classmates.

Ellen Richards, the Tech woman's primary mentor, opposed women's suffrage and reinforced traditional feminine roles. In 1910, among her last public appearances before her death a year later, she gave a series of six lectures on euthenics, the so-called science of better living, a term that she coined. The press, identifying her inaccurately as professor of chemistry at the Institute (her title began and remained instructor in sanitary chemistry for more than thirty years), covered her advice about baked beans being "digestible if cooked properly" along with her hints on "how to reproduce the delicious dish made famous by our grandmothers in their brick ovens." Ruth Thomas arrived on staff in 1910 as a research assistant in organic chemistry, joined by Florence Sargent in 1911. Ida Loring was appointed an assistant in architecture in 1915, but Maclaurin had trouble keeping her there. A cryptic remark that he made in 1916—"Something will have to be done in regard to Miss Loring, but it will be necessary for me to look into the matter before settling it"—implied that there were tensions over her presence in the department. Maclaurin managed to diffuse these and Loring served as an assistant for more than a decade (her duties remained, however, mostly in the library rather than on the instructional side). Loring and Ruth Thomas were the only women to remain longer than a year on staff—Thomas stayed ten years, 1910–20, and was promoted to research associate in 1916—but a few others served as assistants for a year each: Helen Vincent in chemistry of sanitation, 1916–17; Louise Johnson in organic chemistry and Amy

Walker in chemistry of foods, 1917–18; and Charlotte Alling in chemistry, 1919–20. Maclaurin campaigned unsuccessfully in 1916 for the admission of women to Harvard Medical School "as a war measure." The MIT women's association went on record in 1920 as grateful for his strong convictions concerning "the adequate education of women for advanced scientific work."

But Maclaurin's vision for a diverse, inclusive Institute played itself out mostly on the international stage. The number of foreign students rose at a steady clip, from 102 in 1910 to 205 in 1919. In 1910, Maclaurin appointed Jasper Whiting ('89) a special commissioner to increase Tech's appeal to foreign students and to survey what types of courses best met their needs. Because the Far East impressed Maclaurin as the field of greatest opportunity, Whiting spent several months there in 1910 and 1911. He passed through Tokyo, Yokohama, Shanghai, and Beijing before heading to Calcutta and other places further west. Several Asian alumni, notably Takuma Dan ('78) and Stejiro Fukuzawa ('88), helped gain access for him to academic and government leaders. Early evidence of Whiting's success showed in August 1911, when the Chinese government signed up four students for Tech's course in naval construction. These four, Whiting assured Maclaurin, were "but the nucleus of a large number that will be sent from China . . . during subsequent years." And his instincts were correct. In each year of Maclaurin's presidency, Chinese students outnumbered every other foreign student group. A total of 49 registered in 1915 alone. While Japan sent smaller numbers, its constituency jumped from 1 in 1910 to 15 in 1918. Outreach efforts to Latin America on a more modest scale—select schools in Santiago, Valparaiso, Concepción, Punta Arenas, and other cities were canvassed—also produced results. "If the present tendency be maintained," Maclaurin observed in 1913, "it would appear that Technology is likely to move still further ahead of other institutions as regards the number of men who come from foreign parts." The threat of global conflict, meanwhile, reinforced his devotion to the spirit of internationalism.

Whatever joy people felt at war's end was short-lived. A host of worries persisted—some of them a direct consequence of the war—and the least of these, in many minds, was how to deal with some vague, future conflict. The here-and-now was trouble enough; look to *that* first. The war had transformed the world into an overlapping, interdependent sphere. Nations would have to manage not only their own crises, but everyone else's as well—the very result that isolationists had feared and used, effectively, to keep America out of war until it could stay out no longer. Immigration came to be viewed, more and more, as a problem rather than an opportunity. With growing labor unrest, agitation for workers' rights, and the rise of the trade-union movement, warnings that MIT's own Francis Walker had voiced decades earlier—about undesirable elements, revolution, socialism, other foreign-born ideologies—came back to haunt, with a vengeance. Add to this the war's disruptions to industry and commerce, to the entire national fabric, and everyone's burning questions were: what now, what next, where do we go from here?

Maclaurin's immediate concerns had to do with managing the vast influx of students, new ones as well as returnees from war service, and how to solve chronic funding shortfalls. Expenses far outpaced revenues. Maintenance and operation costs were up, the state's contribution of $100,000 a year would soon be cut off (with no chance of renewal, as a state constitutional convention in 1917 prohibited further grants of this sort to private institutions), the collapse of the Harvard-Technology agreement meant no more income from the McKay trust, tuition fees covered a small portion (about one-third) of the cost of educating each student, and faculty salaries would need to be hiked if the best-qualified scientists and engineers were to remain in education rather than move into lucrative positions in commerce and industry.

As the "war to end all wars" drew to a close in 1918, Maclaurin drafted a series of talking points to push the new MIT forward and to leave Boston Tech behind. He presented his plan on December 11, 1918. There was not a moment to lose, he told the Corporation. The Institute must raise an endowment of at least $7 million or face an uncertain, possibly bleak future. The Corporation agreed, but the mood there was glum. Postwar dilemmas—how to recast commerce and industry for civilian needs, how to preserve or rebuild personal fortunes at risk—kept individual donors from stepping forward, willingly. Maclaurin worried that the private sources MIT had come to rely on were tapped out. Funds given by Mr. Smith approached $20 million (around a quarter billion dollars in 2010 terms) and had stimulated, through matching-gift proposals, large donations from alumni.

By mid-1919, there was little more to show. But Mr. Smith stepped in once again, this time with an offer of $200,000 a year from his own private securities, equivalent to a $4 million capital gift. To close the deal, however, MIT must raise $3 million by year's close. "Success would be splendid for the Institute and the industries that it serves," Maclaurin told his funding prospects, "but failure would be disastrous." The goal, unreachable via conventional means, had to be supplemented in some other way.

Maclaurin turned toward self-support: how to capitalize on the Institute's own assets, market its skills, earn money by services proffered, rendered, and paid for. This idea laid the framework for the Technology Plan, managed by a new unit, the Division of Industrial Cooperation and Research (DICR). Launched late in 1919, with William Walker in charge, the plan aimed to attract and streamline industry sponsorship. In return for an annual fee paid up-front, a company was entitled to technical advice from faculty and staff, consulting services, access to alumni records, a variety of quid pro quos. Walker, who had led Tech's highly regarded experiment in cooperative education (the school for chemical engineering practice), employed an aggressive marketing strategy to seal contracts, totaling almost a million dollars, with 189 firms inside two months. He wanted the program viewed as a kind of exclusive club, one that members boasted of belonging to, rather than as a charity to help pull the Institute out of a financial hole. Firms were offered bang for their buck, so to speak, and got it, by and large. "We want to meet the industries more than halfway," Walker told the local press. "There could be no more legitimate way for a great scientific school to seek support than by being paid for the service it can render in supplying special knowledge where it is needed. . . . Manufacturers may come to us with problems of every kind, be they scientific, simple, technical or foolish. We shall handle each seriously, giving the best the institute has at its disposal." Some complained that MIT, in setting itself up as an industrial consulting service, had placed itself in competition with its own alumni. But Coleman du Pont swatted this fear aside—"If any Tech man has made such a criticism," he said, "he must be a poor specimen of the breed, for the real Tech man has no fear of competition."

There were concerns, too, among some faculty that the Technology Plan would push MIT so far into the commercial orbit that parts of its mission would be compromised: teaching, lines of independent research, investigations on the pure rather than the applied side, work tied neither to instant results nor to monetary reward. Eric Hodgins ('22), a student at the time, recalled that "a good deal of hell broke loose" over this. "The arguments *for* were," he wrote, "in addition to revenue, that the Institute would gain by being brought closer to Industry's problems, to the benefit of professor and student alike; that Industry in its turn would learn that there were ways of getting benefits from the Institute other than by raiding its faculty and luring its most gifted men away for good—in those days a real and vexing problem. The arguments *against* were that this sort of contact with Industry would be a soiling and corrupting experience for a university and that, once sanctioned, there could be no telling where it would all end. Education was Education and Business was Business, and no matter how 'practical' the ends the education might eventually serve, there had to be a Zone of No Contact." Hodgins recalled being approached at a social event by someone, either a junior faculty member or a graduate student, who asked him, "What do *you* think of this idea of 'selling Tech'?" This man and half the room viewed the trend as "an idea reared by the devil himself." But opponents found themselves in a vocal minority. Many stayed neutral, as Hodgins did, choosing not to press the issue so that Maclaurin might have room to maneuver the Institute into a position of financial strength.

The minority, however, included powerful figures such as Arthur Noyes and George Hale who saw 1919 as a watershed in Institute history. A flurry of angry catharses boiled over that spring. Hale met with Maclaurin for five hours in late March or early April, laying out his view of why MIT was "going to the dogs." "We are doing no research," he told a dismayed Maclaurin, and "the way the Institute is going its best graduates are losing interest in it." Noyes and Walker, meanwhile, went at each other's throats while Maclaurin played referee.

Walker called Noyes's science hostile to "horse sense," and horse sense, he said, would emerge victorious every time. He threatened to resign unless Maclaurin took Noyes out of his role in key policy decisions affecting the chemistry programs. "I can see no reason," Walker told Maclaurin in March 1919, "why Noyes should assume a superior attitude and feel that he is privileged to remain away from the Institute for months at a time, and yet be able to blow in with a lot of half-fledged immature ideas and have him insist that they be adopted." Noyes had spent part of each year in California since 1916, working with Hale and Robert Millikan to build science programs at Throop College of Technology, which in 1920 became the California Institute of Technology. Maclaurin, anxious to retain Walker at all costs, tried to pacify Noyes by suggesting that the "sinister influences" attributed to Walker by Noyes were "figments of the imagination." Still, Maclaurin eased Noyes out of policy as Walker demanded, which led to Noyes's resignation in November 1919 and his permanent departure for Caltech a month later.

In opting for Walker rather than Noyes, Maclaurin chose entrepreneurial smarts over scientific temperament. But he tried to clarify the difference between choices made for expediency's sake—short-term benefits—and longer-term values that he promised MIT would get back to when the time was ripe, once its financial affairs were in better order. Maclaurin's own convictions, he confided to Noyes, lay closer to Noyes's than to Walker's:

> My instincts and my training combine to make me thoroughly sympathetic with your desire to build up a strong Research Laboratory here and to develop the research spirit throughout the Institute. You will, I think, realize that I have had many unusual difficulties to overcome since I (perhaps unwisely) accepted the presidency of the Institute. I have pursued a policy of doing one thing at a time and this has perhaps given rise to some misunderstanding as to my ultimate aims. In 1917 I hoped that as I had overcome the great difficulties of moving the Institute to a new site and securing for it an endowment sufficient to carry on the work to which it was already committed (and how great those difficulties were very few can realize just as none can know the personal sacrifices that I made in staying on here so as to overcome them) I might turn with greater interest and enthusiasm to what I particularly wanted to do, namely, to strengthen the Institute as an instrument of research. Just as I was turning myself to

> this problem the war loomed on the horizon and soon we were in the eye of the storm and we are not yet recovered from its effects.

Noyes carried through on his threat to leave, regardless. He found it difficult to reconcile Maclaurin's nuanced position, privately expressed, with what seemed otherwise like support for unbridled academic-industry ties. As Maclaurin told Irenée du Pont in August 1919: "The opportunity now presented to Technology to place itself in an unrivaled position as a servant of industry is an extraordinary one." Noyes suspected that this was the true Maclaurin.

Christmas Eve 1919—fair skies, mild for that time of year. Maclaurin was in fine spirits. He could relax a bit. The $3 million mark, reached on December 16, met Eastman's terms, with money still coming in. The Technology Plan promised a steady flow of income and the alumni had rallied. The Institute was in better shape, financially, than at any point in its history. Independence, stability, and security all looked certain. It was a time for celebration. Maclaurin joined in the boisterous fun around the Christmas tree as he, Alice, and their boys hosted students stranded this holiday. Good cheer, plenty of food. The high point was an ad-lib production of St. George and the Dragon, complete with makeshift costumes dug out of closets in the president's house. St. George paraded around in Maclaurin's scarlet robe, a souvenir from his doctoral ceremony at Cambridge in 1904.

The only other thing on Maclaurin's mind was a speech to alumni a couple of weeks away. This wasn't just any speech; this, he anticipated, would be his most dramatic. He had two huge announcements to make: the success of the fundraising campaign and, for a grand climax, Mr. Smith's identity revealed after eight long years of mystery, gossip, wilder and wilder speculation. George Eastman was required, under tax laws, to disclose his latest gift, so the cover of anonymity had to be dropped, whether or not he liked it. Maclaurin kept tinkering with the words. He wanted to draw out suspense, tease his audience, poke fun, much as he enjoyed playing with his guests this festive Christmas Eve. One line switched a common gender stereotype. "I am often asked a recipe for keeping a secret," he planned to tell the gathering about his efforts to guard Mr. Smith's identity. "It is, after all, very simple. Tell it to no man and to very few women. I have told it to two—my secretary, Miss Miller, and my wife."

On January 9, 1920, the day before the big event, Maclaurin left the office early. He had been fighting off a cold all week and suddenly it was much worse. The next day, he contracted pneumonia. The dinner went ahead anyway, as planned, in Walker Memorial, a couple of hundred yards from the house where Maclaurin lay ill. No one felt like celebrating, but Alice appeared, stiff upper lip firmly set, and assured the crowd that that her husband was on the mend.

It was a jam-packed program. After dinner Coleman du Pont, master of ceremonies, outlined the Institute's success in this latest campaign. The $3 million mark had been passed on December 16; four-million-plus by January 9, over and above Eastman's four. The total (and counting), $8,052,681. "We have gone further over the top than we expected we would," du Pont beamed. "I think this is the only college that has gone over the top. It . . . shows that when big men undertake such a thing they do it well and quickly." William Walker and Charles Stone spoke in defense of the Technology Plan. It was new, they said, and untested, but they were confident that its controversial side, its uncertainties, would evaporate once it went into full swing. "What we offer," Walker promised, "is to bring to the industries a great mass of knowledge which we have in our organization—to show them that by applying science to their problems they can be all the more quickly solved." One bitter, discordant note was sounded by John Wesley Hill, conservative cleric and historian, who warned of fissures in the nation's social fabric, dangers emerging out of postwar chaos and reconstruction. "Bolshevism," he said, "is conspiring against us and trying to overthrow our institutions. Just now we are deporting some of them but a little shooting might be mixed with the program to advantage. I see no wisdom in driving these wild beasts into our neighbor's garden. But if the shipping must proceed, then I think it time to sift out the people of this country and separate from the body of intelligent, patriotic and courageous citizens every man and every woman who loves the black flag, the red flag or any other flag, and load these in a ship of stone with the wrath of God for a gale and hell for the nearest port." While many of those present shared Hill's fears, they did not want this meeting—a celebration—to devolve into poisonous political diatribes. There was a time and place for that, and this was not it.

Everyone looked forward to Maclaurin's speech, the one he had worked so hard to perfect. William Sedgwick read it aloud for him. It recapped the history of that first anonymous gift through the latest campaign and milked the punch line, what everyone had come to find out—who *was* Mr Smith?—with coordinated images, decoys really, of suspects like Henry Ford, John D. Rockefeller Jr., Charles Schwab, Thomas Edison, and Andrew Carnegie flashed onto a screen. When the speech ended, still no answers. It was up to Coleman du Pont to rise and fill in the blanks. As du Pont intoned Eastman's name, one more image flashed—finally—this time of the real culprit, known to all as the Kodak king. The gathering, more than a thousand strong, erupted in cheers. They would have to be content with this screen image, however, as the object of their attention, publicity-shy, had declined to appear in the flesh. To show up, Eastman said, "would have been vain and foolish."

Maclaurin drifted in and out of consciousness for several days. Midweek his physicians, William Robbins and Harvard Medical School dean David Edsall, sounded optimistic. Alice sat by him, whispering news of what had gone on at the dinner, looking for some sign of recognition. How much registered she could not tell. To her he seemed spent, weary, resigned—"It has been an arduous life," he muttered at one point, "but well supported from the first." He died on Thursday, January 15, aged 49. It was an unforgiving day, brutal, one of the coldest in recent memory. The next day MIT awoke to drawn shades, flags at half-staff, business suspended till Monday. On Sunday young Rupert, just twelve years old, represented the family at the funeral. Alice, grief-stricken and down with a cold, stayed home to mind little Colin.

The Maclaurins' ties to the Institute remained strong. Alice and the boys lived in the president's house until spring 1921, when they moved to a Beacon Street townhouse whose rear windows looked out across the river toward Tech's Great Court. Although as students Rupert and Colin both chose Harvard over MIT, their subsequent careers played out mostly at the Institute. Rupert joined the economics faculty in

1936 and helped found, then lead, his department's industrial relations section. After several years as an engineer in private industry, then in the U.S. Navy during the Second World War, Colin joined MIT's staff as a personnel officer in 1946. He became director of general services in 1954 and served a year, 1958–59, as assistant to chancellor Julius Stratton. Like their father, both men died young—Colin at 44 on May 4, 1959, Rupert at 52 on August 17, 1959. Alice Maclaurin passed away in Boston on May 31, 1951.

"Not the man for us"

Ernest Fox Nichols, 1869–1924

Richard Maclaurin's death sparked confusion. Its suddenness, and the sense of loss, left the MIT community wondering which way to turn. In February 1920, a three-man administrative committee consisting of Henry Talbot, head of the chemistry department; Edward Miller, head of mechanical engineering; and (as chair) William Walker, head of chemical engineering and the new Division of Industrial Cooperation and Research (DICR), was appointed from the faculty ranks to carry out routine tasks. There never was a plan for an acting president. But when legal counsel advised that diplomas and other official records must be signed by an acting, Elihu Thomson agreed to serve starting on March 20, 1920. Thomson came from his home in Swampscott to sign documents—his only duty, essentially, aside from presiding at official functions.

Walker lasted a month. He and Talbot fought constantly. Talbot, one of Arthur Noyes's close friends and colleagues, and still bitter over Walker's treatment of Noyes, blamed Walker for Noyes's defection to Caltech. He also opposed Walker's campaign to separate chemical engineering from chemistry to form its own department, which took place in 1920 with Walker's protégé, Warren Lewis, in charge. Walker preferred to devote his attention to the DICR and to focus on persuading outside firms to sign up for services under the Technology Plan. Talbot took over as administrative committee chair and a new member, E. B. Wilson, came aboard in early March to fill Walker's slot. Wilson was told that the committee's job was "not to rock the boat but just keep her going." Talbot and Miller deferred daily business to him, the committee's youngest member. Wilson took over the president's office, read and answered mail, conducted job interviews, signed appointment letters, drafted budgets, and arranged dockets for the committee's twice-a-week afternoon meetings. Walker's priorities, meanwhile, continued to shift. He left MIT to return to private industry on January 1, 1921, while retaining his Institute affiliation as nonresident professor and unpaid consultant.

The administrative committee acted cautiously, restricting its decisions to the micro-mundane—whether to install terrazzo-tile or stone flooring in Walker Memorial, how to introduce underground surveying in the summer school, what to do about drainage problems in the new buildings, urging greater care in the editing of *Technology Review*, quashing irresponsible talk about staffing changes, and finding unobtrusive ways to deflect requests for raises and expensive equipment. Talk circulated that each member was more ambitious than the next. "Talbot wanted to be president, Wilson thought he was going to be, and Miller knew damn well he was"—names interchangeable, of course, depending on the source. In fact, they thought of themselves more as placeholders, fulfilling a duty assigned and accepted, not sought. All three longed to return to their regular work and to throw these chores back onto a new chief executive. The best that could be said of them, according to one observer, was that they "carried on in a most devoted manner [but] no temporary body can initiate a policy to be realized during a long term of years."

No serious candidates emerged for several months. When asked in June 1920 for an opinion about Pierre du Pont ('90) and Darragh de Lancey ('90) as presidential material, George Eastman replied that du Pont was too weighed down by company business and that de Lancey, a former Kodak employee, did not manage well under pressure. That fall, talk of recruiting either Henry Suzzallo or Herbert Hoover went nowhere. Suzzallo, president of the University of Washington, spoke at MIT in November, impressing listeners with his perspective on leadership as shared, cooperative, and democratic, rather than the domain of heroic, authoritarian individuals. From MIT's standpoint, Hoover's experience as a mining engineer was attractive. But as a Republican candidate for U.S. president (he lost to Warren Harding in the primaries that year), he had larger ambitions and went on to serve as commerce secretary under presidents Harding and Coolidge before running successfully for president in 1928.

A feeler went out to at least one member of the administrative committee, Wilson, who declined partly because he felt over-burdened as head of the physics department, partly because the problems facing Maclaurin's replacement—low staff morale, unstable funding levels, and shifting priorities—were enormous, and partly because the duties of the presidency itself had worn down Maclaurin and perhaps contributed to his death. But in November 1920, Wilson put forward two candidates of his own. One was Jerome Hunsaker, a Navy man and pioneer in aeronautical studies who had taught at MIT and earned its first doctorate in aeronautical engineering (1916) before returning to Navy service. Hunsaker remained with the Navy after the signing of the armistice in 1918. Wilson worried that his relative youth (he was just 34 years old) might work against him, so he pointed out that Hunsaker was about the same age as Charles Eliot when Eliot became Harvard president in 1869, that he showed the maturity of a man ten years older, and that he had built a fine record in government service. Wilson's other suggestion was Arthur Day. As director of the Carnegie Institution of Washington's geophysical laboratory and vice president of Corning Glass Works, Day bridged the worlds of science and industry. His academic credentials were attractive—Yale graduate, postgraduate training at Berlin's renowned standards laboratory, the Reichsanstalt—and he had served as home secretary of the National Academy of Sciences. Neither Hunsaker nor Day garnered much if any support.

Instead, the executive committee chose Ernest Fox Nichols on March 21, 1921. Everett Morss and Edwin Webster broke the news to Wilson a day or two later, confidentially, over dinner. They offered Wilson the vice presidency—a post not known at MIT since 1869—and asked him to head immediately for Cleveland, where Nichols lived, to wrap up final details before the full Corporation met on March 30 to confirm the nomination. Not a word was said about how the offer to Nichols came about, or what prompted him to accept it. Webster and Morss were vague, too, on Wilson's mission. Was he to look Nichols over, or was Nichols to look him over? Webster and Morss cautioned

Wilson not to breathe a word to anyone, not even to his partners on the administrative committee, Talbot and Miller.

Nichols knew MIT as well as any outsider, and was an old friend of Maclaurin's. The two had met at the University of Cambridge in 1904, when Nichols went over on sabbatical from Columbia University and with Maclaurin there briefly, from New Zealand, to accept the LL.D. degree. Nichols recruited Maclaurin to the Columbia faculty in 1908 and, a year later, helped smooth his path to the MIT presidency. They remained close friends.

Born in Leavenworth, Kansas, on June 1, 1869, Nichols was a delicate child whose parents died when he was fifteen. Cared for by relatives, he attended Kansas State Agricultural College and, in 1888, made his way east to Cornell University. At Cornell he earned the master's degree in physics in 1892, then taught at Colgate University. In 1894 he married Katherine West, daughter of one of Hamilton's wealthiest residents. The couple spent two years in Europe, 1894–96. In Berlin, Nichols helped physicist Heinrich Rubens create a pioneering radiometer for infrared spectrum analysis. On his return to America, he resumed teaching at Colgate before going back to Cornell to complete his doctorate (Sc.D.) in 1898. He was professor of physics at Dartmouth, 1898–1903, then at Columbia, 1903–9. The National Academy of Sciences elected him vice president in 1903. In 1909, the same year that Maclaurin became MIT president, Nichols was appointed president of Dartmouth, a post that he held until 1916 when he became professor of physics at Yale University.

Nichols was highly regarded for his work on radioactivity and electromagnetic theory. Scottish physicist James Clerk Maxwell had conjectured that light beams exert measurable pressure and Peter Lebedev, the Russian physicist, had detected the phenomenon, but Nichols was among the first to quantify it experimentally. During the First World War, he was a founding member of the National Research Council and worked on optical devices as part of a secret U.S. government program. In 1920, he left Yale to direct the pure science program at Nela Park Laboratory in Cleveland, an arm of the National American

Lamp Company, soon to become part of General Electric. At Nela Park, one of the first industrial research complexes in America, Nichols recruited a small army of scientists to work on the science of light and illumination. He preferred experimental physics to administration. As Dartmouth president, he had often talked about wanting to get back to science.

Wilson harbored doubts about Nichols, and felt uncomfortable about the prospect of serving as his vice president. On the train to Cleveland he mulled over something Maclaurin had told him in 1916, the day news of Nichols's Dartmouth resignation reached the newspapers. "Our old friend Nichols is leaving Dartmouth for Yale to return from administration to physics. . . . He never should have let himself be persuaded to take an administrative position; it is contrary to his temperament and he was bound to fail." Then, with Maclaurin's characteristic modesty—"I do not know that I am a success at it, that is very hard to say, but I do know that if I decided I was a failure, I would not try to go back into a professorship of physics but into the business on the legal side." Wilson wondered then, and later, if Maclaurin had confided this to anyone else—someone on MIT's executive committee, perhaps, or to Nichols himself.

Nichols held a different view. He saw his service to science and education as a calling, a deep responsibility. "The strong teacher must ever have the best of the priest about him, in the fervor of his faith, in the healing power of truth." This motto shaped his career, led him to think of executive leadership—whether at Dartmouth or MIT—as a form of self-sacrifice for the greater good. At Dartmouth, he had recruited top-flight faculty, boosted academic standards, encouraged scholarship, and increased the endowment. But conflict had emerged, too. Some complained about his meticulous standards and unreasonable demands, others about his disdain for the extracurricular side of college life, especially sports. Because he liked to weigh different viewpoints and to explore all sides of a question, some saw him as indecisive.

By 1916, anyway, his heart was no longer in it. He longed to return to science and understood that if he did not get back to it soon, he would be unable to catch up. His resignation, effective June 30 that year, left little doubt as to where his revised priorities lay. "There seems . . . no compelling reason why I may not ask you to let me go back to my earlier work, the duties and recompenses of which are in fuller accord with my individual tastes and preferences. . . . The duties of my office, I have found exacting and its high responsibilities, much as you have done to lighten them, a heavy burden. Looking forward to the many years lying before a man of my age, I seriously doubt my ability to hold through to the end and to give the college that vigorous service which its continued welfare requires of its president. Moreover, I feel a growing conviction that the best work it is in me to do for the college is already done."

Echoes of this inner conflict, just five years old, resonated with Nichols when he accepted the MIT presidency and, now, during his conference with Wilson. He struck Wilson as broken, defeated, tentative—a fine, productive physicist, to be sure, but haunted by the Dartmouth experience. "It made me sad, he was so obviously spent and was so widely known to have failed in administration because of inability to make the necessary decisions or to let others make them for him and back them up." Nichols admitted as much, but pronounced himself ready to pick up and try again. He pointed to the case of Daniel Coit Gilman, who had failed as president of the University of California and then gone on to great things as president of Johns Hopkins. Nichols vowed, also, to move equipment to MIT to carry on research in his spare time. Wilson listened quietly, nonplussed by Nichols's naiveté about the demands that the presidency would place on him. But when Nichols asked how the two might share out administrative duties, Wilson spoke up. The vice-presidency offer had come as a total surprise—he wanted to get back to his research and was leaning against accepting. The two men parted, nothing settled. Wilson headed back to Boston, his misgivings reinforced. On March 30, a few days later, the Corporation unanimously confirmed Nichols's appointment.

Wilson sought advice from former mentors Arthur Hadley, Yale University president, and physicist Michael Pupin of Columbia, both of whom urged him to say no. Nichols, so Hadley told Wilson, "may figure that you will run the Institute for him and let him go on with his experiments except for necessary official public appearances." Pupin said that anyone who pushed Nichols for president was doing no favors to either Nichols or the Institute. Wilson, still undecided, asked Nichols for an opinion on a tricky personnel matter. When Nichols wrote back sidestepping the question, Wilson declined the job.

In the wake of all this uncertainty, MIT missed a golden opportunity to trumpet itself as a mainstay of American science and technology. In April and May 1921, Albert Einstein toured America for the first time in a delegation led by Zionist champion Chaim Weizmann to raise funds for the creation of Hebrew University, atop Jerusalem's Mount of Olives. The group stopped in New York, Chicago, Washington, Philadelphia, Hartford, Boston, and other cities. Its political mission was overshadowed, to some extent, by Einstein's stature as the world's most famous scientist, and most of all by public fascination with his seminal, captivating, but thoroughly inaccessible theory of relativity. While hosting Einstein at the White House, Warren Harding "smilingly confessed that he . . . failed to grasp the relativity idea." A whirlwind tour of Boston, May 17–18, found Einstein mobbed by curiosity-seekers at the train station, lionized at receptions hosted by Boston mayor Andrew Peters and Massachusetts governor Channing Cox, wined and dined by Harvard president Lowell, thronged at a sold-out Opera House, honored at the American Academy of Arts and Sciences, private clubs, and other venues.

Einstein showed up just about everywhere, except MIT. The Institute's acting president Elihu Thomson appeared on the guest list for several events, but MIT played no official role and showed little if any interest. While administrative disarray partly explains the apathy, MIT's physicists—most of them on the applied or industrial side—were also to blame. E. B. Wilson's experimental work in vector analysis, aerodynamics, and statistics left him cold to Einstein's abstruse theoretical constructs. "Einstein no sooner had developed the principle of relativity," Wilson had cattily reviewed Einstein's work a few years earlier, "and established it on a sound basis than he went about destroying it, as some would say, or generalizing it, as he says, so as to take account of gravitational phenomena." MIT's great admirer Thomas

Edison, meanwhile, was shocked to learn that Einstein, when asked to recite the speed of sound, replied that he did not care to clutter his brain with facts that could be easily looked up in a textbook. Edison hoped that his son Theodore, an Institute student ('23), would not lapse into an impractical mindset of this sort—"Theodore is a good boy, but his forte is mathematics. I am afraid . . . he may go flying off into the clouds with the fellow Einstein. And if he does . . . I'm afraid he won't work with me."

In late May, about a week before his departure for Boston, Nichols warmly greeted Einstein at Nela Park in Cleveland. They posed together for a photograph, which Nichols passed along to MIT's student newspaper, *The Tech*. The image, in Nichols's mind, captured something of the spirit of research, pure science, and academic excellence that he hoped to nurture at the Institute.

Nichols's appointment was welcomed at MIT, if for no other reason than that it filled a worrisome void. "At one stroke," retiring dean of students Alfred Burton wrote to Everett Morss, "you have secured the confidence of the Scientific and Business World in the future of Technology and you have completely relieved the anxiety of the Alumni and undergraduates." Wilson, Talbot, and Miller, about to head back to their labs and classrooms, were optimistic—Wilson's private doubts aside—that Nichols's "interest in broadening the education of our future engineers and executives" held great promise.

Nichols felt ambivalent about the move, but he was eager to make this, his second try at a presidency, work better than the first. He was alert to the challenge of replacing his late friend Richard Maclaurin. He would have to manage without a vice president, while pushing fresh ideas—a new vision for science, industry, education, and the links between them—at a place that had gone leaderless for more than a year. All this weighed on him as he worked, frantically, to tie up loose ends at Nela Park. A reporter from the *Boston Post* asked him tough, probing questions about science, education, industry, the late war, scientific personnel, women in technical professions, religion, pure versus applied science, sports, hobbies, morality—all this while Nichols stood at his office window, gazing out over downtown Cleveland and Lake Erie. By departure time, he was a bundle of nerves. Katherine Nichols soothed him with talk of her own hopes to match the warmth, grace, and motherliness that Alice Maclaurin had set as the standard for MIT's first lady. The Nichols's only child, Esther, in her mid-twenties, looked forward to helping her mother enliven MIT's social calendar.

The inaugural celebration, held on June 8, was sandwiched between two other events—a banquet for seniors on June 7 and graduation exercises on June 10. Nichols put in an appearance at the banquet, telling his audience that while he would do his best, he could never hope to fill Maclaurin's shoes or to replace him in the minds and hearts of students. "I should like to be called the 'step-president' of Technology. I do not want to come between you and your late president and so I should like to be called that and for the moment I should like to become a step-father and do what they do so well—give a little advice." He then offered up sensible but boilerplate counsel about milestones in life, continuing to learn, giving one's all to one's job, working cooperatively, projecting good cheer, striving for success.

It was a tentative start, redeemed the next day by a stirring, carefully crafted speech that laid out his vision for the Institute—a speech that he had already submitted to *Science*, and arranged for publication in the magazine's June 10 issue. Event organizers, eager to show off the Institute's technical innovativeness, had set up amplifiers on the steps of Walker Memorial so that the proceedings could be heard outside, on the riverbank, by those without hall access—a novelty, wrote the *Globe*, befitting the Institute's "position in the world of applied science." For the evening events, the dome and building facade would be lit by shifting, flashing rainbow colors, while an enormous army searchlight on the Boston side swept back and forth over the river.

The morning ceremony opened with words of welcome from Massachusetts governor Cox (a former student of Nichols's at Dartmouth), Harvard president Lowell, and faculty chairman Henry Talbot. Nichols then stepped forward and read his prepared text. He spoke of how the dynamic between education, science, industry, commerce, and politics had transformed the modern age in unexpected ways. But certain values must remain constant, he argued, even as the relationship evolved further. The best education was not about training professionals to carry out this or that function, but to stimulate intellect, to spark imagination and creativity, to build critical, independent approaches, and to encourage reflection. This concept had been part of the Institute's mission since the earliest days, and had found its strongest advocates in Francis Walker, Henry Pritchett, and acting president Noyes. But Nichols sensed that for a decade it had begun to slip as economic and space pressures mounted and as commerce and industry became primary sources of funding. Nichols cited Maclaurin's view that "a technical school was not doing its whole duty unless it kept in the closest touch with industry." While he intended no criticism of his predecessor—Maclaurin had, after all, guided MIT to an unprecedented level of financial well-being—Nichols contended that now was not the time to stall in service-to-industry mode, but to reassert core values. He pushed for a renewed commitment to pure research, reminding his audience that applied science, the Institute's traditional focus, had always found its best raw material in the pure science laboratories, and that this explained why progressive firms like General Electric and AT&T had built such fine research programs of their own. Applied science simply could not move forward if pure science were shunted aside.

The MIT community had heard this message before, although not recently from Maclaurin. With the Institute's survival on the line, threatened by deep budget and infrastructure shortfalls, Maclaurin had postponed broad-range goals for more immediate, practical ends. The gap between Nichols's vision and the path down which Maclaurin had led MIT connected, too, with tensions that had simmered between the Institute's pure and applied camps, symbolized in the recent victory of William Walker (applied) over Arthur Noyes (pure). Some audience members listened in discomfort, others with hope, as Nichols mapped out a policy where pure and applied science would proceed together, but where, if push came to shove, pure science took preference. And, in yet another departure from the status quo, Nichols laid out a liberal position on labor-management relations that stressed human rather than economic factors, the value of owner-worker mutual respect and cooperation. Maclaurin, in contrast, had taken a hard line after the Seattle dockworkers' strike and red (Bolshevik) scare of 1919, conferring with local financiers on contingency plans—a pool of nonunionized skilled workers, for example—to protect firms and industries in the event of strike activity in Boston. Nichols also referred twice to women, who had been tolerated at MIT but never accepted as coequals. All in all, he laid out a broadminded, progressive program.

Nichols felt worn out by the time he got to the alumni banquet that evening. He spoke off the cuff about his pride in being chosen president, about MIT's purposeful, motivated students and alumni. But then he began to ramble, his thoughts racing, his cadence a bit breathless. "It is going to be my very pleasing privilege . . . to try as hard as ever I can from time to time, as we gather together, to make some of the details a little clearer, a little more definite, without taking the poetry out of the vision. We must go through the forest and cut out the dead wood in order that the live wood may develop and progress and the forest become more beautiful. But these hot-headed men, and the more hot-headed the more silly they are, cry out for a clean slate on which they may draw more beautiful designs. Some of these men would even burn the forest in order that they may plant it anew. . . . A clean slate and a new forest are things we cannot have in this world. We have got to begin where we are, and it has got to be more than a voice crying in the wilderness, more than a child crying in the night. It has got to be faith. We need strong men, capable of doing fourteen hours' work a day for altruistic purposes. We have got to reshape the world as it stands in order that there may be built the foundations of justice, righteousness and peace. The world is calling on the Institute and it shall not call in vain." Nichols's audience, distracted by good food and fellowship, may not have noticed how sharply these words contrasted with the coherence of his earlier speech. After dinner, Arthur D. Little ('85) proposed pressuring city authorities to rename the Massachusetts Avenue bridge over the Charles River, "now known unwarrantedly as the Harvard Bridge," *Technology Bridge.* Alumni whooped their approval as they wended their way over that very bridge to attend Tech Night at Pops in Symphony Hall.

Not quite 52 years old, Nichols looked poised to break MIT's streak of rotten luck in which four out of six presidents had either died in office or been forced into retirement on account of illness. But right after the inaugural, possibly the next day, he fell ill and remained in seclusion for several months. He failed to show up for graduation on June 10. Mrs. Nichols watched over him, keeping house at various locations outside Boston—in Winchester toward the end of June and, by November, at Drabbington Lodge, Kendal Green, in Weston. His diagnosis was shrouded in secrecy, but hints emerged of cardiac problems exacerbated by nervous prostration. He went on total bed rest. With a brave face, he wrote MIT officials at the start of the academic year in October that his doctors were confident of a speedy return to business as usual.

Within weeks, these same doctors concluded that the strain of the presidency would do him in. Nichols composed and submitted his resignation on November 3:

> A sufficient time has now elapsed since the onset of a severe illness, which followed immediately upon my inauguration, to enable my physicians to estimate consequences. They assure me certain physical limitations, some of them probably permanent, have resulted. These, they agree, make it decidedly inadvisable for the Institute or for me that I should attempt to discharge the manifold duties of president. Indeed, they hold it would be especially unwise for me to assume the grave responsibilities, to attempt to withstand the inevitable stresses and strains of office, or to take on that share in the open discussion of matters of public interest and concern inseparable from the broader activities of educational leadership.
>
> As my recuperation is still in progress I have contended earnestly with my doctors for a lighter judgment. I feel more than willing to take a personal risk, but they know better than I, and they stand firm in their conclusions.
>
> The success of the Institute is of such profound importance to our national welfare, to the advancement of science and the useful arts, that no insufficient or inadequate leadership is sufferable. Personal hopes and wishes must stand aside.
>
> It is therefore with deep personal regret but with the conviction that it is best for all concerned, that I tender you my resignation of the presidency of the Institute and urge you to accept it without hesitation.
>
> To you who have shown me such staunch and generous friendship it is pleasant to add that in the judgment of my physicians the physical disqualifications for the exigencies of educational administration are such as need not restrict my activities in the simpler, untroubled, methodical life of scientific investigation to which I was bred. It is to the research laboratory, therefore, that I ask your leave to return.

Nichols was granted two months' paid leave, his resignation not effective until January 4, 1922. Talbot, Miller, and Wilson resumed oversight duty, all three still chafing to return to their academic work. Elihu Thomson became acting president for the second time on the understanding, as before, that he would carry out legal requirements and nothing more. Nichols went back to his research at Nela Park, even as colleagues there worried about his ability "to shoulder . . . duties on account of his physical breakdown." He and J. D. Tear worked on electromagnetism and the infrared, developing experimental approaches to the basic problem—"what *is* light?" Nichols died of heart failure on April 29, 1924, while speaking on this topic before the National Academy of Sciences in a building, newly erected for the Academy, that he had helped raise funds to build.

"Shaping things in orderly fashion"

Samuel Wesley Stratton, 1861–1931

A new presidential search began two or three weeks after Ernest Fox Nichols resigned on November 3, 1921. E. B. Wilson and Francis Hart nominated candidates that some viewed as bold choices, others as desperate last resorts. Hart went to bat for Sir Auckland Geddes, British ambassador to the United States, who had withdrawn as principal of McGill University in order to join the diplomatic service. Wilson thought Robert John Strutt, 4th Baron Rayleigh, worth considering—"a physicist by no means so eminent as his father, but of entire respectability," who the year before had resigned his faculty post at Imperial College, London. When neither candidate gained traction, Wilson and Hart met to hash out what personal qualities MIT should look for in a president. Wilson sensed a growing preference for managerial savvy over academic credentials. "I imagine," he confided to Yale University's new president James Angell, "our Corporation has not yet decided whether they will put an educator in charge of this institution or a business man. You can understand how strong must be the pressure from some of our graduates for the selection of a business man. Faculties, particularly Eastern faculties, are so inert that it seems to me the better chance is to commit our fortunes to some one who has shown that he is able to get some results."

Wilson remained active in the search process even after pulling out as a member of the administrative committee in February 1922, to be replaced by Charles Norton. In March, Wilson wrote on Gano Dunn's behalf. Dunn, president of the J. G. White Engineering Corporation of New York, was a member of the National Academy of Sciences and had served during wartime on the National Research Council. Most impressive, in Wilson's view, were his elegant manner, savoir faire, and "exceptionally well poised bearing." When he failed to attract Corporation support, Wilson suspected that his candidacy was doomed by his ties to a commercial competitor of Stone & Webster, the company founded by Tech classmates ('88) Charles Stone and Edwin Webster.

Then came a stroke of remarkable luck. Hart sailed for Europe in June 1922. In the ship's lounge one evening, he struck up a conversation with two men. One of them was in his late fifties or early sixties, a bit portly, mustachioed, with piercing blue eyes and a fine head of graying hair neatly flattened square to his forehead. The other was slight and clean-shaven, about thirty years younger. The older man introduced himself as Samuel Stratton, the younger as his assistant Morris Parris.

While Hart had never met Stratton before, he knew many who had had dealings with him as founding director of the National Bureau of Standards. Because of the Bureau's broad role in technology assessment, Stratton often crossed paths with MIT faculty and staff. Henry Pritchett had helped guide him through some of Washington's political minefields and remained a close confidant. Stratton went to Boston at least twice in 1901 to seek Pritchett's advice. Bureau board members with MIT connections included Henry Howe ('71), and a sizeable portion of the staff were Institute graduates. George Burgess ('96), who would succeed Stratton as Bureau director, worked in the heat and thermometry division, Albert Merrill ('00) and Franklin Hunt ('09) in instruments, Frederick Grover ('99) in inductance and capacity, and Francis Cady ('01) in resistance and electromotive force. Then there was the wartime research group—Jerome Hunsaker ('16) and others, pioneers in aviation and wind tunnel research—that Stratton had worked with as a member of the National Advisory Committee for Aeronautics (NACA), along with Willis Whitney ('90) of the Naval Consulting Board and George Hale ('90) of the National Research Council. Even Stratton's old college chum, Frank Vanderlip, although not an alumnus himself, was already six years into what turned into a two-decade stretch on the MIT Corporation.

When the two met shipboard, Hart was as well acquainted with Stratton as one could be without ever having met him in person. They took a liking to each other and, over the course of the week-long Atlantic crossing, spent many hours deep in conversation. The voyage was the start of Stratton's month-long escape from Washington's summer heat, a tradition going back to his earliest days at the Bureau. Morris Parris usually accompanied him, juggling responsibilities ranging from secretary, protocol adviser, and troubleshooter to valet, caretaker, and confidant. Stratton's immersion in his work did not allow for a complete getaway, so the annual break—passed mostly in Paris—involved paperwork and attendance at the conference of the International Bureau of Weights and Measures held annually at the Pavillon de Breteuil in Sèvres, just outside the city. Stratton was the permanent American delegate and one of the longest-serving members of the organizing committee. Over the years he had influenced the bureau's decision to take on electrical as well as mechanical standards and, in particular, to promote international norms for electricity. But Stratton enjoyed Paris for its culture, too, and frequented museums, bookstores, concert halls, and antique shops there.

Hart talked to Stratton about science, technology, economics, international affairs, and the way those worlds overlapped. But he was more eager to draw him out on another matter. Without letting on just how precarious the leadership gap at MIT had grown, he sought Stratton's advice on possible choices for a new president. Stratton said he would think it over and get back to him. Neither considered Stratton himself a possibility. He was content where he was and, at nearly 61 years old, getting toward retirement age. He was in the midst of remodeling an old house in Washington's mostly black Georgetown district, the first home he had ever owned. In those days, government employees often rented rather than bought—a sign that they did not expect to stay long—and Stratton had lived since about 1901 in an apartment at the Farragut, corner of 17th and I Streets. But he was ready to trade these quarters for a home with a garden, a place where he could live in uncramped comfort. He may or may not have recalled that Samuel Woodbridge ('79) had approached him fifteen years earlier for advice on a successor to Pritchett. "I would rather be president of the Massachusetts Institute of Technology than of the United States," Woodbridge recalled Stratton telling him at the time.

Late that summer, Stratton and Parris set out on a road trip through New England. When they stopped in Springfield, Vermont, to meet with governor James Hartness, someone handed Stratton a telegram. "Can you come to Cambridge for lunch day after tomorrow?"—signed Francis Hart, on behalf of MIT's executive committee. Stratton wired his consent.

He half expected talk of a new invention, a technical problem that Bureau staff might help out with, a new or modified industrial process that called for a second opinion. Instead, he was blindsided by an offer—almost a plea—that he come to Cambridge as MIT's new president. Hart, on his return from Europe, had talked Stratton up and convinced his colleagues that a better candidate—someone whose experience ran the gamut from higher education to industry, engineering, technology policy, and much in between—would be hard to find. Stratton said he would think it over and talk with his boss, U.S. secretary of commerce Herbert Hoover.

The news shocked Hoover, who had taken it for granted that Stratton would keep the Bureau's affairs humming along indefinitely. But he promised not to stand in Stratton's way, or MIT's, even as he railed about the loss landing a blow from which the Bureau might never recover. Hoover lambasted the government for failing to keep its choice personnel from drifting, sometimes fleeing, out of public service into the private sector. "Dr. Stratton," Hoover told the press, "has repeatedly refused large offers before"—in 1916, for example, Columbia's president Nicholas Murray Butler had tried to recruit him as professor of physics at a salary well above the rest of the department, and with the promise of a substantial pension—"but the inability of scientific men and the government to properly support themselves and their families under the living conditions in Washington makes it impossible for any responsible department head to secure such men for public service at government salaries." Money, however, rarely if ever factored into Stratton's plans. What appealed to him about the MIT offer, as

Morris Parris put it, was the "thought of being among the young men and . . . a part of the great structure which was molding their future."

Stratton got the executive committee's nod on September 19, 1922, confirmed by the Corporation on October 11. According to Hart, "not . . . one word in opposition" was voiced. The larger MIT community sensed stability in the air after three years of not-so-coherent management. "I know of no better man in this country for the place," young Julius Stratton ('23), just entering his senior year, wrote home. "Tech has been without a president for nearly three years and though the administration has carried on well under the circumstances it requires a single head to co-ordinate such an institution. It is just this that I believe Dr. Stratton is well qualified to do. Though you may perhaps never have heard of him you have at least known of the United States Bureau of Standards as one of the greatest and best organized research laboratories in the world. He is in great measure responsible for its existence and later progress. . . . What Tech needs now more than anything else is an organizing genius who will smooth out the running of the internal machinery by a tactful adjustment of inter-departmental difficulties and what is equally important, of bringing to the attention of the engineering world the work which is and can be done here."

Stratton's arrival signaled an end to what one outsider called "the sad . . . almost tragic wanderings of Technology in search of a leader." Students roamed about singing "Samuel Stratton forever," to the tune of "Marching through Georgia."

Old Tech's been looking for a man of caliber and zest
To take the wheel and steer her straight for everything that's best
And now they've found the very one who stands the acid test
Doctor Samuel Stratton forever.

Hurrah, Hurrah, we've got him cinched at last
Hurrah, Hurrah, we'll soon be moving fast,
For he's the man who surely can just lash 'em to the mast
Doctor Samuel Stratton forever.

Tech Alumni ask the Faculty to keep the students crammed
And they answer back that Tech is Hell and Dorms are over jammed.
Sam Stratton is our Beaver and conditions must be damned
Doctor Samuel Stratton forever.

Technique editors dedicated their upcoming yearbook to him "as an expression of . . . confidence that through his efforts there will arise a greater Technology."

Stratton promised to settle in by January 1, 1923, rather than finish out the fiscal year at the Bureau. He and Parris went back to Boston in November, staying with Francis Hart and his wife at their Beacon Street townhouse. Everett Morss, who had succeeded Hart as Institute treasurer, hosted a reception on November 4 at his home on Commonwealth Avenue. Hart's son and daughter took Parris, about their age, to dinner and the theater while the older folks talked MIT business and grew better acquainted. Stratton toured the president's house, which struck him as a fine dwelling equal to or better than his (now abortive) Georgetown dream home. A month later, December 15–16, the alumni hosted a reception in New York patriotically entitled "Get behind President Stratton." Robert Richards ('68), who had known every Institute president from Rogers through Maclaurin, surmised that Stratton would prove the best of them all. "Technology has been hungry for team work all along the line . . . for a long time, and we feel that you have given us assurance that this is coming to us now. Of all the presidents we have had . . . we have never had one who filled our hearts with such confidence as you have."

Stratton arrived for work on January 2, 1923, a day late because a flustered Hoover wanted to spend New Year's Day with him going over last-minute transition details. Parris reached Cambridge a week ahead to set the president's house in order. Stratton's housekeeper and cook, Cordelia Pannell, went with him. Parris hired Postello Jones, an experienced chauffeur, to drive and care for Stratton's Cadillac coupe. In due course, an acquaintance of Jones's—Alonzo Fields—joined the staff. The two may have known each other as fellow congregants in one of Boston's thriving black churches. Fields, an aspiring musician, helped pay for the cost of his lessons at the New England Conservatory by working as Stratton's live-in butler. "I confess," he later recalled, "that I didn't relish the thought of being a house servant, but Mr. Parris pointed out that, if I ever did reach the heights as a concert singer, these conventions he was teaching me would give me a background of good breeding." Fields enjoyed the job so much that he gave up his musical ambitions for a career in butlering.

Stratton traced his ancestry back to several waves of English immigrants who settled in Massachusetts between 1628 and 1659, then dispersed through at least nine of the thirteen original colonies. His own branch laid roots as farmers in Virginia by the mid-eighteenth century. But Robertson Stratton, his grandfather, pushed westward in the early nineteenth century in search of cheaper land near the Ohio and Mississippi Rivers. He settled near Keysburg in Logan County, Kentucky, where he owned and ran a farm. When he died in 1836, his wife Nancy (Miles) packed up the household—including their five children—and moved to Macoupin County, Illinois. One of these children was Samuel Stratton's father, also Samuel, who joined the California gold rush in the 1850s, married Mary Bowman (Webster), and farmed stock near the village of Litchfield in Montgomery County, Illinois, about fifty miles northeast of St. Louis, Missouri. He was among the first farmers to import pedigreed animals to Illinois, including a prized Jersey herd, Shetland ponies, and Brahma fowl.

Young Samuel was born in Litchfield on July 18, 1861. He and his siblings—an older half-brother, William Phillips (his mother's son), and three sisters, Matie (born 1863), Olive (1865), and Lucy (1874)—grew up on the farm. Two other siblings, Oliver Otis and Essie, died in infancy. Sam showed a special talent for machinery, which remained with him throughout life (in his sixties, he listed tool work and physics as his two favorite pastimes). He loved to tinker. When there was a wheel to be repaired or replaced, or a gear to be lubed, all eyes turned to him. He invented, too. One of his early devices, an automatic stirrer, reduced old-fashioned elbow grease needed to mix apple-butter in a vast outdoor iron vat. When his half-brother William, a student at Illinois Industrial University (known as the University of Illinois after 1885), brought home stories of marvelous machine shops, Sam decided he had to go there, or somewhere like it, for a technical education. The best programs—Cornell, Rensselaer Polytechnic, and MIT—were out east. MIT's reputation had spread along with its graduates, many of whom had helped complete the transcontinental railroad, from New York through Illinois, Kansas, Nebraska, Utah, and out to the west coast. But Sam settled for his state school, where tuition was cheaper.

To pay for room and board, he hired himself out as a mechanic and carpenter on full-time night shift.

Sam entered Illinois Industrial in the fall of 1880, intending to study machinery (his term) and become a high-grade mechanic. This plan evolved into a formal program in mechanical engineering led by Selim Peabody, whose son Cecil was an MIT graduate ('77) and faculty member. Sam's perspective on mechanics broadened and deepened, its ties to industry, science, research, and other opportunities growing sharper under Peabody's mentorship. Peabody balanced his duties as department head with multiple roles as university regent (president), registrar, recorder, and secretary. Sam grew close to the Peabody family and joined the household as an all-purpose assistant, helping with chores but also, in exchange for room and board, as personal secretary to Selim Peabody. Sam played a leadership role, too, in campus military affairs. The army impressed him as machine-like, "a beautifully organized, coordinated piece of living human machinery," and in June 1884 the governor commissioned him captain in the state military reserves. Sam joined a campus literary group, the Philomathean, but this was not his forte—"he was always diffident on his feet," a friend remarked, "and never a good speaker." One classmate recalled him as neither showy nor brilliant, but diligent—"a youth . . . clean, healthy, hearty, and human without special talents other than character, personality, good sense, and habits of industry . . . attracting and inspiring men both above and below him." Sam would watch from the sidelines as undergraduate pranksters hoisted cows into the belfry or farm wagons atop the main building. He co-planned and got away with a raid or two of his own, however, on the horticulture department's treasured stock of Bartlett pears.

Stratton's lifelong affinity for male bonding—he never married—dated at least as far back as his undergraduate days. Before moving in with the Peabodys, he roomed in an old university building with Frank Vanderlip, keeping house in a way that would "never have been countenanced by . . . wives." Peabody, citing Stratton's ability to get along with men, appointed him math tutor and housemaster in the university's all-boys preparatory department. Stratton saw this role less as proctor, spy, or disciplinarian than as "chief of clan" fostering a spirit of "fine comradeship." He liked working with the boys, and they looked up to him. Their bond intensified to the point where Peabody's daughter wondered if he suffered from "what modern psychologists would call a father complex." Stratton carried these extra duties while trying to complete requirements for the bachelor's degree. His graduation was delayed past the anticipated date (1884) because he did not complete his final thesis—"A design for a heliostat"—until the spring of 1886.

Stratton took a break in 1887–88 to visit members of his family at their retirement home in Pasadena, California. There he worked in real estate and, pursuing his passion for the military, joined the California Protective Cadets, a branch of the state militia. Back at the University of Illinois in the fall of 1888, he served as an assistant to professor of botany Thomas Burrill, a pioneer in plant-disease research, developing a microphotographic technique to study bacterial organisms. The following summer he went to Europe with a student, Frank Harris—"the first display of that wanderlust," as one friend called it, which remained with him for life. On his return, Selim Peabody appointed him assistant professor of physics. Stratton roomed with Glenn Hobbs, a student who became his assistant, friend, and later brother-in-law (Hobbs would marry Stratton's sister, Lucy, in 1899). In 1890, Peabody asked him to create and take charge of an electrical-engineering course. Stratton traveled east that summer to see what other institutions were doing along this line. With Hobbs along for the ride, the two made a whirlwind tour of several cities, looking over programs at the University of Michigan, Cornell, Lehigh, Lafayette, Columbia, Harvard, and MIT. They bought apparatus in Philadelphia and New York. The Boston excursion included visits to historic sites—Bunker Hill, Concord, Lexington—and their final stop before heading home, Washington, D.C., proved equally educational. "Stratton was a wonderful traveler," Hobbs recalled, "and thoughtfully worked out from his Baedeker . . . the things which should be seen so as to make the best use of our time."

Stratton spent two years building the electrical-engineering course at Illinois and consulting, meanwhile, for railroad entrepreneur and later U.S. senator from Illinois William B. McKinley (not to be confused with U.S. president William McKinley Jr.) on plans for an interurban railway to replace the old mule-car line between Champaign and

Urbana. In 1892, he left the university to become assistant professor of physics at the newly reorganized University of Chicago. Under its president William Rainey Harper, Chicago belonged to a small group of American institutions—Johns Hopkins was a leader in this movement—that sought to reshape American higher education in the rigorous image of German intellectual and scholastic traditions. Stratton was brought to Harper's notice by Nathaniel Butler, professor of English at Illinois, a friend of Stratton's whom Harper also recruited for the new faculty. Hobbs went along as Stratton's assistant. Stratton and Hobbs roomed together for two years, 1892–94, as part of the Butler household before taking an apartment of their own, where Lucy Stratton (later Mrs. Hobbs) joined them as a student at the University of Chicago.

The prized group of scientists that Harper assembled—A. A. Michelson in physics, John Nef in chemistry, C. O. Whitman in zoology, Henry Donaldson in neurology, Jacques Loeb in physiology, John Coulter in botany, MIT alumnus Edwin O. Jordan ('88) in bacteriology—worked at first under primitive conditions out of a dilapidated commercial building several blocks north of the main campus. Stratton ran the physics department for two years, with Michelson on leave in Europe. Michelson relied on him to plan and oversee construction of the Ryerson Physical Laboratory, completed in 1894. Stratton took advantage of contacts made at the World's Fair, mounted in Chicago in 1893, to stock the right equipment and to brainstorm with builders and manufacturers visiting from Europe. Hobbs recalled how he and Stratton wandered the fair grounds "with the enthusiasm of youth," listening to talks by celebrated scientists and engineers—Lord Kelvin, J. J. Thomson, Sylvanus Thompson, Werner Siemens—and marveling at Nikola Tesla's high-tension apparatus, the soon-to-be-famous Tesla coil.

In 1895, a year after Michelson's return, Stratton was promoted to associate professor as reward for creating in Ryerson a truly state-of-the-art facility. He and Michelson worked together on light velocity and harmonic wave analysis. A harmonic analyzer that they devised proved a valuable tool in Michelson's spectroscopic research, which helped earn him the Nobel prize in 1907. Stratton's technical expertise complemented his boss's scientific insights. As their young assistant

Hobbs saw the relationship: "While Michelson was a very constructive physicist, he disliked . . . the mechanical side of the development of his problems. His fertile mind thought out many scientific problems, and with Stratton's mechanical genius and Michelson's theoretical knowledge, they made a very good team."

Stratton went to Europe for the second time in the summer of 1895, accompanied not by Hobbs but by another protégé, Webster Smith, who shared Stratton's love for the military. Meanwhile, he grew friendly with a young naval officer—Lieutenant William Wilson, head of the hydrographic office in Chicago—who shifted his interest to the navy. Stratton joined the Illinois naval reserves on December 30, 1895, and quickly rose through the ranks from ensign, lieutenant junior grade, and lieutenant, to lieutenant commander in May 1898. During the Spanish-American War, he led a naval detachment to Mobile (Ala.) and Key West (Fla.), saw duty aboard the station-ship U.S.S. *Lancaster*, then served on the warship *Texas* as it made its way north after the battle of Santiago. "My naval service was devoid of the excitement of battle," Stratton later recalled, "but filled with hard work of the sort that is essential to that service in time of war . . . [training men] how to run practically all machinery on the ship." He was honorably discharged on May 24, 1899.

The Stratton family held a grand reunion that summer to celebrate Lucy's marriage to Glenn Hobbs. Stratton rented Jacques Loeb's sprawling house near the University of Chicago—Loeb was off doing research at the Marine Biological Laboratory in Woods Hole, Massachusetts—and everyone congregated there. The elder Samuel, Mary, and Matie came out from Pasadena, Olive and her husband Dr. A. T. Newcomb from Baltimore. That fall, Stratton went to Washington to invite Admiral George Dewey, hero of the Battle of Manila Bay, and U.S. treasury secretary Lyman Gage to speak at Chicago-area campuses. The mission was sidetracked, however, when Frank Vanderlip, Gage's right-hand man at the treasury department, joined U.S. Coast and Geodetic Survey head Henry Pritchett in urging Gage to recruit Stratton to promote national standards in technology and industry. In October Gage hired Stratton onto temporary, part-time duty as head of the weights-and-measures office to carry out a survey and prepare a report. Pritchett, soon to become MIT's president, hoped that Stratton would confirm his own belief that the nation must have a standards bureau.

Stratton presented his findings quickly and appended a piece of nicely crafted legislation for Congress to consider. He fielded tough questioning in open session so well that the bill passed with little opposition and was signed into law by President McKinley on March 3, 1901. One congressman wondered if Stratton, who became known for having his way with House committees, used hypnosis to push his agendas through. Congress's concern about bloating the federal bureaucracy was trumped by Stratton's argument that the Bureau would save money in the long run and reduce American dependence on foreign powers, especially with military technology. Up to that point army ordnance and navy construction divisions had had little choice but to send apparatus to Berlin for calibration in the Reichsanstalt, Germany's arbiter of scientific standards and instruments. The bill's ambitious wording endowed the Bureau with considerable authority, flexibility, and range, granting power to tackle larger problems in science and technology as well as to carry out routine testing. Stratton agreed to stay on as director, effective March 1901. His Chicago colleagues were hardly surprised, as for over a year he had spent much of his time in the nation's capital and done little academic work.

Government agencies were accustomed to a slower pace than that which Stratton set for the Bureau. After two years, he had 60 staff instead of the four he started out with. The first two laboratories went up in near-record time on a 10-acre cow pasture in northwest Washington. By 1922, twelve more buildings were erected on 39 additional acres and the staff had mushroomed to 850–900 scientists, engineers, and technicians. In recognition of his administrative work, a stream of honorary doctorates came Stratton's way—University of Illinois and University of Pittsburgh, 1903; University of Cambridge, 1908; Yale University, 1919; Harvard University, 1923; Rensselaer Polytechnic Institute, 1924.

Stratton never played by narrow rules, working around whatever constraints might impede his agency's growth. He urged Congress to expand its vision. "If we are to advance," he told a House committee bluntly in 1902, "we have to create original things." He insisted that the Bureau steer clear of favoritism and bias, whether political or commercial, and make all results public. In 1903, he had the Bureau transferred from treasury to the newly established commerce and labor department, convinced that there he would find more opportunity to move beyond product screening into creative research. His circle of well-connected friends and colleagues, always growing, helped him leverage support. Commerce secretaries regardless of party affiliation—William Redfield on the Democratic side, Herbert Hoover on the Republican side—held him in high esteem. Redfield, asked to comment on a Stratton request, once told a congressional committee: "I can only say that anything that Dr. Stratton wants I back up."

Stratton's leadership style combined personal interest in his staff—the boys of the Bureau, as he called them—with a clear focus, insistence on high standards, and a policy of rewarding achievement. He would roam the Bureau's offices and laboratories with the energy of a man half his age. A hands-on manager, he knew each employee personally and kept up with what went on, down to the smallest detail. "How is that work coming along?" "When did you get your big pot made?" "Did so-and-so send that package to you as he promised?" "Is blank company perfectly satisfied with that?" "Have you heard anything more about how such-and-such turned out?" Some meeting him for the first time felt intimidated by his steely gaze, but behind that gaze, his friends and colleagues discovered, lay a desire to get at the heart of a matter.

Stratton's personal life overlapped in key ways with his professional. During his first two years in Washington, 1899–1901, he rented a room at Mrs. O'Toole's—a boarding house at 18th and I Streets N.W., frequented by government employees of varying levels, mostly bachelors—where he made lifelong friends. In 1901, after signing on to head the Bureau, he moved to an apartment at the Farragut and entertained there with the help of Cordelia Pannell, whom he enticed away from the service of his friend Herbert Putnam, Librarian of Congress. Stratton bought a boat, a 25-footer, with Louis Fisher and spent many a weekday evening and Sunday picnicking while sailing up and down the Potomac. He kept up his naval interests, meanwhile, and on November 18, 1904, was commissioned a commander in D.C.'s naval battalion in charge of the monitor *Puritan* and the steam yacht *Oneida*. Another friend, James Courts, had a place at Arundel-on-the-Bay, just below Annapolis, which Stratton visited regularly. Courts, longtime clerk of the House appropriations committee, helped him shepherd bills through Congress.

Stratton's close-knit circle included assistant commerce secretary Edwin Sweet, Smithsonian Institution secretary Charles Walcott, railway executive and army man George Harries, House clerk Asher Hinds, and Joseph (Uncle Joe) Cannon, colorful, outspoken Republican speaker of the house and Theodore Roosevelt's perennial nemesis. Stratton, Courts, and Hinds went to Europe together in the spring of 1905, hiring a gentleman attendant, George Washington Randolph Lee Green, to accompany them. Green's diary kept during the trip refers to the group as "three gentlemen of Washington, good looking and of proper style," with Stratton further described as "of medium height, rectangular in figure, of an innocent countenance, illuminated by soft gray eyes and a blonde mustache . . . Chief of the Bureau of Standards, a place where they make right the milk quarts and the girth chains used in measuring watermelons." Green, who was black, evidently enjoyed toying with racial stereotypes.

Stratton's attachment to English Walling, a former student of his at the University of Chicago, suggests that politics—left, right, or center—rarely if ever played a role in his friendships. Walling, a socialist and avid supporter of feminist, labor, racial-equality, and other progressive causes, stood about as far-left as Uncle Joe Cannon stood far-right. One evening in 1902, Walling and his brother Willoughby took Stratton to the Parris family home at 3022 P Street, in the city's Georgetown section. There to greet them were Albion Parris, his wife Bessie, and their four children—Albion, aged 19; Worden, 16; Morris, 11; and Juliet, 9. The elder Albion, a well-to-do banker, had family ties to the distinguished nineteenth-century Maine politician and jurist of the same name who had served in both the U.S. House and Senate, as state governor, and as justice of the state supreme court.

Stratton got to know the Parrises well and became a close friend, almost a surrogate father, to both Worden and Morris. He treated all the Parris children to picnics, railroad excursions, and unusual pets—a baby pig one day, a large white goose the next. In 1903, Stratton took Worden camping in Yellowstone Park along with some other bachelor friends. This trip, Worden wrote, "proved to be the realization of a dream to the Doctor's 17 year old companion, who said that the Doctor was a great Pal when he could be gotten away from Desks, Colleges, and Dinner parties." About three years later Morris, aged 16, and a young friend joined Stratton and Uncle Joe Cannon on a sail to Old Point Comfort, Virginia, with instructions to keep autograph- and curiosity-seekers at bay by telling "the most awful lies" about Uncle Joe's whereabouts. Worden went into the army, worked in the Detroit auto industry, and until his marriage to Bernice DeMosh in October 1922, remained Stratton's closest friend. Young Albion and Morris joined the Bureau staff, Morris serving as Stratton's private secretary, all-purpose assistant, and live-in manager at the Farragut apartment. The elder Albion wrote warmly to Stratton in October 1922, just as Stratton and Morris were preparing to move to Boston. "I will miss you greatly and ever recollect the never to be forgotten days in my own home—and . . . I will never fail to recognize the heavy debt of gratitude my children and myself are under to you for the great consideration and multiplied favors and kindnesses, generously and loyally bestowed without limit on the Parris family—their kin and friends."

Stratton, known affectionately around the Bureau as "the old man," was a father-substitute for many there, too. A party that he organized around 1902 for children of staff turned into an annual tradition, replete with hand organs, monkeys, merry-go-rounds, ponies, balloons, and ice cream. On the professional side, he often played protector to guarantee fairness. Where his own collaboration was critical to a project, he would decline credit so that junior, less experienced men might benefit. He aimed to chip away at what he considered a blot on modern science, the presumptive right of senior scientists to take credit for work done by students or staff. Early in his own career, he had been hurt in this way by a senior collaborator—never named, but probably Michelson—and vowed not to fall into that trap himself.

Stratton's name appears on few of the dozens of papers, reports, and circulars published by the Bureau. He saved his byline for annual reports to the commerce secretary and for the driest of position papers, those least likely to reach the public. This posed a problem when his name came up for election to the National Academy of Sciences in 1914. Stratton had just one important scientific paper to his name—"A new harmonic analyzer," coauthored with Michelson and published simultaneously with minor variations in January 1898 in both *American Journal of Science* and *London, Edinburgh, and Dublin Philosophical Magazine and Journal of Science.* "Unfortunately," as he explained to Carnegie Institution of Washington president Robert Woodward, "the circumstances under which I have found it necessary to work have precluded much in the way of publication. . . . It seemed to me that I could perhaps help the cause of science far more by contributing my time and energy toward the establishment of an institution so badly needed by the scientific workers of the country than by any specific investigation in which I might engage personally, although the latter would have been my choice had I considered my own interests alone." The Academy ended up not only electing him to membership but also, three years later, awarding him its gold medal for contributions to science in the public interest.

Stratton's departure from the Bureau resonated on a personal level with many staffers. "I always feel," one employee's wife wrote in December 1922, "that you brought more happiness to Mack than any other one person, for Mack's work was his real life, and always when he wanted to start new work or was discouraged in what he was attempting he would go to you and be sure of understanding and helpful guidance. . . . You have been a father to the Bureau carefully watching over it and putting your whole life into it." August Flegel, molder-technician Division VIII, a recent immigrant, expressed his sadness in broken English—"It makes me very sorry I am losing a good and the best of all superiors you was the firt man in this country which was so humanly and have helped me to a living position." Some reactions leaned toward shock, anger, and worry over the agency's future. "*But*, what, in the name of all efficiencies, is to become of the Bureau?" blurted Herbert Putnam. "But what will the poor old Bureau do?" wrote Edwin Sweet. "I must confess," another correspondent sighed, "to something of that mournful feeling which led King Arthur to bemoan the passing of his Round Table." Others, echoing Herbert Hoover, vented rage at the government's misplaced priorities. "It serves the Department of Commerce perfectly right that a man of your ability is lost to them, for they will pay a lawyer $100,000 to conduct a prosecution in a law case, but they will not pay a man of science . . . a living wage." To many, Stratton's departure represented a tragic blow that could well sink the Bureau.

He arrived at MIT in high spirits, energized by the sight of eager, fresh-faced youngsters milling around. He talked to students about feeling like a schoolboy again, of arriving "to register for the term along with the rest of us." "I am not going to turn the institute upside down. . . . I am going to study it and then become better acquainted with you, its students. I hope to meet personally every man in the institute, and you are all to feel at liberty in my office whenever the time is appropriate." Listeners felt drawn by the open-minded, flexible manner that had impressed his alumni hosts in New York a couple of weeks earlier. "We like very much," Robert Richards wrote, "the quiet way in which you are going to start in not hunting up something big to do right away, but in waiting quietly until the big things that are to be done shape themselves and come along in orderly fashion."

In the weeks that followed Stratton showed up everywhere, unannounced, just as he had done in corridors, labs, and offices at the Bureau. He seemed probing, proactive, alert. "He is learning," young Julius Stratton observed, "in a way which is making the department heads wake up with a start and get down to business. He is not making pre-arranged trips of inspection with complete staff. He just drops around at most unexpected times and places and wants to know the reasons for everything." The inertia that had weighed MIT down since Maclaurin's death—that awkward, rudderless period from 1920 to 1923—looked about to be replaced by something new, vigorous, fresh. A consensus grew that Stratton was a better choice than the ailing Ernest Fox Nichols. Nichols, wrote Julius Stratton, "may have been a fine physicist but he was not the man for us. This man is not only a scientist but an executive, and that is far more important to us at this time than anything else."

Stratton's arrival coincided with near-panic over publication of *The Goose-Step*, Upton Sinclair's scathing indictment of American higher education. The first edition, which reached bookstores in February 1923, sold out within a month. Based on hundreds of interviews nationwide, Sinclair argued that America's college-age youth—over half a million strong—were being corrupted by corporate values and taught, "deliberately and of set purpose, not wisdom but folly, not justice but greed, not freedom but slavery, not love but hate." Sinclair flung insults left and right. He tagged Columbia's president Butler "a climber and a toady," while Harvard's Lowell "could not be a climber, because he was born on a mountaintop, and there was no place to climb to—he could only stay where he was or descend." Stratton sidestepped Sinclair's notice because he was new, and had no prior record in academic administration. Sinclair turned his wrath, instead, on the MIT Corporation. "This is one of the most marvelous collections of plutocrats ever assembled in the world; it includes the president of the Powder Trust, and his cousin Mr. Coleman du Pont, who is emperor of the State of Delaware; also Mr. Eastman, the kodak king; two of our greatest international bankers, Mr. Otto Kahn and Mr. Frank Vanderlip; Mr. Howard Elliot, chairman of the New Haven, Mr. Elisha Lee, vice-president of the Pennsylvania; both members of the firm of Stone and Webster, with all of its enormous electrical interests; also nine other electrical bankers, two officials of the General Electric Company, one big electrical manufacturer, and six others who are interested in electric railways." As some wished a short shelf-life for Sinclair's book (it went into four editions within six months), MIT's new president sailed above the fray.

Stratton put his nose to the grindstone partly out of habit, but also to set an example. On June 11, 1923, his late-morning, early-afternoon inaugural ceremony began with a procession across the Harvard Bridge—students and faculty decked out in academic regalia, for the first time ever at MIT—and ended with speeches at Symphony Hall from governor Channing Cox, major-general George Squier, faculty chairman Edward Miller, ex-president Charles Eliot of Harvard, MIT alumnus Charles-Edward Winslow ('98), Carnegie Institution of Washington president John Merriam, and Stratton himself. Stratton headed to his office afterward, oblivious to advice that he rest up for the inaugural banquet that evening. As he settled in to plow through a pile of correspondence, Thomas Edison showed up at the door with two of his sons, Thomas Alva Jr. (Dash) and Theodore, and Dash's wife. The Edisons were in Cambridge to watch Theodore get his MIT diploma the next day. Stratton and Edison had known each other a long time, thirty or more years. The world of electrical research—wave theory, electromagnetism, quantum mechanics—had leapt ahead and left them relics of an age when all an electrical engineer-inventor needed to stay on top of his game was ingenuity and hard work. They talked electricity for an hour or two, until Stratton pulled himself away to dress for the Algonquin Club affair, where he played the combined roles of honoree and master of ceremonies. Among the speakers were Harvard president Lowell, Merriam again, and many others. Lowell teased about MIT growing exponentially, leaving "no room for Harvard in Cambridge . . . probably no room for anybody else in the city of Boston at all." Stratton enjoyed the joke. The next day, he handed out diplomas to 173 graduate students and 545 undergraduates—Theodore Edison among them—and, this grueling ritual behind him, back he went to his office to finish the chores that old man Edison had interrupted the day before.

Expectations ran high on all sides. Engineers and applied scientists considered Stratton a natural ally, but so did pure scientists, humanists, and other groups, including those who yearned for the old balance between pure and applied that had shifted toward applied under Maclaurin. George Haven, professor of machine design, hoped for greater emphasis on textile technology and allied industries. Professor of electrical measurements Frank Laws saw in Stratton "one trained in administration, with a high conception of the service science should render to the public," while General Electric chairman Owen Young perceived the appointment as a harbinger of closer ties between MIT and the industrial world. Others considered Stratton's "opportunities to help industry" better at MIT than at the Bureau.

Pure scientists, meanwhile, many of whom had felt increasingly marginalized at MIT, sensed a new era in the making. "It is a great relief to me (& many other of my colleagues)," wrote physics professor Harry Goodwin, "to have as President a man with a full appreciation of the value of pure science as a basis for technical training for there is always a real danger in a technical school that 'applied science' alone receives encouragement." Goodwin was relieved that the Corporation had opted for a scientist, rather than what some had feared—"a purely business executive." Another endorsement came from George Hale:

> For years I have hoped for restoration and development of the original policy of President Rogers, in which equal weight was accorded to pure and applied science, but the applications have steadily gained, and pure science has been allowed to fall behind. [Arthur] Noyes's Research Laboratory of Physical Chemistry was an important step in the right direction, but the preponderating tendency of the Chemistry Department was against him, and this was a large factor in his removal to California. Maclaurin recognized the need of much greater support for pure science, but all his effort had to go first into the provision of new buildings and endowment, though he fully intended to move in the right direction as soon as his financial problems could be solved. Finally, [E. B.] Wilson did all he could to stir up interest in science and research, but he could make no progress against conservatism, and resigned to accept his present position. Nichols's plans were excellent, but his health broke down. So you face one of the greatest opportunities ever presented to a man of science.
>
> At the Bureau of Standards you have built up a great institution out of nothing and you have given pure physics and chemistry their right place, in spite of all pressure to subordinate them to their applications. I am therefore perfectly confident that you will do likewise at the Institute, and I feel as though a great load of doubt has been lifted from my shoulders.

Harvard faculty members Theodore Richards, Augustus Trowbridge, George Swain, and legendary astronomer Harlow Shapley looked for far-reaching change. Swain, formerly of MIT, voiced cautious optimism—"There is much to be done, and I hope you will be given the power to do it, although a very popular attitude is that there is nothing to be done and that the Institute is nearly perfect." Shapley, who redefined modern conceptions of galaxy and universe size, had confidence in Stratton's ability to reshape MIT, easing fears of those who "look at science from the research side and who have allowed ourselves to worry a bit concerning the ideals of a powerful scientific institution." "Under your guidance," wrote another astronomer, "we expect to see the M.I.T. take a wider and more intelligent interest in scientific and industrial problems, and . . . become a more truly national asset."

Others pointed to Stratton's special way with young people. For this group the scientific and technical sides of the job were important, but outweighed by a larger responsibility—"the molding of character in the student corps." Someone had watched Stratton work with a high-school-aged group in 1919, bringing to life a rather dry subject—the metric system—as if with "a magic wand." Someone else called him a fine choice to "make . . . the best of America's young manhood." One MIT graduate ('88) was pleased to learn of Stratton's interest in extra-curricular activities, as "in the old days there was not enough of this sort of thing." Bassett Jones ('99) viewed Stratton as an ideal choice to reverse the unfortunate trend among technical schools "to turn out engineers rather than human beings."

Stratton met many of these expectations during his term as president, which lasted only seven years. The applied science departments ventured into new technologies and kept abreast of shifts in older ones. As options in the engineering curriculum expanded, offerings also grew more refined, specialized, focused on the particular rather than the general.

Mechanical engineering strengthened its programs in automotive work, engine design, and textiles. Alfred P. Sloan Jr. ('95), of General Motors, donated funds to build an internal combustion laboratory in 1929. Civil engineering created a new hydroelectric option, while introducing coursework in aerial surveying, aerial mapping, and soil mechanics, as well as pushing for an option in river and harbor engineering. In 1927, civil added a five-year cooperative program in railroad operation with help from the Boston & Maine Railroad. Another option, in geodesy and seismology, was added in 1929. Biology and public health offered two options—public health and industrial biology, the latter focused on fisheries and food technology—and in 1928 introduced another in public health engineering. Electrical engineering shifted some resources toward communications—radio, telegraphy, and telephone—while reinforcing existing programs in the more traditional power-transmission fields. An undergraduate option in communications was added in 1923. Also that year the first set of bachelor's degrees—13 in all—was awarded in architectural engineering. Chemical engineering developed subjects in fuel and gas engineering, furnace design, heat exchange, and petroleum refining. A master's-level program in fuel and gas engineering was offered for the first time in 1925. Coursework in chemical warfare was added in 1926, attended by designated army personnel—captains John McLaughlin, Carl Marriott, and Maurice Barker to begin with—from the U.S. Chemical Warfare Service. A bachelor's degree in military engineering was first awarded in 1927. Mining and metallurgy added subjects in optical ceramics, while naval architecture crafted a new option in ship operation. Because of his navy background, Stratton took a special interest in naval architecture and in building both the Institute's Pratt School of Naval Architecture, which had opened in 1921 after a decade of delays, and its associated nautical museum, created in 1922 and eventually named for Francis Hart. In 1926, aeronautical engineering broke away from physics to form Course XVI and offered bachelor's degrees for the first time, the first two undergraduates earning their diplomas the following year. Also in 1926 the mining, metallurgy, and geology department was split in two with mining-metallurgy (William Hutchinson, head) and geology (Waldemar Lindgren, head) going their separate ways. In 1928, aeronautics added coursework in meteorology.

Stratton's efforts, beginning in 1925, to establish a degree-granting program in graphic arts never caught on. But a course in building construction (Course XVII) admitted its first group of students in February 1927, with the first set of bachelor's degrees—nine in all—awarded in 1929. Those who wanted to slow this proliferation of courses found themselves in a distinct minority. Course XVII, however, raised more than a few eyebrows. "Those that had heard of it," the *Technique* editorialized in 1927, "expressed great surprise that even bricklaying and ditch-digging had become part of a college education."

The rich, dizzying array of offerings kept growing, and the same held true on the research side. Dozens of projects continued under the so-called Technology Plan, the Division of Industrial Cooperation and Research (DICR) serving as intermediary, negotiator, and manager of contracts between faculty members, departments, and outside firms. Staff attached to the Laboratory of Industrial Physics operated under several such contracts, in a more or less self-supporting way. So did the Research Laboratory of Applied Chemistry, with projects carried on simultaneously in oil refining, iron and steel corrosion, paper waterproofing, automobile fuels, combustion reactions, rubber, leather, lubrication, and metallic oxidation. At one point the lab consulted on textile cleansing for the Laundry Owners of New England. Several members of the chemical engineering group—Warren Lewis,

William McAdams, Clark Robinson, and John Ward—worked as consultants for fuel, oil, and gas companies. The electrical engineering laboratories held contracts for research on high-tension cables and the impact of illumination on industrial efficiency. As interest in communications grew toward the mid-1920s, they entered into agreements with Western Electric, New York Telephone, Bell Labs, and AT&T. Projects of this sort included short-wave radio research conducted at Colonel Edward Green's Round Hill Station in South Dartmouth, Massachusetts, which Stratton considered a splendid example of industry's dependence on original scientific research.

Also notable was Vannevar Bush's work on an electromechanical integrating device—product integraph, a form of analog computer—which vastly improved computing capacity, speed, and accuracy. Bush's other research projects included the network analyzer, a power-system simulator sponsored largely by General Electric and completed by 1929, and, a decade later, the differential analyzer, a more advanced version of the integraph developed with support from the Rockefeller Foundation. The biological laboratories sought industry contracts to study nutrition, vitamin values, fermentation, effects of irradiation, and applications of bacteriology in food processes. A Division of Industrial and Municipal Research was established in November 1926 "for the purpose of providing communities with competent impartial advisory service regarding their industrial and other problems . . . and to train and develop men for work in this field." Two years later, Stratton talked about establishing a course in city planning. The Daniel Guggenheim Aeronautical Laboratory, donated by the Guggenheim Fund for the Promotion of Aeronautics, was completed and dedicated on June 4, 1928. Even though the Guggenheim gift had no industrial strings attached, the sense of aviation as the next-great commercial success story in transportation—after ships, railways, and automobiles—helped motivate research in aeronautical engineering.

Key players on the applied side of MIT's research and teaching programs were Edward Warner, head of aeronautical engineering, along with C. Fayette Taylor and Carl-Gustaf Rossby; Samuel Prescott, head of biology and public health, with Clair Turner and John W. M. Bunker; Ross Tucker, in charge of the building construction course; Warren Lewis and later William Ryan, heads of chemical engineering, with Robert Wilson, Robert Haslam, John Ward, and Robert Russell; Charles Spofford, head of civil and sanitary engineering, with Charles Breed, George Hosmer, and Charles Terzaghi; Dugald Jackson, head of electrical engineering, with Frank Laws, Vannevar Bush, William Timbie, Edward Bowles, Carlton Tucker, Karl Wildes, Ernst Guillemin, Harold Hazen, Parry Moon, Samuel Caldwell, Harold Edgerton, and Wilmer Barrow; Hoyt Hottel in fuel and gas engineering; Edward Miller, head of mechanical engineering, with Allyne Merrill, Charles Fuller, George Haven, Charles Berry, and Harrison Hayward; William Hutchinson, head of mining, with George Waterhouse, Robert Williams, Charles Locke, Carle Hayward, and Edward Bugbee; and James Jack, head of naval architecture and marine engineering, with William Hovgaard, Henry Keith, and George Owen.

The incentive underlying many of MIT's research programs was earning power and service to industry. Department heads in the engineering disciplines argued that the Institute accrued substantial benefits from industry's need for help with problem-solving. "The Institute," wrote one, "true to its traditions, should occupy the first place for the training of leaders for these industries." But by the mid-1920s, Stratton and some engineering faculty began to question this focus. Stratton appreciated funding boosts—the Research Laboratory of Applied Chemistry brought in $171,880 in outside contracts in 1927–28, while its poorer, pure-science sister, the Research Laboratory of Physical Chemistry, brought in only $25,483—as well as the professional opportunities that came students' way because of MIT's ties to commerce and industry. But these often turned into mixed blessings. Robert Haslam, director of the applied chemistry research lab, warned in 1924 that reliance on industry for six-sevenths of the lab's support had prevented the program from branching out in creative ways. "While our outside relations are peculiarly fortunate and happy, this work is of necessity carried on under a certain pressure that interferes with the productive capacity of the Department in its contributions to general science. The Department would regret to lose all this outside work, but if the proportion of it could be reduced, the contributions of the Department to the prestige of the Institute and to the development of the profession could be greatly increased." A year later, Dugald Jackson wrote with respect to electrical engineering that "our staff should consist of highly analytical men with a strong leaning to the purely scientific aspects, but with full recognition of the importance of applications."

Stratton was troubled by evidence that the Institute's reputation as a center for both pure and applied research had waned. The 1910 edition of *American Men of Science* ranked MIT first in the nation in chemistry and seventh in physics, but by 1927 the Institute had dropped to third in chemistry and fallen out of contention in physics. More than twice as many National Research Council fellows in chemistry chose the University of California, Harvard, or Caltech over MIT in the 1920s. The Commonwealth Fund, in 1925, did not list MIT among accredited institutions where the Fund's fellows and grantees were permitted to study or to carry on research. One MIT alumnus, George Hooper ('91), heard a talk by Robert Millikan at Caltech in 1926, in which Millikan ignored MIT while comparing Caltech to other well-known eastern colleges and universities. When Hooper asked why, Millikan told him that MIT aimed to "turn out a fair grade of man suitable to average industrial work, but in no way competent to carry on research." Stratton tried to reassure Hooper—"I know of no other institution where more [advanced research work] is going on, not excepting the institution in southern California"—but he also recognized that perceptions were difficult to change. Hooper suspected that Millikan's view was colored by Arthur Noyes, whose disillusionment with MIT had led to his departure for Caltech in 1919. George Hale, meanwhile, insisted that MIT had more to learn from Caltech than Stratton cared to admit. "The weakest point at the Institute, in my opinion, is the lack of funds for research in fundamental physics. . . . A specific accomplishment of industrial research is a great thing, even though it solves but one problem. A fundamental discovery, applicable . . . to the solution of thousands of problems is a much greater thing. Millikan has found it easy to build up . . . a physics department which draws the best graduate students from all parts of the country and also from the leading laboratories of Europe. I wish Tech would do the same thing." Hale's classmate, Willis Whitney ('90), looked for ways to make the "spirit of research pervade the instruction in science" at MIT.

Like Hale and Whitney, Stratton wanted less time devoted to tasks and more to issues, "working out problems of fundamental importance to the industry as a whole." There were also prickly proprietary concerns—patent rights, restrictions on publication, control over data—and personnel attrition to contend with. In 1926, Stratton complained how hard it was to retain capable staff when firms offered salaries and benefits on a scale way beyond MIT's means. And, when the Institute lost three of its principal chemical engineering faculty—Robert Haslam, Robert Russell, and John Ward—to industry in 1927 and 1928 (Haslam agreed, however, to stay on as a nonresident professor), Stratton wrung his hands over the competitive disadvantage that MIT faced. Industry's penchant, he said, for "cutting the throats of schools" was in no one's best interests, least of all its own.

In response to such problems, Stratton proposed reforming MIT's view of industrial research and shifting resources toward basic-science fields. In 1923, he pushed to establish an MIT chapter of Sigma Xi, the national honor society of research scientists and engineers. The effort was unsuccessful—a chapter would not appear until 1934—but it was a symbolic gesture that turned into a clear articulation of policy by June 1926:

> It is more and more difficult to make a distinction between pure and applied science. Many of the most difficult theoretical and experimental problems that the mathematician, physicist, and chemist are called upon to solve arise because of the need of information as to fundamental laws. Progress in what we have designated as the field of technology is almost entirely dependent upon progress in the fields of science. . . . The Institute should participate in the search for information upon which its work in technology depends. This is equally true of the technical departments. They must be producers of information if our instruction is to be kept in the forefront. Otherwise we shall be followers only.

Stratton considered $2 million a reasonable starter goal for an endowment to support basic research in the pure and applied sciences.

But funding of this magnitude was difficult to come by, for several reasons. Older alumni, who had given generously to Maclaurin's ventures the decade before, felt both tapped out and wary of drastic shifts in MIT's mission. Private foundations resisted giving funds on grounds that the Institute was an engineering school whose natural base of support lay with industry. In 1929, when Stratton applied for a large grant to build "fields of pure science on the very highest plane possible," Rockefeller Foundation president Max Mason told him that MIT faced an uphill battle because it was so widely viewed as "a technical school only." Stratton tried to persuade Mason otherwise. "Personally, I believe the Institute has long since passed that stage, and that as far as pure science is concerned, the Institute is in as favorable a position for development as any of the large universities, and perhaps in some cases more so." But Mason remained unconvinced. Stratton wondered if the word *technology* was itself a problem, confusing MIT with lower-level trade schools and obfuscating its mission. At one point he suggested, half seriously, that MIT change its name to Institute of *Science and* Technology.

Stratton tried to shift both the perception of MIT and the reality that *was* MIT, looking toward quantum mechanics and other recent developments in science. He and Charles Norton, head of physics and the DICR, agreed that the physics department—or at least part of it—must move in a new direction. In 1924, Norton earmarked a portion of MIT's proceeds from contracts with General Electric and Victor X-Ray Corporation to encourage pure science research, not only in physics but in chemistry and biology as well. Stratton and Norton also proposed a new unit, the Laboratory of Theoretical Physics, which the executive committee approved on March 4, 1924. Early staffers included Paul Heymans, Manuel Vallarta, Walter Dehlinger, and William Allis. Even Vannevar Bush, whose experience lay on the applied side, was excited about prospects for fruitful give-and-take between this strong group, mathematics faculty such as Norbert Wiener and Henry Phillips, and electrical-engineering faculty and staff such as Gustav Dahl, Gleason Kenrick, and Julius Stratton. Wiener would help originate, define, and systematize the multidisciplinary field of cybernetics.

A stream of notables trickled into MIT from Europe's foremost labs and universities, for periods ranging from a few days to two or three months. One alumnus, Calvin Rice ('90), congratulated Stratton on bringing to campus the "world's greatest men," as "the one thing that engineers need more than anything else is *world vision*." Physicists

Otto Oldenberg of Göttingen and Charles Fabry of the Sorbonne, and Belgian mathematician Charles de la Vallée-Poussin, arrived in the fall of 1924. Oldenberg lectured twice in October on phosphorescence and fluorescence. Vallée-Poussin gave four lectures in November. Fabry spent the semester, teaching a graduate-level course in optics and offering a series of public lectures. He had visited MIT once before, in 1910, at the old Boylston Street campus, and was delighted to find a new spirit, "a technical institution [with] men interested in pure science, because without pure science, no progress in industry is possible." Peter Debye of the Eidgenössische Techniche Hochschule (ETH) Zurich lectured during the 1924–25 academic year. Max Born, Théophile de Donder, and Abram Joffe came in 1925–26. Erwin Schrödinger, Joffe (again), Victor Henri, and Arthur Haass came in 1926–27. William Lawrence Bragg spent part of 1927–28 and Werner Heisenberg was in residence for much of the spring semester, 1929. "It will be a great pleasure," Heisenberg told Stratton, "for me to make the acquaintance of the Physicists working in your Institute and to see a new part of scientific life in America." Heisenberg's lectures on quantum theory filled MIT's largest lecture hall, 10-250. Distinguished American physicists came for shorter visits, one or two days. William Meggers, a member of the staff that Stratton had assembled at the Bureau of Standards, visited in 1925, as did Nobel laureate Michelson, Stratton's old boss and collaborator from the University of Chicago. International visitors from fields other than physics included chemists Ernst Hauser, Herbert Freundlich (director of the Kaiser Wilhelm Institute for Physical Chemistry in Berlin), Edouard Urbain, and Jean Piccard; gas and fluid dynamics authority Ludwig Prandtl; electric power expert Reinhold Rüdenberg; and hydraulics specialists Theodor Rehbock, Wilhelm Spannhake, and Dieter Thoma.

Whereas Einstein's visit to Boston four years earlier had gone largely unnoticed at MIT, Born's three-month stay as a special lecturer played out quite publicly. He had spent a brief period in 1912 at Chicago's Ryerson Physical Laboratory, but this was his first visit to America since he and his research group had made huge strides in systematizing quantum theory. Born shared Paul Heymans's office at MIT and got to know the community well. He gave ten lectures in all—the first on November 14, 1925, the last on January 22, 1926—divided into two series, "The structure of the atom" and "The lattice theory of rigid bodies." In introducing Born, Stratton pointed to the surging importance of pure physics to engineering. Born composed his lectures in German but delivered them in English, with translation assistance from physicists in MIT's emerging theoretical group: William Allis, Hans Müller (Peter Debye's former assistant at ETH Zurich), Francis Sears, Nathaniel Frank, and Manuel Vallarta. Physicists from Harvard and elsewhere flocked over to listen. The lectures were published by MIT in 1926 as *Problems in Atomic Dynamics*, the first of Born's works to appear in English. Also in 1926, the *Journal of Mathematics and Physics*, an innovative interdisciplinary periodical launched at MIT in 1921, published a much-talked-about paper on quantum mechanics coauthored by Max Born and Norbert Wiener. While Stratton was unable to persuade him to join MIT's permanent faculty, Born agreed to continue his affiliation as a nonresident member and to return occasionally to give more lectures. De Donder's lectures were also published by MIT, in 1927, as *The Mathematical Theory of Relativity*. "Frankly," Charles Norton confided to Stratton, "it does not look very interesting to me, but some of our mathematical friends seem to get quite excited over it."

As Stratton grew intrigued by MIT's potential role in scientific publication, he worked with Corporation member John Ripley Freeman ('76) and the new, young managing editor of *Technology Review*, James Killian ('26), to move the idea along. Stratton and Killian went down by train to Providence, Rhode Island, for a meeting in Freeman's office sometime in 1927. Stratton agreed that MIT, with financial support from Freeman, would oversee translation and publication of select German monographs. He asked Killian to coordinate the project. Killian prepared a report, which doubled as his undergraduate thesis (coauthored with John Crawford '27) in September 1928. Entitled "The advisability of establishing a press at the Massachusetts Institute of Technology," the report argued convincingly that "the University Press has proved its usefulness at other institutions and that such an organization would be valuable here." A press emerged at MIT in due course, semiofficially, then officially, in 1932, under the imprint Technology Press. It published eight titles in its first five years, with Killian at the helm working concurrently on *Technology Review* and at one point as acting secretary-treasurer of the alumni association. "I think," Stratton reassured Freeman about Killian, "you will find him a very competent man to handle the routine work of our press."

Meanwhile, he probed foreign guests about the state-of-research at MIT. Bragg offered his view a month after returning to England in 1928. "I take it . . . that you wish a good deal of this work to be on fundamental problems rather than on the problems sent in by firms which require a quick solution to some difficulty that has been encountered. . . . I think you need one or two men in the laboratory who know the theory of the analysis thoroughly, and who are free to follow to its end any interesting clue which turns up, without the feeling that they must produce practical results as soon as possible. Of course the man who has the power of producing the practical results is just as important and valuable, but he draws his ideas largely from the other fellow who is doing more fundamental work." Stratton also sought advice from Ernest Rutherford, director of the Cavendish Laboratory, University of Cambridge, with respect to recruiting promising physicists. In 1930, he coaxed Bragg's equally distinguished father, William Henry Bragg (the two had shared the 1915 Nobel physics prize), into giving the keynote address at MIT's commencement.

While the theoretical group made significant strides during the 1920s, MIT's core physics faculty—Charles Norton, John Norton (Charles's son), William Drisko, Newell Page, Gordon Wilkes, William Barss, Frederick Norton, and Louis Young—remained attached primarily to industrial work. Chemistry was not so bifurcated, but Frederick Keyes, who took over as department head on July 1, 1923, sought to shift as far as possible from the industrial side. "One can easily observe," he wrote to Stratton in 1925, "that most of the younger members of the Institute staff are dissipating the most formative portion of their lives in working out in their spare time various small applied science problems for pecuniary reward, instead of devoting themselves to the pursuit of pure science." Keyes hoped to replace professor of organic chemistry F. Jewett Moore, who died in 1926, with one of Harvard's academic stars, James B. Conant. Harvard retained Conant, who would succeed Lowell as Harvard president in 1933, and Keyes's search turned up empty in spite of an all-out campaign to reach "every organic chemist in the country of any pretensions." Keyes's success with recruitment was limited—he brought in George Scatchard in 1924, Louis Harris in 1928—but he did nurture a fine group, already in place, that included Augustus Gill, James Norris, Henry Monmouth Smith, Miles Sherrill, Samuel Mulliken, Earl Millard, Arthur Blanchard, Joseph Phelan, Tenney Davis, Louis Gillespie, Walter Schumb, James Beattie, Avery Morton, and Ernest Huntress. Mathematics, under Harry Tyler, shared out teaching or service functions among its faculty while promoting periodic leaves for research men such as Norbert Wiener. Wiener spent time in Göttingen and Copenhagen in 1926, and he was instrumental in recruiting Dirk Struik, a rising star in differential geometry, from the University of Utrecht. Wiener advised Stratton that Struik, a committed Marxist, promised that while at MIT he would "keep clear of American politics." Among the other outstanding faculty in mathematics were Frederick Woods, Clarence L. E. Moore, Henry Phillips, and Philip Franklin (married to Wiener's sister, Constance).

Stratton was eager, as he put it in 1923, to ensure that teaching and research in the sciences did not grow "stale and behind the times. . . . I am going to do all that I can toward encouraging the selection of exceptional men to fill the junior ranks of our scientific staff, and to provide them with the time and facilities for keeping up with their professions which to the exceptional man is often more of an inducement than salary." Besides Wiener, Stratton encouraged other faculty to expand their intellectual horizons. John Drisko studied hydraulic developments in Germany, Sweden, and Russia between 1927 and 1929. Manuel Vallarta spent those same two years in Berlin and Leipzig working under Einstein, Max Planck, Schrödinger, Heisenberg, and other luminaries. When physics instructor Nathaniel Frank's application for a National Research Council fellowship was turned down in 1929, Stratton gave him a paid leave to visit Heisenberg's laboratory in Leipzig and Arnold Sommerfeld's in Munich. In 1927, he arranged for X-ray specialist John Norton to visit laboratories in Germany, Switzerland, England, Holland, and Sweden. Wilmer Barrow, whom Stratton and some others considered "the most brilliant man" in electrical engineering at MIT, went to Munich for postgraduate studies in 1929.

One of Stratton's favorite protégés was Julius Stratton. At first the relationship was personal—did they, the elder Stratton wondered, share a common ancestor?—but it evolved on close professional terms, too, as the president recognized special promise in this young scientist. Julius Stratton spent two years in Europe, 1926–28, with the president's watchful support. They corresponded regularly. Julius kept the elder Stratton posted on his program and where he hoped his career might lead, the president wrote back with encouragement and thoughtful advice. In July 1926, not long after his arrival in Zurich to work under Peter Debye, Julius wrote the president:

> I am learning, learning a tremendous lot. I feel already that my year is going to be a most profitable one. As you told me, the resources for individual experimental work are most limited. The library system, if it may be termed such, is medieval. But I shall have time and opportunity for experimental work later. What I need now is a thorough theoretical grounding and I think I shall get it. At all events, I have never enjoyed my work more, nor felt that I was getting more out of every day; and I am sure that in preparing to enter the field of mathematical physics as I am, I am choosing wisely.

After earning his doctorate at ETH Zurich in 1928, Julius Stratton returned to MIT as assistant professor of theory of electricity and magnetism. He started in the electrical engineering department, but the president insisted on flexibility there. "My own wish is that you shall join the Department of Physics as one of a group of mathematical physicists of which I have often spoken to you, and which I hope will be second to none in the country." Two years later, the appointment was transferred to physics.

President Stratton's immersion in science and engineering left two other areas—architecture and the humanities—to carry on essentially under their own steam. This worked well for architecture, which reasserted itself in the 1920s under the leadership of William Emerson after a slower period the decade before. Besides Emerson, key faculty included William Lawrence and Henry Gardner. In 1924, eleven of the department's alumni were serving as heads of architecture programs at schools elsewhere—Edmund Campbell ('06), Armour

Institute of Technology; John Howard ('86), University of California; Frederick Mann ('94), University of Minnesota; Henry McGoodwin ('94), Carnegie Institute of Technology; Allen Kimball ('11), Iowa State College; Cecil Baker ('07), University of Cincinnati; A. Lawrence Kocher ('13), Pennsylvania State University; Frederick Giesecke ('04), Texas State University; Emil Lorch ('93), University of Michigan; Ellis Lawrence ('01), University of Oregon; and Robert Taylor ('92), possibly MIT's first black graduate, at Tuskegee Institute. This pool of talent continued to grow under Emerson, even though the department did not assume in Stratton's eyes the same priority levels as physics, say, or chemistry. Because architecture remained in the old Rogers Building on Boylston Street, it also did not feel quite integrated with the rest of the MIT community.

Humanities continued to function as a service area to broaden students' cultural exposure—to languages, literature, history—and to hone skills useful in professional life, especially written and oral expression. "I would stress public speaking and ability to put over ideas in a clean-cut and convincing way," read a comment quoted in a 1924 visiting committee report. "Forceful speech opens the doorway to business and success; the lack of it has doomed many to disappointment." Henry Pearson, who headed the English and history department, sometimes appealed for more staff by underscoring the student body's "level of semi-illiteracy." Among Pearson's core faculty were Archer Robinson, Henry Seaver, Matthew Copithorne, and the universally popular Robert (Tubby) Rogers. In 1929, when Rogers sparked controversy in a speech to graduating seniors—"Be a snob, marry the boss's daughter instead of his stenographer, dress, speak, and act like a gentleman, and you'll be surprised at the amount of 'murder' you can get away with"—Stratton resisted calls to fire him and helped quell what turned into a public-relations nightmare. Frank Vogel and Herman Kurrelmeyer taught German, while Ernest Langley headed Romance languages with a small staff that taught French, mostly, with occasional forays into Spanish and Italian. On the social sciences side, Davis Dewey headed a department—Economics and Statistics—that covered economics, political science, and industrial research, as well as engineering administration (Course XV) under Erwin Schell. In 1928 efforts were made to add applied psychology, "a knowledge of how the mind of man works," to the Course XV curriculum. A special lecturer, Johnson O'Connor, spoke from time to time on "human engineering."

Stratton promoted other kinds of broadening experiences. The Aldred lectures, sponsored by New York banker and power-company executive John Aldred, sought to "give . . . students not only technical information, but a picture of the problems to be met in practice, and especially the necessity for common sense in engineering." Subjects in "humanics" were introduced in 1928 as "a prototype for courses . . . to prepare men to better understand the human relations met with in industry and in the practice of an engineering profession," with support from William Nickerson ('76), Gillette Company executive and co-inventor of the disposable razor. Charles Gow became MIT's first professor of humanics, followed by F. Alexander Magoun in 1930. When someone suggested that poet Robert Frost be invited to MIT in 1926, Stratton doubted he could arrange this but agreed that "it would be a good thing for our students to have a better contact with men of prominence in reference to literature and philosophy." Two years earlier he helped the Technology Christian Association coordinate an innovative series called "Sex factors in human life." The speaker, F. N. Seerley, dean of the YMCA College in Springfield, Massachusetts, was said to "treat sex not as something vulgar and bad, but as a great constructive force which properly conceived and used aligns man most closely with the eternally creative heart of God." Identical talks were given to male and female students, but in separate rooms.

Under Stratton, overall student numbers remained fairly constant. The mid-1920s saw a slide—3,505 students in 1921–22, only 2,671 in 1926–27—but by 1929–30 the number recovered to 3,066. A marked shift, however, occurred in students' chosen areas of study. In 1920–21, about 90 percent of students opted for engineering, 5 percent for science, 2 percent for architecture, and the rest unclassified. A decade later, 1929–30, about 78 percent opted for engineering, 11 percent for science, 7 percent for architecture, and the rest unclassified. The doubling of the proportion in science mirrored reforms pressed by Stratton. Why the proportion in architecture more than tripled is another matter—a trend somehow triggered by the department, as Stratton paid relatively little attention to this area. The number of faculty rose from 139 in 1920 to 220 in 1929, with supporting staff—instructors and assistants—holding steady at 188 and 184, respectively. The first three undergraduate degrees in mathematics were awarded in 1923, to John Hinds, George McReynolds, and Richard Ovenshine.

But it was with graduate degrees that the curriculum shifted most dramatically. In the 1920s, the number of master's degrees soared more than fivefold. Between 1910 and 1920, 291 master's degrees were awarded, while 1,609 were awarded between 1921 and 1930. Several new master's programs were launched during this latter period. The first master's in architecture (M.Arch.) went to Walter Church, Arthur Stanton, and Amory Williams in 1921. The first set of master's degrees (S.M.) in chemical engineering practice, 32 in all, were awarded in 1922. The first S.M. in metallurgy went to Haig Solakian in 1923, in engineering administration to Percy Bentley in 1925, in fuel and gas engineering to Francis Ford and Jennings Hamblen in 1927, and in sanitary and municipal engineering to Roland Hutchings in 1929.

The number of doctorates (Ph.D., Sc.D., or D.P.H.) increased more than fourfold from the previous decade. While 39 doctorates were awarded between 1910 and 1920, that number jumped to 165 between 1921 and 1930. The largest contingent was in chemistry (59) followed by chemical engineering (23), metallurgy (14), electrical engineering (12), geology (12), physics (11), biology (10), mathematics (8), aeronautical engineering (4), public health (4), mechanical engineering (4), civil engineering (2), electrochemical engineering (1), and naval architecture (1). There were 91 Ph.D.'s and 74 ScD.'s awarded. The first Ph.D. in mathematics went to William Fitch Cheney Jr. in 1927. The first Sc.D. in chemistry went to Ram Prasad in 1922, in metallurgy to William Frazer in 1923, in physics to Paul Heymans and Bailey Townshend in 1923, in chemical engineering to Charles Holmes Herty Jr. and John Lewis Keats in 1924, in civil engineering to Alberto Ortenblad in 1926, in electrochemical engineering to Donald Stockbarger in 1926, in mathematics to Gleason Kenrick in 1927, in mechanical engineering to Clifford Duell in 1927, and in naval architecture to Leo Jürgenson in 1929. The first D.P.H. (Doctor of Public Health) was awarded to Edwin Maynard in 1924. To help handle these expanding graduate needs, Stratton appointed Harry Goodwin dean of graduate students in 1926. The title dean of the graduate school was also considered, but since there was no graduate school as such—each department admitted students and ran its own programs autonomously—dean of graduate students seemed more appropriate. Goodwin coordinated policy, while departments continued to self-manage.

The growing emphasis on research and graduate education led some to wonder if "the Institute now regards itself as primarily a research institution where a faculty man's hope of promotion varies distinctly with his publication output [and where] classrooms become discouraging affairs, supervised by men who do not, in general, dare to allow teaching to become a primary interest." Stratton's position was that without the research mindset, faculty members risked becoming "mere routine men" out of touch with their fields. But some corporate interests worried, too, about MIT losing sight of the ordinary engineer. One executive told Stratton that he wanted to support "some university that is not trying to make science abstruse and difficult to understand, but one that either believes as I do, or one to whom I can sell the idea that the greatest contribution that all of us can make to science is to make it so simple that it can be understood by the mass of our people and its benefits applied by all kinds of people and in everyday problems." Stratton assured him, and others, that there were no plans to abandon MIT's traditional mission, simply to broaden it in ways that would better serve the interests of science, industry, and higher education.

Stratton was sensitive to lingering anxieties about overwork, about students' ability to "stand up under the strain" and to avoid turning into "physical and nervous wrecks." Corporation member William Kales ('92) raised the issue in 1924, and again in 1930. His wife, he said, harbored a grudge against MIT for the way he turned out—"she always attributes every flaw in my character to the years I served in a certain penal institution in Massachusetts"—and both wondered whether their son Morris ('33) should be "condemned to do time" under the Institute's "unfeeling jailers." Visiting committees urged more "social and out of hour life" for students. Stratton worked hard to show, through a rich, systematic network of extracurricular activities, that MIT had no "plan to over-work the boys."

A rebirth of athletics began before he took office as president, and continued afterward. The gym team and hockey club reorganized around 1920 after several years of inactivity, while boxing and soccer teams were first mentioned around then. Basketball, baseball, and golf teams were revived in the early 1920s. In 1926 polo, the king of sports, drew interest with about thirty riders on the charter team. Formal lacrosse and squash teams emerged around 1930, although Stratton talked of installing up to eight squash courts as early as 1926. Sports with unbroken traditions—crew, track, fencing, rifle, swimming, wrestling, and tennis—thrived throughout this period. The first varsity crew team was organized in 1920 and, to accommodate its needs, MIT acquired the Boston Athletic Association's boathouse in 1922. The crew won its first major regatta in 1924, defeating Cornell's team at Ithaca. That same year MIT's junior varsity edged out Harvard on the Charles River. An honorary society for crew champions, the Tech Boat Club, grew active in 1927. Crew was Stratton's favorite college sport, its value tied more to healthful exercise than to victory, defeat, or public show. As he wrote to Corporation member Charles Hayden ('90) in 1930: "We have consistently looked with disfavor upon the establishment of football and baseball teams because of their interference with the regular activities of the Institute, but rowing can be, and is, participated in by a very large number of our student body. . . . I am a firm believer in the attitude of the Greeks, who encouraged athletics for their value from the point of view of physical exercise, whereas the Romans favored it to amuse the populace. I am thankful to say that the Institute has carefully avoided the latter course." Even so, some complained that MIT needed huskier students to build teams in heavyweight wrestling, shot-put, and the hammer. "It seems either that Technology is no place for a fat man, or else the huskies who do come . . . are too lazy or too indifferent to be interested in sport."

Stratton focused less on sports, in fact, than on residential facilities, social life, community services, and cultural activities. Fraternities continued to provide housing options in the absence of adequate dormitory space. New chapters appeared, however, at a slower pace than in prior decades. After a brief flurry in the early 1920s—Phi Beta Delta, in 1920; Alpha Mu Sigma and Phi Lambda Alpha, in 1921; Phi Mu Delta, Psi Delta, Sigma Nu, and Sigma Omega Psi, all in 1922—no others appeared until 1924, when Kappa Eta Kappa emerged followed by Alpha Phi Delta in 1928 and Alpha Kappa Pi in 1929. Stratton's push to build on-campus housing gradually relieved pressures on the fraternity groups. In addition, the Technology Christian Association worked hard to identify clean, respectable lodging for students in the Boston and Cambridge area, offering "service to anyone, anywhere, anyway, anytime, regardless of creed or color." In 1922, around 1,300 students found rooms through TCA. Other useful TCA services included an orientation handbook (the Tech bible, as it was popularly known), employment bureau, book exchange, freshman advisers, special assistance to foreign students, and church referrals. A dormitory first called 93 Hall—named for the Class of 1893, which gathered funds for the project as part of their 30th-anniversary celebration—and later known as East Campus was ready for occupancy by 1924, accommodating about 80 men, with a late-night eatery known as the owl nest in the basement after a year or two. Also in 1924, Stratton spearheaded the purchase of about 30 acres of land on the western side of Massachusetts Avenue with funds raised by subscription from alumni. What MIT would do with this land was unclear, but Stratton floated ideas for more dormitories, labs, a gymnasium, an infirmary, possibly a new home for architecture to replace the Rogers Building on Boylston Street. A few more dorms went up adjacent to the '93 group in 1927, big enough for 136 men. But this still meant that only a small fraction of the student body could live on campus. Stratton, unhappy that students were "forced to cultivate manners and morals and the social graces in boarding houses," vowed fundamental change. The Homberg Memorial Infirmary, a state-of-the-art medical facility, began servicing the MIT community in 1928.

Musical clubs found new life under Stratton. "We have too long been looked upon," students complained about a year before he arrived, "as a body . . . interested in nothing that does not pertain to science or engineering [and] it is the function of our musical organizations . . . to dispel this idea." A tour organized in the early 1920s included a performance by MIT's recently formed jazz band, along with its glee, banjo, and mandolin clubs. A choral society, all-male, was first mentioned in 1923, while a dance orchestra appeared in 1924. When Stephen Townsend was hired that year as director of choral music, the *Boston Globe* joked about MIT perhaps changing its name to the Technology Conservatory of Music or the Massachusetts Conservatory of Technology. In 1926 the jazz band, known by then as the Techtonians, hired itself out for local weddings and other functions. The Instrumental Club emerged by 1926, the Saxophone Quintet by 1927. For several years Stratton supported a concert series, organized by pianist Arthur Whiting, to expose the MIT community to fine performance. A "really high hat musical treat" in April 1927 presented Whiting and his group playing Beethoven's Kreutzer Sonata and Brahms's Horn Trio before a large audience of students and faculty. Yet John Burchard ('23) recalled being berated for ducking out of class one Friday afternoon to hear the Boston Symphony. "We have to run a tight ship here," his professor warned, "and symphony concerts are not part of MIT's program." Stratton disapproved of such rigidity, which seemed likely to reinforce the Institute's image as factory-like, or as obsessed with work.

Dramashop, a new theatrical group, was formed in May 1927 as an alternative to the slapstick shenanigans of the Tech Show. It began work in October 1927 and staged its first production, Eugene O'Neill's *The Hairy Ape*, that December. Its second production, Charles Brooks's *At the Sign of the Greedy Pig*, took place in May 1928, and its third, George Kelly's *The Show Off*, in December. Dramashop used the common room of the Rogers Building on Boylston Street for its performances and, unlike the Tech Show, recruited real women—rather than "beskirted, lip-sticked and silk-stockinged young men"—to perform female roles.

Groups like Dramashop were part of an effort on Stratton's part to elevate cultural tone and to counteract a rising tide of rambunctious student behavior. Complaints about the Tech Show as overly commercialized or "low and unworthy" were not uncommon, and in 1928, when the show poked fun at federal alcohol prohibition, some wondered about the wisdom of MIT appearing to flout the law. Students went to creative lengths to improvise their drink of choice—some imbibed wood alcohol, along with a foul-tasting concoction known as Zepp's hair tonic—or to manufacture their own cocktails. Bathtub gin, a staple for alcohol-deprived partiers in the roaring twenties, made the rounds at MIT as everywhere else. "We bought sweet spirits of niter," recalled Burchard, "which could be obtained at any drugstore. Then we filled the bathtub with water as hot as the tap would supply. Into this we immersed the flask of sweet spirits of niter to drive off the niter and leave us ethyl alcohol. Dense clouds of dark brown smoke would pour out for a few minutes. At the end of this fractional distillation . . . we diluted the mixture with water to bring the alcoholic content to about 50%. We put in a few drops of oil of juniper, 'aged' the concoction for a day and our gin was ready." Burchard was hardly much of a rebel—he went on to become a respected faculty member in architecture, eventually dean of humanities and social science—and what he described here was typical student behavior for the time. But in 1924, president Raymond Hughes of Miami University of Ohio thought twice about sending his son to MIT when a friend told him that "he had never seen as much liquor anywhere in recent years as was on exhibition and being drunk in the D.K.E. house." Stratton, angry over the fraternity's "flagrant offenses against the Institute's good name," saw to it that the chapter's lease was terminated for code violations, but the group reappeared at a different address.

Even more disruptive forms of conduct were on the upswing. The 1920s saw cars and animals perched precariously on building-tops, madcap pranks that would later be called hacks. The term hack in this sense did not come into common usage until much later, probably in the 1950s, but ironically it showed up in a flamboyant bit of mischief perpetrated by Tech students in the early 1920s: the so-called great hack race, "hack" in the sense of horse-drawn taxicab. This involved a planned brawl at a pageant, normally a staid affair, sponsored by the haute-culture Copley Society and held annually in the old Rogers Building.

The society asked MIT architecture students to play the role of supernumeraries, to arrive decked out in Arabian costume, and, according to Burchard, "to rush in and at command 'salaam' to prostrate ourselves, and then take a quiet position, almost as part of the décor, while the more serious acting went on." The students, however, ring-led by one Louis Metz ('23), came up with a more outrageous plan. "Louis . . . had the idea of hiring all the open hacks in town (about six I would guess). We would then meet at his place which was on Beacon Street just above Massachusetts Avenue. There we would put on our costumes, get at least a little tanked up. . . . The hacks arrived in due time and when we piled into them they were so full that Arabs were literally hanging out over the wheels, all exhorting their drivers to whip on the horses. We tore down Beacon Street all the way to the Public Gardens, turned right on Arlington Street to Boylston and then up Boylston to the steps of the Rogers Building above Berkeley Street. It was never really a race . . . more like a rapid small exotic parade. When we arrived at the Rogers Building we poured up the monumental steps. At their top were two large stone or concrete urns. Only the cooler heads among us dissuaded others from rolling those down the steps in the hope of bowling over the trolley trains which passed at close intervals. Instead we rushed to the exhibition room, entered it with an extemporaneous but unanimous shout, then salaamed and sat down according to plan. . . . As we sat there Julian Berle would inflate a balloon and raise it at a critical moment in the libretto. James Sullivan would prick the balloon with a needle and the loud explosion pop would drown out what was intended to be a sophisticated passage."

Pranks sometimes turned into high-profile disturbances. In November 1925, crowds of MIT students—part of their night-before Field Day celebration—streamed over the Harvard Bridge, stormed the subway, held up traffic, wrecked cars, and provoked a sweep by riot police. Stratton's response—stern lectures, temporary bans on certain activities—helped reduce student misbehavior to the point where Corporation member Frank Jewett ('03) detected a carryover influence on alumni. The tone of one reunion banquet in 1928 impressed Jewett: "I have attended many of these affairs in the past and have usually been rather disgusted with them, either because of the buffoonery which was present or because of some element of boisterousness which did not speak very well for the character of the alumni." But some, even dean of students Henry Talbot, thought that too much was made of student (and alumni) high jinks, that it was healthy for "hard-working boys [to] 'cut loose' and revive the spirit of deviltry."

Options for students on the literary or expressive side grew slowly. For light reading, they liked *Voo Doo* best (because it helped relieve "the sordid aspects of life" at MIT), followed by *La vie parisienne* and the *Tech*. The *Tech Engineering News*, first published as a supplement to the *Tech*, vol. 39 (1919–20)—with the goal of "broadening, extending, and stimulating . . . the engineering instruction given at the Institute"—offered professional food-for-thought but little cultural or entertainment value. In 1920, summer surveying-school students started an annual yearbook known as *Benchmark*, modeled loosely on the *Technique*. Lawrence Conant ('23), self-described as "one who has been through the mill," wrote about the student experience in a favorably reviewed book, *Tackling Tech: Suggestions for the Undergraduate in Technical School and College*, published in 1923. The *Freshman Gray Book*, hot off the press in January 1924, was a guide somewhat like the Tech bible but with first-year students as its primary audience. An underground "razz sheet" known as *Filter Paper* appeared by 1925. *Dorm Rumor*, first published in 1927, served as a gossipy outlet for campus news. Some members of the Speakers Club, tired of that group's static format, formed a Debating Society under the leadership of Arthur Kallet ('24) in the spring of 1923. That December the society met Boston University in a debate over the Paris Peace Conference's proposals for a world court to adjudicate war crimes (MIT argued in the affirmative). MIT students, meanwhile, had played a central role in creating the Intercollegiate Conference on Undergraduate Problems and Government, first held in New York City on December 23, 1920. A Republican Club formed in December 1923 to rally support for candidates in the 1924 elections. The Liberal Club, founded in October 1928, brought in lawyer and orator Clarence Darrow, of Scopes-trial fame, as one of its first guests.

During the 1920s, much effort went into professional groups. Cheops and Triglyph, both honorary architectural societies, were first mentioned in 1921, along with Hexalpha, for undergraduates in electrical engineering. The Radio Society was joined by a chapter of the honorary radio fraternity, Alpha Sigma Delta. The 1923 annual convention of New England radio amateurs was hosted jointly, at MIT, by the Radio Society and the American Radio Relay League. In 1924, Stratton discussed with radio enthusiasts possible sites for a station on campus. The Engineer Unit Association appeared in 1922, and a Mathematics Club by 1924. In 1922, MIT formed a chapter of the national engineering fraternity, Tau Beta Pi. A branch of the American Institute of Electrical Engineers and a chapter of the Unitarian Laymen's League emerged around 1923. A number of military groups drew support from war veterans, including Stratton himself. Among these were the MIT Post-Army Ordnance Association, first mentioned around 1922; Sigma Alpha Beta and the MIT Post Society of American Military Engineers, 1923; Scabbard-and-Blade and the Colonels, 1924; National Coast Artillery Fraternity and Mortar and Ball, 1925. A Flying Club appeared in the fall of 1927. Stratton took a few risks here, what with worried parents besieging him about students engaging in so dangerous an activity. "I fear that in case of accident," he wrote to president Angell of Yale, who was thinking of establishing a similar program, "the Institute would be subjected to severe criticism. . . . On the other hand, aviation is growing by leaps and bounds; fatal accidents do not check its progress in the slightest degree, any more than the enormous loss of life on account of automobiles has checked the growth of that industry." A chapter of Chi Epsilon, the civil engineering fraternity founded at the University of Illinois (Stratton's alma mater) in 1922, emerged in 1928, and by 1928 there were references to a Society of Automotive Engineers.

With respect to places of origin, MIT's student body shifted hardly at all during the 1920s. In 1920, 76 percent of students came from the North Atlantic region—New England states, New York, and Pennsylvania—and by 1929 this proportion had dropped just slightly, to 73 percent. Proportions (and numbers) from other regions also remained constant. The next-largest group, in both 1920 and 1929, came from the North Central region, followed by the South Atlantic, West, South Central, and, finally, offshore territories, districts, and dependencies (Canal Zone, Hawaii, Philippine Islands, and Puerto Rico). International students, however, showed a steady decline, from 267 in 1920 to 187 in 1929, reversing a trend begun in Pritchett's administration and continued under Maclaurin's. Antiforeigner sentiment, on the rise nationwide, contributed to the downturn. Congress passed legislation—the Emergency Quota and Immigration Restriction Acts of 1921, the 1924 Immigration Act folded in with the National Origins and Asian Exclusion Acts—that made port entry difficult not only for those seeking permanent residence, but also for students who wanted short-term permits.

Even before Stratton arrived at MIT, some faculty joined a campaign originating at Columbia University to reduce the impact of this legislation. Stories of students being hauled off ships in New York, detained at Ellis Island, and harassed grew commonplace in the early 1920s. LeBaron Colt, chair of the U.S. Senate immigration committee, offered little hope that the legislation would be reversed in view of public support for restrictions, but immigration authorities promised in 1922 to ease the burden on foreign students by allowing them to post a five-hundred-dollar bond on entry. This was itself a hardship—how many students could come up with that amount?—but the policy allowed some, at least, to get through. Stratton did what he could to encourage admissions from abroad, as in 1925, when he sought to boost scholarship funds for students from India and South America. But there were problems even when students entered legitimately. In 1926, the Coast Guard seized a boat flying a Chinese flag and detained two of its crew, MIT students of Chinese origin, until they could produce passports. In 1929, when theoretical physicist Manuel Vallarta was considered to replace Charles Norton as physics department head, his nationality—Mexican—became an issue. "I felt uncertain about him," wrote one referee, "did not trust my impressions as to his personality and character, until I learned that his mother was English-American." Vallarta, incidentally, served as faculty adviser to MIT's chapter of Phi Lambda Alpha, a Latin-American fraternity.

While racial and ethnic attitudes did not harden, exactly, during the 1920s, they grew more exposed as xenophobic sentiment deepened nationwide. Harvard's president Lowell, a staunch advocate of immigration limits, endorsed racial segregation in Harvard campus housing in 1922 and, in 1926, adopted a covert quota policy that reduced the number of Jewish undergraduates from 27 percent in 1927 to 10 percent by the time he left office in 1933. Stratton, by contrast, favored a more inclusive environment. When Charles Richards, president of Lehigh University, asked him to divulge the methods used by MIT to limit the number of Jewish students, Stratton's response was unequivocal: "The Institute of Technology has never to my knowledge taken any steps to control its attendance as to nationalities. . . . I have never looked into the question of the proportion of Jewish students attending the Institute, but I would say off-hand that it was not very large." In fact, of 20 fraternities in existence at MIT in 1928, as many as five—Alpha Mu Sigma, Phi Beta Delta, Sigma Alpha Mu, Tau Delta Phi, and Tau Epsilon Phi—were all-Jewish, each with a substantial membership. The Institute's Menorah Society counted 42 members in 1928 and 72 in 1929, suggesting that while Harvard, Lehigh, and other institutions were cutting back on Jewish students, the number at MIT was growing.

This did not mean that MIT was free of anti-Semitic bias. "Any man with an Arabian name like Ginsberg will do, pick your own," said the 1923 *Technique* in identifying students whose politics leaned socialist. John Ripley Freeman repeatedly used anti-Semitic slurs about an MIT graduate, Samuel Shulits ('24), whom he had commissioned in 1930 to translate Armin Schoklitsch's *Der Wasserbau*: "Shulits is a Hebrew and asked me a stiff price . . . which I agreed to with insistance [sic] on a strictly first class job"; "Shulits is a Hebrew, and has a well-developed tendency to take care of himself"; "I was forewarned that Shulits was a Jew [but] some of the finest characters of my acquaintance and some of the ablest brains have been possessed by Jews, and I try to keep free from prejudice on racial matters." Stratton's responses routinely ignored such comments. Religious ties ranged widely among MIT students—about 14 percent identified themselves as Congregationalist, 13 percent Catholic, 12 percent Episcopal, 6 percent Jewish, 2 percent Christian Scientist, and the remaining 53 percent either religion-free or subscribers to "varieties of heathen religions." Shulits's translation, incidentally, was published as *Hydraulic Structures*, in 1937, by the American Society of Mechanical Engineers, not by Technology Press.

The number of black students rose during this period—still small, on the order of a couple of dozen all told, but several times the number in previous decades. The impetus came, possibly, from a statewide commission that met in the early and mid-1920s to explore a range of issues in higher education, including racial discrimination. Stratton assigned certain faculty and staff to attend the hearings. In 1927, he asked for a report on black students and assistant dean Harold Lobdell, after scouring the available records, counted 45 since 1892. There were at least 3 blacks in the class of 1923, 5 in 1924, 4 in 1925, 7 in 1926, 3 in 1927, 2 in 1928, and 1 in 1929. Many, but not all, earned degrees. The most popular field for this group was electrical engineering (8), followed by civil engineering (7), mechanical engineering (3), electrochemical engineering and chemical engineering (2 each), and chemistry, architecture, and mining and metallurgy (1 each). When Victor Smith (S.B. '24, S.M. '26) was awarded the Sc.D. in chemical engineering in 1930, he may have been the first black to earn a doctorate at MIT. One black undergraduate, James Evans ('25), belonged to the Institute's debating team. A special fund, the Mary J. Buchanan Scholarship, was created in 1925 to support "deserving and ambitious young men of color, of respectable Northern parentage, in whatever department of physics they may select, as being best adapted to their capabilities." As no black students specialized in physics (Course VIII), Stratton helped modify the agreement so that both electrochemical engineering (Course XIV) students—Marron Fort and Paul Jewell—could benefit. In August 1927, Stratton corresponded about possible financial help for a Liberian student who was described by his recommender as "a very intelligent and energetic Negro."

The number of women students in the 1920s dropped slightly from immediate postwar levels. The peak number (45) in 1921–23 fell to 43 in 1923–25 and 29 in 1925–26, before recovering to slightly over 40 during the remainder of the decade. While women still leaned toward architecture, chemistry, and biology or public health, several ventured into physics and a few into the engineering fields. A pattern observed by the *Boston Globe* in 1921, that at MIT "girls for the most part prefer architectural and other courses where the emphasis is not laid on mathematics," grew a bit fluid. Women could be found in electrical, civil, chemical, electrochemical, metallurgy, engineering administration, and the engineering option in architecture—all the engineering courses, in fact, except for aeronautical, mechanical, and naval architecture and marine engineering. "I just love it," said electrical engineering student Helen Hardy ('24), deftly bridging the gender gap, "and see no reason why women should not be as good electrical experts, plumbers or carpenters as men. . . . I have at times become very much upset over some of the mathematical problems that come in the course of my studies, but if I start to cook or sew my upset will simply fall away." The 1920s also saw women earn master's degrees for the first time in several fields: Bertha Wiener, sister of Norbert Wiener, in chemistry, 1922; Evelyn Clift in physics, also in 1922; and Dorothy Quiggle in chemical engineering, 1927. The first woman to earn a doctorate in chemistry was Elizabeth Gatewood, 1922; in physics, Louisa Eyre, 1924; in metallurgy, Frances Clark, 1926; and in mathematics, Dorothy Weeks, 1930.

Employment opportunities for women at MIT were rare, but greater than in previous decades. Each year from 1921 to 1930 between four and eight women served on staff, mostly as lab or research assistants in physics, chemistry, architecture, and biology and public health. Two women—Dorothy Quiggle and Valerie Schneider—found spots in chemical engineering. Dorothy Weeks, instructor in physics, was the first woman since Ellen Richards to achieve the rank of instructor. There were no women faculty. When the American Association of University Women asked in 1924 about MIT joining that organization, Stratton replied that there would be no point "in view of the fact that the Institute of Technology is engaged in purely technical work and has no women members on its faculty." In 1928, Stratton rejected someone's suggestion for an assistant dean of women as untenable in light of the small number of women students, but he did think it wise—at some future point—to erect a dormitory for women with a suitable matron in charge.

Women found more acceptance in student governance and extracurricular life. Helen Hardy was elected class secretary in her sophomore year. Gertrude Harris ('24) was treasurer of the Math Club, while Dorothy Quiggle ('26) served one year as class treasurer and member of the finance committee. Mary Soroka ('26) contributed to the comedic campus magazine *Voo Doo*. But women still faced gibes from male classmates, like those who in 1926 voted their preference for "the girl who does not let her brains interfere with her pleasures." Access to athletics facilities was a special problem. Allan Winter Rowe ('01), secretary of the advisory council on athletics, appealed in 1926 for help in organizing some sport—possibly fencing—that women could participate in. "I am a sort of step-father," he wrote, "to the forlorn young women who constitute the feminine undergraduate group at Technology." While Stratton refused to allow Rowe use of the third floor of the president's house for women's fencing practice, by the late 1920s at least two women—Ruth Davies ('29) and Constance Sharp ('29)—were competing in that sport.

Stratton's pursuits outside MIT remained quite circumscribed—they revolved, mostly, around his annual trips to Europe—but he did play a role in one of the most contentious public controversies of the 1920s. In June 1927, Massachusetts governor Alvan Fuller asked him to serve on a panel to review evidence, guilty verdicts, and death sentences in the notorious Sacco-Vanzetti case. Two Italian-born, working-class immigrants, Nicola Sacco and Bartolomeo Vanzetti, had been convicted of armed robbery and murder in trials so loaded with anti-immigrant and political bias—the presiding judge, Webster Thayer, was said to have boasted to friends about purging Bolsheviks—that a firestorm of protest erupted, led by intellectuals such as Bertrand Russell, Upton Sinclair, and George Bernard Shaw, and by jurists like Felix Frankfurter concerned for due process and judicial integrity. Governor Fuller organized a last-ditch appeals process that he hoped would mitigate public outrage. Besides Stratton, his three-member review panel consisted of Harvard's president Lowell and retired probate court judge Robert Grant. Lowell was chairman, and the panel came to be known as the Lowell committee.

Stratton agreed to serve reluctantly, torn between his sense of public duty, his inclination to avoid the limelight, and his desire not to upset plans for his trip to Europe. The hearings were scheduled to fill part of June and much of July. By the second week in June he had moved out to Manchester, the picturesque seaside community on Cape Ann where the Fabyan family—Francis Fabyan ('93) had been elected to life membership on the MIT Corporation in 1925—lent him their house as a summer getaway. Stratton loved it there. This was the one residence, aside from his Georgetown dream home, that he always said he would have liked to own. He enjoyed sitting outside with his Boston terrier, Ku, on the terrace that sloped toward the bay. But each day, for several weeks, he had to leave this bucolic outpost to catch the first train into Boston, sit through hours of testimony in a hot hearing room at the State House, work through lunch on legal matters, catch the evening train back to Manchester, and plow through court transcripts into the wee hours.

Stratton had little if any impact on the case's outcome. He said hardly a word, allowing Lowell and Grant to dominate. One of the defendants mistook his quietness for reflection, writing from prison on July 21 that, unlike Grant, Stratton seemed honestly intentioned and not hostile. But on July 27 he joined with Lowell and Grant in declaring that the process had been fair throughout and that justice should now take its course. Governor Fuller denied clemency on August 3 and signed death warrants.

Stratton, meanwhile, was determined to get in his European vacation before the fall semester began, even if it had to be a short one. The steamship company, worried about security, agreed to take him only under strict cover. A friend whisked him by car quietly to portside. Stratton remained hidden from view in the harbormaster's office and boarded at the last minute. His name did not appear on the passenger manifest and he kept a low profile, confined to his cabin for the duration. On August 23, mid-voyage, the captain stopped by to tell him that Sacco and Vanzetti had just gone to the electric chair, news not shared with anyone else on board. "I don't know how the crew feels about this affair, and what they don't know won't hurt them."

Reaction at MIT was muted, at first. Not many on campus discussed it, much less resonated with the visceral anger shared by a few. John Burchard, a beginning instructor in architecture at the time, later reflected on the episode as a sad commentary on Stratton. "His principal claim to fame (or *infamy*) occurred when he was a member of the three-man independent investigatory committee . . . which upheld Governor Fuller's refusal to grant clemency. . . . As a young man I lunched more than once with Judge Thayer and am quite satisfied from what he then said that he was a prejudiced judge. After all the trial was published I read all the evidence carefully . . . and am fully satisfied that at the least Sacco and Vanzetti did not get a fair trial." Eric Hodgins ('22), who had just passed on the editorial reins of *Technology Review* to James Killian and who would become publisher of *Fortune* magazine, vice president of Time Inc., and author of the popular Mr. Blandings novels, also recounted his emotions later. "A more incompetent and inappropriate threesome could not possibly have been chosen, and their eventual decision not to disturb the status quo was scarcely a surprise. Old Judge Grant was senile. Harvard's Abbott Lawrence Lowell was slowly running down with all the outward dignity of an unwound grandfather clock. And as for Samuel Wesley Stratton . . . the less said the better."

These were comments in hindsight. At the time, Stratton's mailbox never piled high with responses either pro or con. A couple of dozen letters came in, about half from MIT-connected people—alumni, Corporation members, faculty, current students—and about half from outsiders. Of the MIT group, an overwhelming majority praised the Lowell committee's conclusions. "I honor you, your associates and Governor Fuller," Harry Jordan ('91) wrote to Stratton, "for your granite courage in standing by your opinions—instead of running away before the clamor of those misguided, sentimental people who slobber over criminals, and crimes." Corporation member Franklin Hobbs ('89) thanked the committee for showing "the world . . . that Massachusetts still stands for law and order and not for mob rule." Miles Sherrill ('99), professor of chemistry, wrote of Stratton's "fine and unselfish service." Even former MIT president Henry Pritchett chimed in, congratulating Stratton on his brave patriotism. But two alumni expressed a different view, more along the lines of what Burchard and Hodgins wrote later. James Ellery ('99) pointed to political prejudice—anticommunist propaganda—as having tarnished the case. Atherton Hastings ('23) denounced Stratton in a letter sent a little over a month after the executions were carried out:

> It is because you have cast an ugly reflection upon M.I.T. that I write this letter. It is because I want you to realize that there are some graduates of the Institute who utterly condemn you and your disgraceful conduct. If you made this decision out of ignorance or as a result of perversive influence I forgive you, but not otherwise. Your standing in the field of science and technical development mark you as one who takes the scientific approach to important questions, but here you were weak and allowed ulterior motives to dominate. . . . If there ever was a case of a man who has prostituted his good offices to white-wash a sordid trial it is yours. . . . You whom I once looked to for guidance and inspiration have fallen lower than a Judas. You have prostituted a high office to serve a frenzied, irrational plutocratic group.

Responses from outside were even more polarized, and polarizing. One expressed relief that "these destructful brutes"—Sacco and Vanzetti—had been liquidated, while another depicted them as martyrs who had "died to set men free." Harold Ickes, a then-obscure Chicago lawyer who would later serve as U.S. interior secretary under Franklin D. Roosevelt, wrote a thoughtful critique in October 1927. "I was particularly gratified when you were appointed a member of the Commission because I knew that you had a scientific mind and to me that means a penetrating and impartial mind. . . . I did think that you would approach this investigation as you would approach an intricate problem in physics; that you would exclude no possible evidence and would examine critically every act and thing adduced in evidence, keeping your mind open and clear and not accepting any conclusion that would not reconcile all the facts. One cannot read your finding in this case and fail to come to the conclusion that for some reason you did not bring your scientific training to bear on the issues. . . . The best that can be urged on your behalf is that you were misled and made a tool to their purposes by shrewder and less scrupulous men." William Howard Taft, chief justice of the U.S. Supreme Court and former U.S. president, wrote from the opposite angle in November, praising Stratton for his heroic, dedicated service to the nation.

Stratton's reputation began to fray after this episode. Already well into his sixties, he seemed to age a decade or more overnight. He lost energy, he looked disengaged and withdrawn. His role on the Lowell committee—as a passive rubber-stamp, essentially, for the views of Lowell and Grant—did not inspire confidence, and motivated many to think of him, now, as compromised, flawed, even expendable. As his obvious faults drew more-than-usual attention, and as his not-so-obvious ones became talking points, few found much good to say about him.

Eric Hodgins remarked on the not-so-obvious many years later. The MIT community had grown used to the president's house as a mainstay of traditional family life—Mrs. Pritchett and Mrs. Maclaurin had been surrogate mothers, and their children were little brothers and sisters—and Stratton's bachelor lifestyle seemed out of place, if not inappropriate. "This was unfortunate enough," wrote Hodgins, "on a job where a Mrs. President is a social necessity; what made matters somewhat worse was that Dr. Stratton's 'confidential secretary' was a moist-eyed young man of full red lips and slightly lisping speech . . . who was his constant companion. No scandal ever broke into the open as the result of all this, but Dr. Stratton soon found himself an almost isolated man at the institution he was supposed to head, with few friends among faculty, trustees, alumni or students." Burchard wrote, also much later, about Stratton living alone with "a fairly effeminate male companion-secretary," and squirreling himself away in the basement of the president's house for hours on end tinkering with machinery, crafting fine furniture, or repairing old gadgets. "Stratton," Vannevar Bush observed retrospectively, "was a bachelor and lived with a sidekick called Morris Parris—a great deal could be said about that, but probably shouldn't be." As ubiquitous secretary, social organizer, community liaison, gatekeeper, and devoted jack-of-all-trades, Parris provoked wild conjectures. Julius Stratton later called the talk more malicious than truthful, but acknowledged its prevalence.

Gossip aside, when custom or etiquette required a lady's attendance as chaperone or tea-pourer—at student dances, say, or receptions—Stratton had to recruit Mrs. James Jack or some other faculty wife. The basement lab struck some as a throwback to the Edisonian-style tinkerer-inventor, now passé. Stratton's loyal black staff—cook Cordelia Pannell, maid Laura Allen, and man-servants Postello Jones and Alonzo Fields—seemed quaint, hyper-elaborate, rather anachronistic. The president's house became known less for its community-friendly feel than for its fancy black-tie affairs. Houseguests included dignitaries like Herbert Hoover, even after his election as U.S. president in 1928. MIT had never gone for this level of exposure or formality, and did not go either for Stratton's tales of rubbing elbows in London with the Queen of Spain and her entourage, or of sitting at dinner between Lady Astor and the Duchess of Sutherland. Stratton, in short, was a purveyor of elegance in a place where elegance did not count for much. Those who had been around long enough pined for Maclaurin's rollicking, informal Christmas get-togethers.

Some complained about Stratton's frequent absences from campus, his forgetfulness, his mental lethargy. When a department head would press him in writing for a decision, he would send the request back with a note in the margin to the effect of "Okay, if nobody objects." On transatlantic crossings he would sit at dinner with German or French scientists, size them up, and offer appointments to those who impressed him. But when he got home he would forget to follow up, to consult department heads, and to organize the necessary paperwork. This caused trouble when an offer was taken seriously, accepted by mail, then ignored or withdrawn.

Some faculty—those who were not his favorites, or special allies—grew annoyed about having to be reintroduced to him over and over. Few made allowances for his age, or for the fact that he was almost blind in one eye and severely myopic in the other. If he was this infirm, what was he doing as president? He hung on, or so it seemed to some, "in the declining years of an active life [whose] crotchets had come to cover more surface area than his adaptability." A milder man might have garnered sympathy. But Stratton could be aggressive, cantankerous at times. Some department heads swore off going anywhere near him, if they could help it. Those who met with him willingly were forceful, irascible characters, men like Dugald Jackson who enjoyed a good fight. When Stratton pounded his desk, Jackson would pound back. But few faculty had the stomach for this. Many stood aside, smoldered in silence, sensed inertia all around, and waited—hoped—for change. "The communications system . . . during S. W. Stratton's regime," one department head later recalled, "consisted of reading the Monday morning paper and finding out what the hell had happened at the Institute. Most of it had happened, not by somebody doing something, but by lack of it." E. B. Wilson, at Harvard, heard grumbling from his former MIT colleagues about Stratton "frittering away the resources on all sorts of new-fangled odds and ends instead of consolidating the position of the really important activities." Horace Ford, MIT's bursar, felt that Stratton was "not accustomed to anything of the sort of work that we were doing." Even Francis Hart, who had pushed Stratton for president after meeting him on board ship in 1922, was said to have "regretted that sail."

Stratton retained some supporters. His staunchest ally—Samuel Prescott, head of biology and public health—never wavered. Prescott felt drawn to Stratton from the moment the new president was introduced to the faculty in 1923. "I shall never forget the first time I saw him, when across a reception room and before I had ever spoken to him, I noted his kindly but searching blue-gray eyes, his illuminating smile, and the firm set of his head. Instantly, I inwardly felt that here was a friendly and sincere man of sterling character, wise but not hasty of judgment, and fine and generous impulses—a man, in our New England country phrase, 'that was safe to tie to.'" And tie to him Prescott did, with an intensity that bordered on hero-worship. Sometimes he puzzled over why he found Stratton so magnetic, in light of how negatively others felt. "Perhaps the similarity of our boyhood experience gave us, at the outset, a common basis of understanding and sympathy . . . and what I may call spiritual response. Perhaps it was an unexpressed faith in each other, and tacit recognition of that nameless something that draws a man to his chief, or men to each other in a relation closer than blood kinship. Of this I cannot say, but I found in him comradeship untainted by favoritism, loyalty which could withstand differences of opinion, a friendship steadfast, sincere, and sure."

Prescott was not alone. Stratton's other loyalists—not a huge group, to be sure—included several faculty who felt beholden to him for redirecting MIT onto a path where serious research and pure science mattered. Edward Bowles testified to being "moved deeply" by the president, to the confidence that he had "instilled and inspired in many of us youngsters." Henry Phillips, of the mathematics department, thanked him for his "efforts to make of this a more scientific institution. . . . The revision of courses you have started is I believe the most fundamental step taken by any administration during my more than twenty years of service." Julius Stratton appreciated his "devotion to all that is best for the Institute," while fellow physicist Manuel Vallarta pointed to his "splendid services to the cause of pure science . . . the rebirth of pure physics at this Institute." Waldemar Lindgren, who had relied on Stratton to help geology achieve independence from mining and metallurgy in the mid-1920s, wrote of how the president had "stoutly upheld the view that technical education must be based on the strong foundation of pure science."

By 1928, critics—and even some allies—wondered if Stratton ought to make way for new leadership. Executive committee member Gerard Swope ('95), of General Electric, took the lead in easing him out, or, if not out, at least to the sidelines. What worried Swope was MIT's increasing tunnel-vision. His own experience as a student there had been broad: economics under Davis Dewey, business law from Louis Brandeis, courses in literature and history, besides his core subjects in electrical engineering. Such offerings remained, but was there, he wondered, a strong enough commitment to them? "The danger that I always foresaw," Swope later said, "that MIT would become too narrow in its preparation of the men for the work that lay ahead of them, occurred in the twenties." Corporation member Frank Jewett shared this concern. "The engineering student must be given more humanities," he wrote in a March 1928 article ("Scientific education: Do we know what we want and can we get it?") in *Technology Review*, "he must be taught fundamentals; he must be imbued with an appreciation of research and the power of the research method; he must be trained for leadership and for engineering management (whatever that may mean)." Was there, Swope and Jewett wondered, enough emphasis at MIT on science, or enough attention to breadth in the rush to churn out workmanlike engineers? These were the same concerns that Stratton had spent his presidency trying to address, but not, evidently, to the satisfaction of some. A couple of his primary achievements—expanding coursework, recruiting distinguished guest lecturers—began to look less like assets and more like liabilities. Swope and Jewett wanted to encourage greater coherence in the academic program (quality over quantity), and a more systematized approach to faculty recruitment (permanence over transience).

Stratton shared the values that Swope and Jewett considered crucial—the president had proven himself open to reform—but they worried that he was too old to build and to sustain momentum. He made it clear, however, that he was not ready to retire. How to ease him out with no hard feelings—this was a tricky question. The executive committee suggested appointing a vice president who would be groomed and elevated to the presidency once Stratton retired. When Stratton expressed little enthusiasm for this idea, and made no move to recruit anyone, Swope suggested that Stratton be made Corporation chairman with a new president to help him run things. Stratton took offense, telling treasurer Everett Morss that the proposal was little more than a polite way of pensioning him off.

The executive committee pressed ahead anyway. It was a practical idea, and innovative too; never had this corporate management model been tried at an academic institution. Jewett canvassed his close circle for possible candidates, someone, he said, in his late thirties but no older than fifty or so, "the right type of man with executive capacity among the so-called pure scientists." The core group at Caltech—Robert Millikan, George Hale, and Arthur Noyes—recommended Karl Taylor Compton, of Princeton, in September 1929.

Compton was chosen in March 1930 and Stratton agreed to move up to Corporation chair. "I am sure," Corporation member Joseph Powell soothed Stratton, "the new arrangements by which you are relieved of many of the active duties of Management . . . are going to work out most satisfactorily to everyone. I . . . congratulate you on being the first head of a big Institution to realize that the business has grown so large as to require someone to do the thinking and someone else to do the running!" Stratton, at least, approved of Compton. He knew him personally, got along with him, saw him as quiet, conciliatory, unlikely to throw his weight around. "As long as it's Karl Compton," Prescott recalled Stratton saying, "I'm perfectly happy to yield the place, but I would get out in a minute if it was Arthur [Compton] or anybody else." Stratton opened up sooner than he was supposed to, to Julius Stratton. One day the two sat talking in the back of the president's limo, Postello Jones at the wheel. "I've got something very good to tell you," the elder Stratton said. "It will please you very, very much. And I'm very pleased, too. It is that Karl Compton—you know him, don't you?—has accepted the presidency of MIT. He will make a big change."

Stratton was in good health, in better health, he felt, than at any other time in the past decade. The Corporation chairmanship agreed with him. He had more time to travel, yet felt of some value to MIT.

Compton often sought his advice. It was a fine partnership, Stratton told George Hale in 1931:

> The new arrangement here is enabling us to build up our scientific work. The new laboratory for graduate work and research in physics and chemistry is well under way, also the new spectroscopic laboratory which is expected to be the best of its kind from the point of view of stability, temperature control, etc. We have some very excellent young men on our physics staff, and the new quarters will be a great encouragement to them.
>
> It has been very difficult to get some of our engineering departments to understand the value of creative work, and to do the things that will promote the interests of the Institute, their profession and themselves.
>
> Since coming to the Institute I have devoted a great deal of energy toward encouraging the spirit of research in the engineering departments, and I think, with some success, as we now have some very important investigations going on in all these departments and quite a number of young instructors who are going to be heard from, notably in hydraulics, soil physics, automotive engineering, aeronautics, physical metallurgy and seismology, as well as the newer branches of electrical engineering.
>
> Among our engineering students there are generally quite a number of brilliant men fitted by nature to undertake research.

Stratton's successor as director of the Bureau of Standards, George Burgess, captured the new regime in nautical terms: "Under Admiral Stratton and Captain Compton the good ship Technology will weather the seas of technical education with the broom lashed to the mast as a signal that it has no rivals."

From his home at 370 Beacon Street, Boston, Stratton would gaze out across the river at the Great Dome that crowned Tech's sprawling campus, his other home. Now his third home, actually. His second was a place up in farm country—West Springfield, New Hampshire, not far from Dartmouth College—that brought back memories of his younger days in the Midwest. There, at his 70th birthday celebration in July 1931, he talked about living to a grand old age and of taking on a massive twenty-year research program that would make a younger man wilt: hydraulic field studies, models for measuring water flow, data to help tame rivers, control floods, harness currents, and predict silt shifts. Stratton hoped to avoid the mistake made by a neighbor he remembered from his boyhood days. When this man reached 70, he retired, drifted, sat around, waited to die—then proceeded to live for twenty more years, still waiting and accomplishing nothing. "He would have been far happier," Stratton declared, "had he lived every day as though he had long years ahead of him."

Whether this meant that Stratton was poised for a late-career surge, there was no time to find out. When his old friend Thomas Edison passed away in the fall, on October 18, 1931, the media chased him down for a statement that same day. "It seldom has fallen to any one man to be of such service to humanity," Stratton told a reporter from the Associated Press. But other reporters he waved off—all except for one, an MIT undergraduate moonlighting for the *Post*, Boston's popular daily. Stratton invited him in, sat him down, asked about his journalistic interests, his coursework, his life at MIT. "My associations with Mr. Edison extend over a period of more than thirty-five years." As Stratton reminisced, he reached over to pat his blind bulldog. "When his son graduated, that was when he came to Tech last. And oh, yes! Have you written it already? About his accomplishments which will be the inspiration of the generations to come? His interest—"

Suddenly, he slumped to the floor. The young reporter yelled out. In rushed Morris Parris, who called for Dr. William Robbins over on Marlborough Street, a few blocks away, the same physician who had attended Maclaurin in his final days. But it was too late—a massive coronary occlusion, Robbins said. Parris choked back tears as the dog nuzzled Stratton's cold cheek.

The household dispersed quietly. Cordelia Pannell, getting on in years, may have retired back to the Washington, D.C., area. Postello Jones lived in West Medford, a Boston suburb, with his wife Irene and daughter Charlotte, although whether or not he found other chauffeuring work during the hard economic times of the 1930s is anyone's guess. Alonzo Fields moved out and up. Through a close aide to Herbert Hoover, Parris arranged for him to join the White House staff. Fields served there for more than twenty years—in the Hoover, Roosevelt, Truman, and Eisenhower administrations—rising through the ranks from butler to chief butler and, finally, to maître d'hôtel. None of these faithful retainers was mentioned in Stratton's will, which itemized a sizable store of antiques, silver, china, and other collectibles for sharing between Parris and Stratton's three sisters. One-third of the rest of the estate went to Parris, two-thirds to the sisters. MIT was left a large Sheffield urn and Stratton's portrait painted by famous society artist Philip de Laszlo, whom Stratton and Parris had befriended on one of their many European trips. Stratton's remains were buried in the family plot in Altadena, California.

Parris stayed on to help Samuel Prescott gather material for a planned, but never finished, biography of Stratton. Prescott abandoned the project to focus on writing a history of MIT, published many years later (in 1954) by Technology Press as *When M.I.T. Was "Boston Tech."* Parris, meanwhile, fell into a slump from which he hoped to recover by joining his sister, Juliet, and her children on a trip to Europe. "Morris . . . is all prepared to sail," Stratton's brother-in-law Glenn Hobbs wrote to Prescott in June 1932. "The itinerary that he recited . . . reads like the activities of a society bud, but that is Morris' life, and if he can be on the go with parties and other things of that sort, he is supremely happy." Prescott agreed. "I joked him about being a social butterfly, flitting from flower to flower. He loves the social life, has had it for years and has carried it on with marvelous skill during the long period when Doctor [Stratton] had to be host and do much entertaining. Doctor loved it, too, but it would have burdened him to attend to all the details which Morris arranged so well. I certainly hope that Morris will find his niche where he can apply his talents and be happy." Some years later Parris married the daughter of old friends of Stratton's from France, and the couple settled in Marshall, Virginia.

"All knowledge his sphere"

Karl Taylor Compton, 1887–1954

By the fall of 1929, with support for president Samuel Stratton dwindling, a small MIT Corporation group—Gerard Swope, Everett Morss, and Edwin Webster—set out in search of a replacement. A long list of candidates was quickly winnowed down and the brothers Karl and Arthur Compton, both well-known physicists, rose to the top of the short list. Karl came highly recommended by Corporation member Frank Jewett and by the core group at Caltech—Robert Millikan, George Hale, and Arthur Noyes—who shared Karl's views on the role of university science. In 1927 Karl had written a piece for *Science* magazine, "Specialization and cooperation in scientific research," that mirrored the Caltech group's position on basic research as vital to the academic mission.

For some, Arthur's Nobel prize awarded in 1927 made him a special draw. "Dr. A. Compton . . . better than Karl," so one outside source advised Swope. But another preferred Karl, as an administrator if not as a physicist. One of Karl's own colleagues, someone in the classics department at Princeton University, had no idea there were two Compton physicists. He had always conflated them. This man placed Karl Compton "on a very high level," jotted Swope. "He says he is one of the three most distinguished physicists in America, a winner of the Nobel prize, and at the head of a huge research organization. On the personal side he is tactful, modest, and most agreeable to meet. He is an administrator of the first order, and a great teacher. . . . Mrs. Compton is said to be a very charming and cultivated woman and a good neighbor."

Like Stratton, Karl Compton came from America's heartland. Born September 14, 1887, in Wooster, Ohio, he grew up on or around the campus of the College of Wooster. His forebears on his father Elias Compton's side, originally of English stock, had joined the population push westward and settled in Ohio around 1817. Elias taught Latin, English, philosophy, and psychology at Wooster, then served as its first dean. The college, steeped in Presbyterian traditions, was noted for training ministers and missionaries. But it was nonsectarian, too, offering a high-quality, liberal education regardless of religious affiliation. Elias had intended to train as a missionary, but after studying at Wooster, followed by a brief period at Western Theological Seminary in Pittsburgh, he went into teaching instead. In 1889 he earned a doctorate in psychology from Clark University, by thesis in absentia. Karl's mother, Otelia (Augspurger) Compton, was from a Mennonite family of German origin, mostly farmers. Her father went into the mill business, however, and was not strictly observant by his family's standards—singing, square dancing, and picnics were permitted, even encouraged, in his household.

While Elias and Otelia brought up their children in the Christian faith, the household was much like the college itself: tolerant, open to ideas, rigorously thoughtful. "As I try to evaluate the character of the old guard as I knew them at Wooster," Karl later reflected, "I think of them in two categories: (1) those who were passionate and sure in their advocacy of Christian and ethical principles and of their stand on all current issues, and (2) those who were judicial and scientific in their attempt to study and evaluate the facts of life and the issues of the day. The two groups worked well together, and their combination developed in their students a respect for scholarship, an adherence to high ideals, and within these ideals an objective guide to action." Karl and his siblings—brothers Wilson and Arthur, sister Mary—grew up and remained, with little evident inner conflict, both devoutly Christian and broadly intellectual. Karl was the eldest, followed by Mary (born 1889), Wilson (1890), and Arthur (1892).

As children, they read widely. Besides the usual young-folks' fare, Elias introduced them to Homer, Greek and Roman mythology,

Norse and German legends, the novels of Walter Scott. Their earliest exposure to science came via the Darwin controversy, which continued to rage, especially on campuses with strong religious traditions, decades after the initial publication of *Origin of Species* in 1859. Elias and another faculty member, biologist Horace Mateer, were among the lone voices at Wooster raised in defense of evolutionary theory. Elias was drawn to many currents in modern thought, life, and industry—from inventions like the automobile, the telephone, the Ferris wheel at the Chicago World's Fair of 1893, to the latest discoveries and theories on atomic structure, geometry of space (Henri Poincaré), X-rays (Wilhelm Roentgen), radioactivity (Henri Becquerel), electrons (Joseph Thomson), polonium and radium (Marie and Pierre Curie). Under Elias's influence, Karl read John Trowbridge's *Three Boys on an Electrical Boat,* an adventure story whose plot twists revolved around electric batteries, and Agnes Giberne's popular *The Story of the Sun, Moon and Stars: Astronomy for Beginners,* which posed questions about where space begins and ends, what light is and how it travels. These books were Karl's introductions to physical science. Trowbridge had taught at MIT in its early years, 1868–70, before heading to Harvard where, as director of the Jefferson Physical Laboratory, he helped modernize physics teaching away from lecture-demonstrations toward laboratory experimentation.

When an older acquaintance of Karl's, Paul Swartz, talked about heading east to train as a chemical engineer, Karl told him about wanting to become a scientist. Swartz graduated in chemical engineering at MIT in 1906, while Karl attended the College of Wooster and leaned, first, toward biology. Horace Mateer hired him as a lab assistant in his sophomore year, but when funding fell through he went over to physics instead, under George Bacon. Karl was responsible for setting up lecture and lab experiments, and for equipment maintenance. In class he excelled in algebra, trigonometry, and analytical geometry. He was a fine athlete as well, a valued member of the baseball team and captain of the football team in his senior year. He was remembered, too, for a spectacular bit of mischief—the type of prank that MIT students would come to call a hack—pulled in the first few days of his freshman year. "One sunny autumn morning," the story went, he "climbed up the inside of [the heating plant's] 150-foot brick smokestack, carrying with him a flexible steel hoop of carefully calculated diameter to which his class colors and numerals had been attached. When he got to the top he dropped the hoop down over the outside of the tapered stack, whose widening girth held it firmly in place and effectively out of reach from above or below. On this glorious morning the numerals '08 were to be seen waving in the breeze for all to admire, and there was no getting them down. The ingenuity, agility, courage, and enthusiasm of the young hero remained visibly confirmed for some weeks." He was viewed on campus as a natural, instinctive leader, remarkably well rounded.

Karl graduated Ph.B. (bachelor of philosophy) cum laude, in 1908, and announced his engagement to classmate Rowena Rayman. In 1909 he earned a master's degree in physics under George Bacon in preparation, possibly, to teach in a missionary college overseas. As Bacon's assistant, Karl discovered how much he enjoyed teaching and how good he was at it. But he also grew excited about research, especially anything concerned with the new electron theory. His master's thesis, which explored the operating principle behind the Wehnelt interrupter, a relatively new device in the emerging field of radiotelegraphy, was published in the February 1910 number of *Physical Review*.

Karl took long Sunday afternoon walks with his father, who kept him from falling for science's cruder, mechanistic certainties. "One time," Karl recalled, "he had been quoting the ideas of the older philosophers all the way back to the Greeks. I made the comment that it must be discouraging to be working in a field in which the answers to the problems were never finally established and I tried to argue that scientific pursuits were more satisfying because it was possible to get clear cut answers to which everyone would have to agree. Father then proceeded to point out that in the entire realm of human values and human understanding the real value of study of these problems lay not so much in finding the answers that would convince everybody as in developing a pattern of thought and action which would be satisfying and unifying to the individual himself who made the study." Such home truths kept Karl grounded at a time when physics, through quantum theory and relativity, opened new windows to the scientific imagination. He was thirteen when Max Planck published the quantum hypothesis in 1900, and a college sophomore when Albert

Einstein explained the mechanism controlling emission of electrons by atoms struck by light waves—both followed, in due course, by Niels Bohr's explanation of atomic structure in 1913, Louis de Broglie's particle wave theory in 1924, Werner Heisenberg's matrix mechanics in 1925, and Erwin Schrödinger's wave mechanics in 1926. But thrilling as each of these discoveries proved, they were always balanced for Karl by his father's sensible, holistic, down-to-earth views on life and the world.

A job offer from a missionary college in Korea, what Karl had half-aimed for, materialized in the spring of 1909. But he was not sure he wanted this. Elias reinforced his doubts, urging him instead to go on for a doctorate at some prestigious university. "I had such faith in my father's good judgment," Karl wrote, "that I decided to follow his advice as quickly as ways and means could be found." Karl chose Princeton by browsing through university catalogs that a college friend, Robert Caldwell, had lying about the house. What drew his attention was the presence at Princeton of two rising stars, J. H. Jeans and O. W. Richardson, products of the famed Cavendish Laboratory at the University of Cambridge. Jeans returned to England soon after Karl arrived in 1910, but Richardson appealed to him more anyway. Jeans's work was purely theoretical. Richardson was an experimenter with strong theoretical interests, and Karl liked laboratory work.

Karl adapted well to life at Princeton. He attended chapel regularly and wrote to his family about sermons he heard on Sundays. According to fellow Richardson student C. J. Davisson, Karl was "quiet and pleasant, and serious and reasonable in what he said and did. He had already acquired that kindly subdued smile which all who knew him remember well." But Richardson's ways took some getting used to. When Karl arrived, Richardson led him to the basement, pointed to a nook where he could set up shop, and paid him little mind until his thesis was done. Karl's doctoral research fed into a growing body of evidence that confirmed Richardson's theory of thermionic emission, the removal of electrons from matter, for which Richardson would receive the Nobel prize in 1928. Karl earned the Ph.D. in physics, summa cum laude, in 1912. The summa, a source of quiet pride to him and his family, was a rare distinction at Princeton, awarded no more than once or twice in a generation. It pointed, too, to a possible career for Karl on the cutting-edge of physics: electromagnetic theory, light waves, laws of nature whose paradigms were in flux, even upheaval.

In June 1913, on completing a one-year postdoctoral fellowship at Princeton, Karl married his fiancée Rowena, by now a high-school history teacher in Canton, Ohio. The couple honeymooned in the Canadian Rockies on their way to Portland, Oregon, where Karl had accepted an appointment as instructor in physics at Reed College. Reed, a coeducational, religious but nonsectarian institution founded in 1908 (its first classes were held in 1911), prided itself on offering not the socially clubby atmosphere of an east-coast establishment institution but a flexible, individualized approach focusing on intellectual growth. Karl took charge of physics instruction and most aspects of departmental and laboratory development. He also carried on experiments, introduced undergraduates to research methods, and published papers in *Physical Review*. A college-level textbook on electricity and magnetism that he drafted remained unpublished because he was never satisfied with it.

Reed College, his first post, offered Karl a huge opportunity—"while the ship he commanded was more skiff than liner, he was the captain bold as well as the cook." But he did not stay long. The salary was too low to support a family. Rowena was expecting, and their daughter, Mary Evelyn, arrived late in December 1914. O. W. Richardson recommended him for a job at Yale. But Princeton was eager to have him back, too, so in 1915 he returned there as an assistant professor.

Karl, often known as K.T. to close friends and colleagues, energized Princeton's teaching program. Richardson's pedagogical skills were notoriously lacking—Arthur Compton once remarked that his lectures on electron theory were among the worst ever—and Karl boosted student morale by laying out clear, goal-oriented ways to tackle research problems. His brothers, meanwhile, had followed him to Princeton. Wilson (Ph.D. economics '15) was now teaching at Dartmouth. Arthur, still a year away from earning his doctorate in physics—he, too, would make summa cum laude—did some collaborative research with Karl. They co-invented a device known as the Compton electrometer, for measuring voltage sensitivity. Karl declined a job offer from Willis Whitney, director of research at General Electric, but agreed to serve as an outside consultant, an arrangement that lasted many years. Scientists at GE's Schenectady laboratories—isolated geographically, sometimes intellectually, from university life—grew to rely on his expertise and to appreciate his personable manner.

During the First World War, Compton saw service locally, then with the U.S. Signal Corps in Washington, D.C.—"Research work under Gov. auspices," he broadly described it—and finally in Paris as associate scientific attaché at the U.S. embassy and as an officer in the Research Information Service. At Princeton in the summer of 1917 he joined fellow physicist Augustus Trowbridge in a study of sound-ranging methods to trace enemy artillery, then became part of a team of scientists at the Edison Laboratories in Menlo Park, New Jersey, investigating torpedo propulsion and submarine detection. That fall, until December 17, he served under Robert Millikan in Washington, D.C. Through the National Research Council, Millikan had taken charge of mobilizing scientific personnel for national defense. Assigned to aviation systems, Compton called himself an aeronautical engineer. But he could not escape the physicist label so easily. When General John Pershing reached France, he telegraphed back to Washington—"Send forty physicists at once." "What in blazes *is* a physicist?" asked Millikan's superior officer. "A physicist," Millikan replied, "is a particular kind of nut—I can supply you with forty at once." Compton was among the nuts shipped over. In Paris he continued the work he had begun with the Signal Corps, preparing reports on artillery-sound and submarine detection. From December 1917 to January 1919, he was also an adviser to the Council of National Defense and to military and naval intelligence on inventions classified as tools of war.

Back at Princeton, Compton resumed his routine gradually. Students were slow to return from war service, and for a while his only graduate student was Henry Smyth, who would later chair the Princeton physics department and author the famous Smyth Report for the U.S. Atomic Energy Commission following the Second World War. The lag gave Compton more freedom to move on his own research. He churned out a dozen or more papers in 1919, meanwhile sharpening his focus to a single area—electron collisions—which set the stage for his research during the 1920s. He became a pioneer in experimental work on electrical plasmas, aggregates of subatomic particles that in the approaching nuclear age would prove a key to generating controlled fusion reactions from hydrogen.

Rowena was expecting another child when she died suddenly, in October 1919, of acute peritonitis from a ruptured appendix. The shock laid Compton low for many months, but he remarried in July 1921. His second wife, Margaret Hutchinson, was the sister-in-law of a Princeton classics professor, Henry van Heusen, who helped Compton through the difficult period following Rowena's death. Van Heusen and his wife, Ruth, Margaret's sister, often looked after Mary Evelyn. Ruth and Margaret's father, J. Corrin Hutchinson, was a minister-turned-polymath who taught classics, Biblical analysis, and mathematics at the University of Minnesota. Margaret, herself a University of Minnesota graduate, taught school for a year and then went into YWCA field work. The newlyweds honeymooned on the lakes above Duluth, canoeing two or three hundred miles over three weeks. While they longed to do what another young Minnesotan, Eric Sevareid, had just achieved—row from the Upper Peninsula all the way to Hudson's Bay—there was no time as Compton, just promoted to full professor, had to get back to Princeton for the start of the fall term. The family grew to five in due course. Jean Corrin was born in 1924, Charles Arthur (usually referred to as Arthur, after his uncle) in 1927.

Compton's reputation in the world of physics continued to grow. In 1923, he was elected vice president and chair of the physics section of the American Association for the Advancement of Science (AAAS). He was a member of the AAAS executive committee for several years and would later, in 1935, become president. The American Physical Society elected him vice president in 1925 and president in 1927, both two-year terms. He chaired, 1927–30, the National Academy of Sciences physics division and the National Research Council's physics subcommittee charged with providing expert advice to organizers of the 1933 Chicago World's Fair. "These jobs," he wrote to his parents, "involve chiefly making some speeches," but they were also a mark of the high esteem in which he was held by scientists nationwide. Offers from other universities flowed in and one, in particular, he almost accepted. In 1922 the University of Chicago tried to recruit him at a high salary, with a number of attractive perks thrown in. Margaret supported the move, mostly so that she could live closer to her family in

Minneapolis. Compton was tempted by the opportunity to work with A. A. Michelson, one of the grand old men of physics, now nearing retirement. In the end, Princeton kept him by matching Chicago's offer and promising a fast track toward maximum salary, along with a reduced teaching load.

But Chicago persisted. Another offer came in 1924, this one more difficult to refuse. Arthur Compton, now established at Chicago after brief periods at the University of Minnesota, Westinghouse Lamp Company, and Washington University, St. Louis, suggested a package deal, that is, bringing the brothers together—these two rising stars of American physics—to help sustain Chicago's eminence, post-Michelson, in the world of academic science. Arthur's work on X-ray and gamma-ray scattering, the so-called Compton effect, for which he would be awarded the Nobel prize in 1927, had edged ahead of Karl's in its creativity and momentum. He was earning a reputation, too, as more brilliant but less steady than his older brother. Karl went so far as to sign a contract with Chicago. But Princeton, not to be outdone, countered with a plan of its own: hefty salaries, along with fringe benefits and special research funds, for Karl to stay and for Arthur to come to Princeton. This contest ended amicably, with the brothers staying where they were. "A feeling of loyalty and obligation," Karl wrote to University of Chicago president Ernest Burton, "together with faith in the future, lead my brother to wish to remain with Chicago at this time, while the same feelings lead me to desire to remain at Princeton."

Karl assumed that Princeton was where he would live out his career. A generation of physicists learned vacuum and gas-discharge measurement techniques under his guidance, while a stream of publications on thermionics, ultraviolet spectroscopy, and the electrons of gases flowed from his lab. In 1926 he spent part of a sabbatical at the University of Göttingen, a mecca for up-and-coming American scientists, where he met, among others, a young American graduate student by the name of J. Robert Oppenheimer—the "quite excellent Mr. Oppenheimer," as Max Born described him to Samuel Stratton. In Göttingen Compton found more than a dozen young Americans working under theoretical physicists Born, James Franck, and Otto Oldenberg. On his return to Princeton in 1927, he was appointed research director of the Palmer Physical Laboratory and given a newly endowed chair, the Cyrus Fogg

Brackett professorship, which freed him from undergraduate teaching to commit full time to graduate work and research. In 1929 he became department head, vowing to spend the rest of his career building Princeton's physics program into the world's best.

But then came the MIT offer. Gerard Swope, Compton's chief advocate on the Corporation, asked him to drop by his New York office for a chat sometime in January 1930. The two had never met, even though Compton had been a consultant to Swope's firm, General Electric, for more than a decade. Swope asked him about GE's research program. What directions should it move in, how might it shift focus? Compton promised to go home, think, and get back to him in writing, which he did a few days later.

Swope invited him in again, and this time got to the point. "What I really want to know," he said, "is whether you would be willing to come up to MIT as president." Compton was stunned. A research physicist, he had no administrative ambitions and no desire for the limelight. Budgets, finances, economics—these, apart from the bare essentials needed to manage his department, were beyond him. He was no good at fundraising and did not want to do any. "I know something about the difference between an electron and a proton," he told a quizzical Swope, "very much less about the difference between a stock and a bond." He did not like public speaking—no talent for it, he said, and, unless physics was the topic, the prospect of giving a speech ruined his peace of mind for weeks in advance. He did not care much for educational theories either, and his ideas about what made for a good teacher were neither well thought out nor particularly creative—someone, essentially, "who lets his student cooperate in some way with him in his investigations."

Swope tried to ease Compton's concerns. Fundraising, he said, was the MIT Corporation's responsibility. The president's job was to get the Institute's educational and research programs on the right track, not to run around, hat in hand, begging for money. Just two speeches a year were required, one to greet incoming freshmen and the other to "bid them Godspeed" at graduation. When Compton pointed out that MIT's budget for physics research was skimpy compared to Princeton's, Swope replied: "If you will be president you can make the physics research budget as big as you wish." Compton agreed to mull the offer over and talk with Frank Jewett, president of Bell Labs, whom he knew well and whose opinion he valued. But he said this out of politeness, only half-intending to follow up.

Margaret sensed something on his mind when he got home. They talked pros and cons for hours, and they were hard-pressed to come up with a single pro. They worried that the children, now at impressionable ages, might find the MIT president's house confining, disruptive, even unnatural. Compton would miss his laboratory, graduate students, colleagues, and well-laid plans for Princeton's future in physics. All signs pointed to no, but Margaret dutifully left the decision to him. "Wheresoever thou goest," she told him, "there will I go also." Biblical aphorisms often captured their basic values, raised as they both were in Christian evangelical households.

Compton arranged to meet with Jewett on a train carrying Jewett from Washington to New York. That morning Compton left home convinced MIT would never work out, that he had best stick to the job he liked and knew he could handle. He boarded at Trenton, an hour or so outside New York—more than enough time, he thought, to persuade Jewett to help him ward off Swope's advances. But when the train pulled into Penn Station, they were still talking. Compton followed Jewett to his office in the Bell Labs building overlooking the Hudson River in lower Manhattan.

Jewett talked, Compton listened. The Institute, he said, had become stuck in neutral or reverse and needed a jump-start. Its vision—outmoded, if not exactly irrelevant—was not meeting industry's needs. MIT's founder, William Barton Rogers, had conceived and built a place that would lead in science as well as in its practical applications. But MIT had fallen behind in the former and its approach to the latter bordered on the obsolete. For decades, industry had relied on MIT to supply personnel trained in certain techniques, lab methods, shop practices, and other routine functions. In turn, MIT "took on a strong slant of immediate practicality." But Jewett's firm (AT&T), Swope's (GE), and others had evolved with such precise, refined, often proprietary technical demands that they quickly began training their own staffs. As this trend grew, MIT's value as a supplier of technical expertise declined. What industry needed from MIT was not so much practical skills as graduates broadly educated in science, mathematics, and basic engineering principles—personnel with minds open to "the anticipation of technological change . . . some contact with the spirit and methods of research, and preferably some experience in it, so that [they] would be prepared to grasp new technological opportunities and either to participate in their development or at least to understand something of the conditions required for such development." This pool, with its more flexible, creative mindset, would help industry take advantage of new opportunities opened up by advances in science. Jewett explained that the opportunity to break with the traditional style of engineering education, and to reinforce the pure sciences, could be realized only under leadership that would move aggressively to build a new type of faculty. There had been some tinkering around the edges under Samuel Stratton, but nothing like the transformation that Jewett was hinting at. Engineering departments, he said, must have more research faculty, whether trained as engineers or as scientists. In his mind the type of leader most likely to make this happen was a research physicist—not an engineer, or a chemist, or a biologist, but a physicist, someone like Compton, whose background and experience lay at the juncture between academic science and technological know-how.

Jewett laid out a vision that Swope had merely hinted at. This was not the MIT that Compton had heard so much about, productive yet satisfied with its traditional mission. He had dismissed the challenge because it felt like giving up a career for something opposite. But the transition was as natural as one could find—"an opportunity," Compton realized, "to draw on this background of experience and my scientific contacts in order to enlarge the scope of their value and influence in the educational and research fields generally."

He understood the part that MIT had played in American science and technology. In just short of seven decades, it had grown from a small experimental school into America's foremost institute for technological education. The first collegiate program in architecture had started there, along with the earliest formal coursework in architectural engineering. The first programs in aeronautical engineering, chemical engineering, food technology, industrial biology, naval architecture, and marine engineering—these were MIT innovations. Jewett, however, talked to Compton not about legacies, traditions, or achievements, but about where MIT stood currently, poised at a crossroads, ripe for change in a world where science and technology were moving in unprecedented directions.

Compton went away convinced. That afternoon, when Margaret picked him up at Princeton Junction, he made just one comment on the way home—"Jewett, who really knows these engineers, tells me I'm what the Institute needs."

The next step, from MIT's angle, was to meet Margaret. Her interview took place over dinner at the Swopes' Manhattan apartment. The Swopes (Mr. and Mrs.) were there, along with other members of the search committee—Edwin Webster, Francis Hart—and their wives. Margaret acquitted herself well, combining "demureness with maturity most effectively." Swope called her lovely, charming, and able. After dinner the gentlemen retreated to the study, where they puffed cigars and talked over Institute affairs. The ladies waited. When the men emerged, Swope led Compton over to Margaret with the words "Let me introduce to you the new president of MIT." The executive committee voted unanimous approval on March 6, 1930, confirmed by the full Corporation on March 12. Also on March 12, Stratton introduced the new president in person at a special meeting of the faculty.

Compton recognized the risks. In the days ahead, he weighed them as a scientist would, logically, against the opportunities, his own goals and abilities, prospects for success or failure. "I admit freely," he wrote to a colleague, "that it is a gamble on which I have staked about all I have. The situation appealed to me in this wise: the Institute needs some things in which I thoroughly believe; the sentiment of the Corporation and of leaders of the Institute is strongly demanding those things; various people whose judgment in other directions I respect have expressed faith in my ability to make a valuable contribution in these directions; if successful, I believe the value of my services to the advancement of science will be greater in the new position than here at Princeton. I should therefore always have felt myself a coward had

I refused to make the gamble. Doubtless the safer policy would have been to demand assurance of success before undertaking the adventure. That is impracticable, but I did convince myself that there is a genuine possibility of making progress."

Some at Princeton thought of Gerard Swope as "a very wise and wily old serpent" for having invaded their Eden and stolen away their man. Most disappointed were members of Compton's own family. Princeton reflected their Presbyterian values, while MIT had no church affiliation and professed no religious faith. Arthur worried that Karl, by accepting the MIT presidency, was sacrificing his career as a research physicist. When Karl told him that he would bring assistants and continue his work, Arthur's instincts told him that this would not work. "If Karl had known," a close colleague remarked some years later, "that he would soon be giving up his beloved research completely and for all time, he would have answered no at once. But he had the pleasant idea that he could set aside a couple of half-days a week for carrying on his researches."

Compton did not lack for people eager to hand out advice. His friend Henry Gale, dean of the Ogden Graduate School of Science at the University of Chicago, cautioned him not to talk too much, at least at the start. He had seen many a new president jump at the chance to speak before having much if anything to say, then become convinced that "the half-baked ideas which he had pulled out of thin air were a great educational doctrine, and from that time on his career as a really great educator was ruined." Some pointed to the MIT faculty's fear of change, how difficult it would be to win their support for adjustments on any level. Then there was the lingering impact of three shocks—Maclaurin's death, the misguided appointment and resignation of Ernest Fox Nichols, and the advent of aging, befuddled Samuel Stratton—that called for urgent attention from the standpoint of morale.

As the MIT community awaited his arrival, Compton prepared a brief, preinaugural greeting for *Technology Review*. The *Review*'s young editor, James Killian ('26), had suggested the idea. "I am looking forward," Compton wrote, "to joining you in the work of the Institute. . . . I join you in this program with enthusiasm for its importance and confident of your support in every effort to make the Institute still more effective in the pursuit of knowledge and the practical applications of knowledge to life." Gerard Swope, meanwhile, wandered the campus beaming about MIT's new acquisition. One day he ran into professor of electric power transmission Vannevar Bush. "Did we make any mistake this time?" he asked. "You certainly didn't," Bush replied, "but you made a honey the time before." Swope went on his way, resolved that nothing—not even Bush's sour remark about the former president—would dampen his confidence about what Tech's future held in store.

Killian sat down with Compton in April 1930, a few weeks after the appointment was announced, using the opportunity to garner for *Technology Review* readers a sense of the human qualities that Compton would bring to the presidency. What struck him most was Compton's personal warmth, his approachability. "The manner of his greeting is so unaffectedly cordial, his smile so disarming, his speech so decisive and unequivocal, that the interviewer is impressed by the exceptional hospitality with which he is received, and more, by the dispatch with which Dr. Compton disposes of his questions." Also conspicuous was Compton's reluctance to talk about himself. Killian waved a bundle of news clippings at him, bits and pieces about his younger years that sources in Wooster, Ohio, had sent around: his fondness for date pie, for Homer's *Iliad* and Walter Scott's novels, for football and baseball—his skill with the spitball. When Compton dismissed it all as hype, Killian probed further. Compton opened up enough to give his interviewer a glimpse of himself, and Killian walked away pleased with his candor, his grasp of issues in science and technology, and his conception of MIT. "Dr. Compton," he wrote, "will be a President who can give voice to the Institute's ideals, who can carry its philosophy to the outside world, who can himself be a symbol of those things for which the Institute stands."

The inaugural was set for June 6, 1930. On the way up by train, Compton and Margaret sat with Gerard Swope and GE's board chairman Owen Young, also en route for the event. Young went over a list of key executive goals. Number one, guarantee continuity. Nurture someone, Young said, to fill in in an emergency, and over time to take over. Compton took this advice to heart. At MIT he would cultivate a small, effective inner circle.

Several thousand guests, about half of them alumni, gathered in the Great Court. Academic hoods and gowns, a wash of colors, blended with the pink, white, and blood-red rhododendron blooms lining the perimeter. The crowd sat hushed, expectant, while inside a nervous Compton confessed to MIT's public-relations officer John Rowlands, "Frankly, I'm terrified to go up there in front of that group. I've never talked to an audience of this kind before." He was used to small groups of physicists, not this vast sea of amorphous humanity. But once on the podium, he settled down. His relative youth—next to the grandfatherly Stratton, who spoke right before—underscored the vigor, the energy, the vitality that many hoped he would bring to MIT. "There is every indication," Compton told his audience, "that only a beginning has thus far been made in the science of discovering and understanding nature and in the art of usefully applying this knowledge." While his vision would require a greater emphasis on basic science, he reassured traditionalists—MIT's engineers—that their interests and concerns were also vital. His message of change, in other words, meant rebalance rather than transform, evolution not revolution.

Karl spent much of the summer with Arthur at the University of Chicago's Ryerson Laboratory, and he corresponded regularly with Stratton about the fall transition. Margaret and the children stayed with her family in Minneapolis, awaiting completion of renovations to the president's house, from bachelor-mode to family-style. She worried about hiring a good household staff and was grateful when Stratton's assistant, Morris Parris, offered to lend "one or two of their faithful helpers" to get her started. Toward summer's end the Comptons canoed for two weeks near the Canadian border, unwinding before the crush of activity that fall.

Compton also began recruiting new faculty. This was a tough but exciting challenge: how to draw the most distinguished scientists away from institutions like Princeton, Harvard, wherever the best happened to be. John Slater, a young theoretical physicist with a growing international reputation, agreed to come over from Harvard to replace Charles Norton as head of the physics department. Compton had tried to get him for Princeton two or three years earlier, but Slater did not like the small-town atmosphere there. Compton's MIT offer, however, was difficult to refuse: department head, full professorship—Slater was then associate professor at Harvard—and a salary of $10,000 a year ("rather unheard-of for those days," Slater remarked). Slater's boss at Harvard, Percy Bridgman, was stunned. He had joked with others about Compton taking on the MIT presidency—"crazy," they said, "another good physicist gone wrong"—but poaching from Harvard's territory raised a different kind of alarm.

Compton and Slater brainstormed about other prospects, in chemistry as well as physics. They hoped, for example, to lure physical chemist Linus Pauling from Caltech as the main draw for a new research laboratory of physical chemistry. Compton had met Pauling in Munich, where one of the pioneers in quantum and atomic physics, Arnold Sommerfeld, raved about him as the best young mind he had worked with in many years. MIT's chemistry department head, Frederick Keyes, was as excited as Compton and Slater, but more skeptical about the Institute's chances. Pauling had rebuffed Harvard, Keyes warned, and recruiting him would be difficult for several reasons. "I believe," Keyes told Compton, "Mrs. Pauling is 'wedded' to California as well as to Pauling. Also our friends A. A. Noyes and Millikan are prepared to offer considerable inducements." As Keyes suspected, Caltech did whatever was necessary to keep Pauling there. Pauling did come to MIT briefly, as a visiting lecturer in 1931–32, but the Institute would have to look elsewhere to boost its long-term prospects in physical chemistry and its coursework in atomic and molecular theory. Both areas, Compton recognized, were essential to the interdisciplinary alliances that must be forged if MIT were to reshape itself as a leader in the world of modern physical chemistry.

The fall of 1930 began with Compton appearing at freshman camp, convocation, a Corporation meeting, and an alumni gathering in Rochester, New York. In October, he gave a speech ("The influence of scientific on religious thought") at a church in Boston, talked to the Alumni Council and the Technology Club of Philadelphia, and appeared on a CBS radio program in New York—all this besides getting the office up and running, making a start on strengthening departments, and raising money, the very thing that Swope had promised he would not have to bother himself with. When he asked his secretary to arrange an appointment with a tailor, she told him, "Sorry, you're booked solid for the next three months." So crammed was his schedule that he awoke one morning realizing that he had forgotten an engagement the night before. His staff grew adept at compensating for such lapses—carrying duplicates of speeches, for example, which Compton sometimes left in taxicabs—and soon he was able to relax much, he said, like Queen Victoria, who never needed to check if there was a chair behind or under her to sit on.

Compton still could not imagine giving up research, no matter how many demands were placed on him. When he arrived in Cambridge, he vowed to set aside two afternoons a week for the lab. "The new work here at M.I.T. is proving to be very absorbing and interesting," he informed his mentor Richardson, now back in England, "and the indications are that I am going to enjoy it very much. At present I am keeping on with my research at Princeton with the aid of a couple of research assistants. . . . The work in progress is in the fields of spectroscopy in the extreme ultraviolet and electrical discharges in gases." He took on four graduate students and signed himself onto the staff roster in experimental physics. Some members of his Princeton group—Joseph Boyce and Edward Lamar among them—moved to Cambridge to work with him. When several weeks passed and they had not laid eyes on him, they called the president's office to ask when he might come over. He showed up a week later, in lab coat, and spent part of an afternoon cleaning out a mercury pump. But when a call came in from his office—a distinguished visitor had dropped in unannounced—he raced back to attend to this, his first line of business.

The episode marked an end to Compton's career as an active experimentalist. While he kept his research outfit going for a while, and even coauthored a few papers with members of the group (Boyce and Lamar eventually joined the MIT faculty), his attention was increasingly drawn elsewhere. A book that he had worked on with Irving Langmuir, one of GE's star research scientists and soon-to-be Nobel laureate (1932), would never be completed, although some of the results were published in *Reviews of Modern Physics*. Compton felt guilty about this, harkening back to what his brother Arthur had warned him about abandoning science for a bureaucratic career. He could not manage both, and in the end he chose office over lab. Yet in 1931 he was elected by starred physicists in *American Men of Science*—the top hundred in the nation—as one of six outstanding among their number. The other five were his brother Arthur, Percy Bridgman, Michelson, Millikan, and Robert Wood of Johns Hopkins. In 1933, when organic chemist James Conant succeeded A. Lawrence Lowell as president of Harvard, Compton commiserated with him about difficult choices leading to sacrifices on the one hand, compensatory rewards on the other.

For a while, Compton continued to direct his reformist energy toward physics. If he could make progress there, he felt, everything else would fall into place. It was not that MIT's department was weak, more that its aims suffered from narrowness and tunnel vision, its prime goal being to drill engineering students in professionally relevant principles. The faculty, which was sizeable on account of the large amount of teaching required, had minimal contact with physicists elsewhere, had few scholarly interests, and lagged in new knowledge. Most came from the ranks of MIT graduates, which promoted insularity, further narrowing, a cycle of limited—and limiting—intellectual perspective. Under Compton, former physics department head Charles Norton, an industrial physicist of the old school, made way for Slater and spent most of his time thereafter managing the Division of Industrial Cooperation and Research (the "research" part of this title was dropped in 1932). Others were reassigned to new duties, let go, recommended for positions elsewhere. Compton's emphasis was not so much on recruiting older, established figures as on the value of promising younger men, those who would inject new vigor, a fresh outlook, a capacity for intellectual leadership.

Slater helped set the tone. A core of distinguished theoreticians and experimentalists remained—Manuel Vallarta, Hans Müller, Julius Stratton, Arthur Hardy, Nathaniel Frank, Bertram Warren, and William Allis—supplemented by fresh recruits during the 1930s. George Harrison arrived in 1930, brought over from Stanford

University. In 1931 came Victor Guillemin Jr., Robert Van de Graaff, Philip Morse, and Wayne Nottingham. Van de Graaff had worked under Compton at Princeton. Robley Evans and George Harvey arrived in 1934. Chester and Lester Van Atta, both on staff by the early 1930s, joined the faculty in 1938, William Buechner in 1942. In 1932–33, when six National Research Council fellows, one international fellow, and one Rockefeller Foundation fellow chose to spend time at MIT, it was clear that the department's reputation was in the ascendance. With Compton's help, Slater built the department over the course of the next two decades into one of the powerhouses of academic physics, turning out as many as one in twelve American physics doctorates.

A tradition established by former president Stratton, of bringing in distinguished international visitors for limited periods, continued through the decade. Paul Scherrer came from ETH Zurich in the fall of 1930 to speak on atomic theory. Peter Debye, also of ETH, visited in the spring of 1932. The two were close colleagues—the Debye–Scherrer method for measuring crystal structure was named for them—and both understood from Compton that their primary mission was to bring the nonphysics faculty and staff at MIT up to snuff on new developments in physics. Only once did Scherrer's frustration show. While lecturing on the "Compton effect," he observed his audience of mostly engineering faculty nodding, smiling, nudging each other as if to say, "Look, our new president discovered *that*!" "No, no!," blurted Scherrer in heavily accented English, appalled by their ignorance—"Zees Compton effect, zees is not Kah Tay [K.T.], zees is Ah Hah [A.H.] Compton!" Jewish physicist and Nobel laureate James Franck visited MIT in December 1933, having abandoned his post at Göttingen and fled Germany following the Nazis' rise to power that same year. Léon Brillouin, of the Collège de France, came in 1938–39.

Physics was Compton's test case. While the other science departments did not undergo such a radical transformation, Compton made several key appointments to reinforce the Institute's research-oriented mission. New arrivals in geology included, in 1931, Louis Slichter (a geophysicist "from a truly scientific standpoint"), Warren Mead in 1934, and Robert Shrock in 1937. Among Compton's hires in chemistry were Robert Hockett in 1935 and Clifford Purves in 1936, with

Isadore Amdur promoted to the faculty from a staff position in 1940. For biology, Compton brought in John Loofbourow in 1940, followed by Francis Schmitt, David Waugh, and Richard Bear in 1941, to broaden a departmental profile that, under William Sedgwick and then Samuel Prescott, had stayed confined largely to sanitary engineering, food technology, and industrial microbiology. Schmitt, Compton said, represented a "crowning stage of a long effort to lay out a forward-looking program in biology." Irwin Sizer joined the biology faculty from a staff position in 1939.

Mathematics, already populated by distinguished scholars, hosted several visitors from overseas, including Otto Szasz in 1933. Szasz, a Jew, had just been forced by the Nazis from his position at the University of Frankfurt. Norbert Wiener, who called him "a lovable little Hungarian," arranged a temporary home for Szasz at MIT and later at Brown University, before helping him find a permanent position at the University of Cincinnati in 1936. MIT's math department, according to Compton in 1933, ranked among the top three or four in the country. Eberhard Hopf spent four years there as an assistant professor before returning to Germany in 1936, attracted by the offer of a full professorship at Leipzig when many of his Jewish colleagues (he was not Jewish) fled in the other direction. While at MIT, Hopf and Wiener developed the important Wiener–Hopf equation combining Hopf's ideas on cosmic radiation with Wiener's in prediction theory. Among the department's other hires in the 1930s was Jesse Douglas, who in 1936 was awarded one of the first two Fields medals—math's unofficial counterpart of the Nobel prize—followed by the coveted Bôcher prize, in 1943. Fellow math professor Dirk Struik inferred that Douglas's dismissal in 1938 occurred because in those days MIT did not tolerate, much less appreciate, brilliant eccentrics in the way that they—Wiener, the iconic wandering, absentminded professor among them—would come to be nurtured, even cherished, in years ahead. Douglas, Struik recalled, was "nervous, emotional, and a remarkably sensitive mathematician. . . . He was fond of anecdotes, some quite funny, mixed with his own special little prejudices; my experience, he said, is that geometricians are as a rule nice fellows, and analysts are nasty. He had his own lifestyle, which did not include coming to class on a regular schedule, so that [department head Henry] Phillips, who stuck to the Runkle discipline of conscientious teaching, had to let him go, to my and others' regret." Douglas went on to teach at City College of New York. William Ted Martin and Norman Levinson, both of whom had served on staff, joined the mathematics faculty in 1938 and 1939, respectively.

Compton's bold hiring initiatives spilled over, as planned, into the engineering departments. Thomas Camp was appointed in civil engineering in 1930, Roy Carlson in 1934. Manfred Rauscher joined aeronautical engineering in 1930, while Charles Stark Draper was promoted from staff to faculty in 1935. New arrivals in electrical engineering included Ralph Bennett in 1931, Edward Moreland in 1935, and Arthur von Hippel in 1936. John Trump, Wilmer Barrow, William Radford, and Gordon Brown, already on staff, joined the electrical engineering faculty at different times between 1936 and 1941. Brown found good reason to regret a remark that he had made to classmate Ken Germeshausen at their graduation ceremony in June 1931, the first presided over by Compton. "Well, look at this diploma, it's got a strange signature—Karl T. Compton—I wonder if it will be worth anything in five years." Thomas Sherwood joined the chemical engineering faculty from a staff position in 1930, while Walter Whitman arrived in 1934, Edwin Gilliland and Ernst Hauser in 1935 (Hauser had been a visiting lecturer in the 1920s). Jerome Hunsaker replaced Edward Miller as head of mechanical engineering in 1933. New hires in that department included Alfred de Forest and Joseph Keenan in 1934, John Lessells in 1936, and C. Richard Soderberg in 1938, with John Hrones promoted from staff to faculty in 1941. In mining and metallurgy, Compton looked to hire and retain experts in the science of metals—what would later come to be known at MIT as materials science—and, by 1940, to eliminate the program in mining engineering as best left to technical schools in the Midwest. Among the distinguished metallurgists brought in to refocus the program were Francis Bitter in 1934, John Chipman in 1937, and Antoine Gaudin, Carl Floe, and John Wulff all in 1939. Wulff moved over from physics, where he had been hired in 1931 to work on the physics of metals. Compton recruited Frank Lewis for marine engineering in 1936 and, to reinforce the program in meteorology begun by Carl-Gustaf Rossby in 1928, Sverre Petterssen in 1939 and Bernhard Haurwitz in 1941.

Architecture and parts of the humanities, meanwhile, which had coasted if not languished under Stratton, found renewed vigor under Compton. For architecture, Compton brought in John Burchard via the Albert Farwell Bemis Foundation in 1938, Walter MacCornack in 1939 (to replace the retiring William Emerson as dean), and Alvar Aalto in 1940, while a related innovative program—city planning, begun in 1933—found leadership under Frederick Adams. Compton made little effort to bolster the softer humanistic disciplines—English, history, and modern languages—but he worked hard to strengthen economics and business administration. Among key hires in general economics were Ralph Freeman in 1931 and Paul Samuelson, a former student of E. B. Wilson, in 1940. "Samuelson . . . is one of the ablest young fellows I have ever met," Wilson told Compton. "I am sure he will have a distinguished career whether he stays with you or goes elsewhere. It seems to me that it is particularly appropriate for MIT to have in its department of economics persons who understand science and mathematics." Sociologist Edwin Burdell joined the economics group in 1934, marketing expert Ross Cunningham in 1937, and social anthropologist Conrad Arensberg in 1938. Compton's interest in the emergent field of industrial relations led him to recruit Rupert Maclaurin, son of former president Richard Maclaurin, in 1936 to develop a program in this area, assisted in due course by Douglass Brown, hired in 1938, and Paul Pigors in 1941. Charles Myers was promoted to the faculty from a staff position in 1941. To capture its broadened scope, economics and statistics was renamed economics and social science in 1934. In February that year, responding to what he called "a general feeling in the air," Compton urged more emphasis on the "social or human point of view" in economics and hoped that this would set a trend with respect to humanistic studies at MIT.

Compton's laissez-faire approach to the softer humanities probably did not influence many students, a number of whom, especially the aspiring scientists, were already inclined to think of these disciplines as irrelevant fluff. Physics student Richard Feynman ('39), later a Nobel laureate, put it this way: "When I was a student at MIT I was interested only in science; I was no good at anything else. But at MIT there was a rule: You have to take some humanities courses to get more 'culture.' Besides the English classes required were two electives, so I looked through the list, and right away I found astronomy—as a *humanities* course! So that year I escaped with astronomy. Then next year I looked further down the list, past French literature and courses like that, and found philosophy. It was the closest thing to science I could find." But many students continued to appreciate MIT's efforts to broaden the curriculum, and in so doing to dodge the twin pitfalls of overspecialization and technical narrowness.

In 1933, Compton voiced confidence about MIT's research program "adding prestige, contributing to human welfare, and . . . providing a most valuable adjunct to the educational program . . . directly felt by those who participate in the work and . . . by the stimulus given to the entire body of staff and students." His partiality toward the physical sciences showed, sometimes, in the projects that he chose to highlight. Among those he mentioned most often was the research on fog dissipation, high-altitude air masses, and electronic communications carried on by Edward Bowles, Julius Stratton, Henry Houghton, Wilmer Barrow, Carl-Gustaf Rossby, and others at the experimental research station set up on Colonel Green's estate, Round Hill, in Dartmouth, Massachusetts. Also at Round Hill, Robert Van de Graaff built his multimillion-volt electrostatic generator, working with colleagues Chester and Lester Van Atta, John Trump, William Buechner, and others on applications ranging from cancer radiation therapy to electric power transmission over long distances. Often cited among projects on the Cambridge campus were Harold Edgerton's stroboscopic high-speed photography capturing still images of microscopic and ultramicroscopic processes; Stark Draper's engine- and propeller-vibration studies and, in collaboration with Bowles and others, techniques for blind landing and instrument landing of airplanes; Gordon Brown's work on servomechanisms and principles of automatic control; Vannevar Bush's work, with Samuel Caldwell, Harold Hazen, and others, on mechanical and electrical computing machines; Arthur von Hippel's on semiconductors; George Harrison's in spectroscopy, what Compton called science's master key; the work by Julius Stratton and Philip Morse on ellipsoidal wave functions and propagation of ultrahigh frequency waves; Morse's on atomic structure and acoustics; William Allis's on the theory of the electric arc; John Slater's on the electronic structure of metals; Manuel Vallarta's studies of cosmic rays, in collaboration with astronomer Abbé Lemaître; Hans Müller's work on light scattering, Francis Bitter's on atomic arrangement in alloys, and Robley Evans's on radioactivity, with an emerging emphasis on medical applications and development of a cyclotron.

The clearest pattern underlying all these projects was their multi- and cross-disciplinary fluidity, the blurring of lines between science and engineering and between fields in the physical and life sciences. This was as true in chemical engineering, mechanical engineering, civil engineering, aeronautical engineering, chemistry, geology, and metallurgy as in physics and electrical engineering, the two fields with which Compton was most familiar. Besides rubber research and vulcanization of latex, chemical engineers joined forces with geologists to tackle solar radiation and high-temperature colloid chemistry, "a practically virgin field." Mechanical engineers worked in traditional areas—internal combustion engines, for example—but also with metallurgists on surface fatigue and resistance in metals. Civil engineers, much interested in hydraulics, worked with geologists on seismic risk to water supplies. Biologists carried out nutrition and vitamin studies, but also joined physicists in radiation-therapy research, with the concept of biological engineering articulated and undergoing definition for the first time by the mid-1930s. Chemists cooperated with physicists and engineers on studies in the thermodynamic properties of steam and cryogenic (low-temperature) approaches to investigating the heat capacity of salts, rotation of atomic groups in molecules, and coefficients of temperature dilation. In the spring of 1939, Compton conducted an informal census that identified 529 research projects underway at MIT, many of which coalesced into programs that broke across departmental boundaries.

In order to thrive, a research environment of this sort required not only first-rate personnel but also a network of well-equipped workspaces. An early prototype came with the opening, in 1932, of an interdisciplinary facility for physics and chemistry, the George Eastman Research Laboratories, which Compton proudly called "the finest in the world for their purposes." The Eastman group was the first of a series of Compton-inspired improvements to the Institute's physical plant. After persuading the executive committee to set aside just over a million dollars from unassigned capital to fund the project, he worked with architects and select faculty members to plan the layout of each nook and cranny. To accommodate delicate experiments, he helped design a structure free of vibration. The blueprints that he personally approved incorporated lecture rooms, shop facilities, a library, even spaces to foster social interaction between faculty, staff, and students. Richard Maclaurin had chosen a phrase from Pliny the Elder to adorn the mantelpiece in the president's office, and Compton wanted a fitting capstone—also from the classics—for the laboratories' entrance lobby. He came up with a passage from Virgil's *Georgics*, roughly translated: "Happy is he who has been able to learn the causes of things and has cast beneath his feet all fears of inexorable fate and roar of greedy Acheron." Inspired by his father Elias, Compton often sought wisdom in the writings of the ancients. On his office desk sat a favorite quote from Aristotle: "The search for Truth is in one way hard and in another easy. For it is evident that no one can master it fully nor miss it wholly. But each adds a little to our knowledge of Nature, and from all the facts assembled, there arises a certain grandeur."

For Compton, the process of completing the Eastman labs underscored problems in the Institute's management scheme, specifically the power-sharing arrangement with Corporation chairman Samuel Stratton. Early on, when Compton asked him where each might focus their energies, Stratton had replied: "Well, I will look after the things that I'm interested in, and you do the same." This, Compton realized, was a recipe for administrative paralysis. But he remained deferential to Stratton in public, in private polite yet firm, careful to preserve the older man's dignity while retaining his own hold on policy. "In carrying forward this program," he had told the alumni just before taking office, "you will still have the leadership of Dr. Stratton, whose experience, judgment and intimate knowledge of the problems of the Institute are invaluable. I have been happily associated with him for many years in various scientific societies. His desire to have me cooperate with him in the Administration was a major consideration in my acceptance, and I am convinced that the new administrative arrangement will be a personally happy one, just as I believe that it is in effect a wise one." When

Stratton vetoed plans for two shops in the Eastman laboratories—one for physics, one for chemistry—Compton applied his flair for diplomacy. "Dr. Stratton has given orders that there is to be a single shop," Compton told disgruntled physicists, "but I don't see any objection to having rows of shelves down the middle of one large shop area, to separate the chemistry and physics sections." The shelves were installed as Compton suggested, and Stratton was never the wiser for it.

While it lasted, the "two-headed operation" complicated matters for Compton. But he accepted it, made it work, established a good balance, and carved out a cordial, trouble-free relationship with Stratton. Following Stratton's death in October 1931, Compton moved quickly to build a new model. He revived the concept of a vice president—two-headedness, still, but with a difference—realizing that he would need a backup, someone to fill in during emergencies and who would qualify, in time, as a successor. He sought Vannevar Bush's advice. Intrigued by Bush's famously blunt personality, he wanted to sound out his opinion on who might be worthwhile. Bush suggested himself. "But I thought you wanted to be a research professor," said Compton. "No," Bush replied, "I've done some research and I like it, but I'm interested in the administrative end of things fully as much." As Compton thought more, he saw that this just might work: two men, opposite personalities, styles that complemented each other. Bush got the job, approved by the executive committee in May 1932.

They had met for the first time just a few months earlier, when Bush charged into the president's office to protest a policy change with respect to faculty consulting. Compton had imposed a one-day-a-week limit on outside consulting plus a 50 percent tax on outside income, the proceeds to go into a special fund, known as the Professors' Fund, to help raise faculty salaries across the board. Bush bristled about the policy as an "ukaz," a decree or edict in the manner of Russian autocrats from czars to religious patriarchs to (most recently) Comintern officials. As no one had talked, much less negotiated, with the faculty, in he marched to confront the president. "I'm leaving MIT," he announced. Compton asked why. "Look," Bush said, "when I joined this outfit, I had an agreement under which I was allowed to do a certain amount of consultation while I carried on my MIT affairs. I just won't work for an outfit that abrogates its obligations unilaterally. And furthermore, I won't stay at a place that has so little sense it tries to clip down to nothing proper consultation on the part of its engineering professors." Compton offered to make an exception for Bush, who only grew more irate. "Your whole damn fool plan," Bush retorted, "will dissolve in a welter of exceptions." To prove his point, someone whom he considered one of MIT's finest teachers—physics professor Louis Young ('15)—had read the new policy, donned his hat, walked out the front door, and never returned (he ended up at the Gillette Company).

Bush stormed out of Compton's office, but he did not follow through on his threat. "I'm essentially a mild chap," he said, with more than a hint of irony, "but I'm occasionally charged with being belligerent." His bluster was meant to be constructive, to get Compton thinking about consequences of actions possibly not well thought through. Compton admired his outspokenness and Bush, once he cooled down, was grateful for Compton's sane, reasonable approach. The encounter did not appear likely to cement good working relations, much less a close friendship, but according to Bush it did both. "The reason was that Karl met that issue in such a kindly manner, and with such obvious willingness to look at all sides of things that I changed my tune completely." In any event, Compton's tax on outside consulting lasted just a few years, abandoned as impractical but also as achieving its primary goal—to establish the core principle that even where consulting brought MIT useful benefits, the faculty's first obligation was to the Institute.

Compton and Bush got along well. Their personalities, disparate yet smoothly coexistent, kept the Institute moving efficiently. One colleague described the relationship as a near-perfect balance of contrasts: "Karl knew that he needed a tough right-handed man, and Bush brought to the administrative center great talents for policy making and detailed operation. His native shrewdness and realistic understanding of human nature excellently supplemented Karl's primarily idealistic approach, and he was not so likely to become overly sympathetic in situations involving the personal feelings of others. Bush, with a somewhat fiery temperament, was much the more practical of the two men, but sometimes rubbed people the wrong way. When a proposed action was discussed he would consider first the workability of a given procedure, and with great acumen would analyze the power realities of a competitive situation. Karl would reluctantly agree with Bush when such matters were pointed out to him, but would still want to give every potential villain the benefit of the doubt, for to him there were no villains."

Compton took to the presidency with ease, poise, and the spirit of a reformer. He moved quickly to group the Institute's diffuse, proliferating string of departments and units into some sensible, cohesive framework. Immediately following Stratton's death in October 1931, Compton notified the executive committee that he would bring a special item of business to the next meeting: a proposal to create three schools—Engineering, Science, and Architecture—each run by a dean, and two divisions—Humanities and the Division of Industrial Cooperation (DIC)—each run by a director. The schools would consist of degree-granting departments while the divisions would oversee activities, including instruction, that were more service-oriented. Deans of the professional schools would be responsible for developing and maintaining strong faculties in their respective departments, for preparing sound budgets, and for building robust teaching and research programs. The humanities group had a special mission to enrich students' educational experience, while the DIC would stimulate problem-solving partnerships with business and industry. Compton proposed another unit, the Graduate School, whose dean would oversee policies with respect to admission and handling of graduate students, examinations, degree requirements, and fellowship awards, but carry no responsibility for curriculum and no direct degree-granting authority. An administrative council consisting of the president, vice president, deans, directors, bursar, and faculty chairman would meet weekly as a sort of cabinet to review and act on matters of policy. These changes, Compton held, would not only streamline, consolidate, and simplify administrative functions but also bring science, architecture, and other areas into parity with engineering, thus helping to define and build "a fine university . . . in which a great engineering school is one important part, but not the whole."

The executive committee eagerly adopted Compton's plan in February 1932, along with the reasoning behind it. The full Corporation followed suit in March. The new layout looked this way:

School of Engineering

Aeronautical engineering
Building engineering and construction
Business and engineering administration
Chemical engineering
Civil and sanitary engineering
Electrical engineering
Electrochemical engineering
Mechanical engineering
Meteorology
Mining and metallurgy
Naval architecture and marine engineering
General science and general engineering

School of Science

Biology and Public Health
Chemistry
Geology
Mathematics
Military science and tactics
Physics

School of Architecture

Architecture
Architectural Engineering
Drawing

Division of Humanics

Economics and statistics
English and history
Modern languages

Even though it overwhelmed the other schools in sheer size, engineering gradually saw itself as a partner with science, architecture, and humanities in reconfiguring the Institute. For Compton the emphasis was on alliance, particularly in view of growing ferment between fields and across departments attached, for administrative convenience, to different schools. Fluid crossovers were to be encouraged, barriers avoided.

The structure was amended, now and then. In 1933, the general science and engineering options shifted from the School of Engineering to the School of Science, city planning was added to the School of Architecture, and military science and tactics moved from the School of Science to the Division of Humanics. In 1934, the Division of Humanics became known as the Division of Humanities (humanities was the term that Compton originally preferred), economics and statistics became economics and social science, and electrochemical engineering moved from physics to mining and metallurgy. In 1936, general studies was added to humanities in order to "make sure that in the effort to perfect the successive steps in our technical training we never lose sight of the need for a corresponding acquaintance with the great field of the humanities." In 1937, mining engineering separated from metallurgy and meteorology combined with aeronautical engineering. Mining engineering was eliminated altogether in 1940, Compton calling it "pretty much in a rut of a routine educational program of the traditional type." The sanitary side of civil and sanitary engineering was dropped in 1940, in line with Compton's view that Harvard, through its School of Public Health, "was doing a better job at teaching sanitary engineering . . . and there were better ways of handling the few students who remain at the Institute in this field."

Compton's choices for the inaugural deanships were Vannevar Bush (engineering), Samuel Prescott (science), and William Emerson (architecture). In announcing the appointments at a faculty meeting, Compton encouraged comment and queries about what the deans might do, or steer clear of, but insisted on his prerogative as chief executive to select deans without debate or consultation. The appointments took effect on July 1, 1932. One surprise was Prescott—generally liked, but viewed by some as an intellectual lightweight. "Just think," observed one faculty member, "appointing a man Dean of

Science who probably doesn't know that there is any such thing as 'higher Mathematics.'" Some expected a physicist or a chemist for this post, but Compton may have strayed for two reasons: first, to broaden the base so that the administration was not weighted so heavily toward the physical sciences; and second, aware of Prescott's close ties to former president Stratton, to retain continuity and a reservoir of good will among Stratton's close allies. Compton granted Bush two portfolios—three at one point, actually, as he was assistant treasurer from January to June 1934 as well as vice president and dean of engineering. The value of Bush's advice outweighed any concern about overlapping responsibilities or potential conflicts of interest. Prescott continued to head biology as dean of science, and Emerson, as dean of architecture, remained head of the architecture department. Harry Goodwin, MIT's first dean of graduate students, became the first graduate school dean.

Compton, apparently in no rush to appoint anyone for humanities, assigned Bush, Prescott, and Emerson to jointly manage that side of the curriculum. Edwin Burdell, a member of the economics faculty as of 1934, agreed to serve as head of humanities in 1937, by which point Compton had decided on dean rather than director as a more appropriate title even though humanities did not have school status. "The country is becoming increasingly faced with human problems," Compton wrote in announcing Burdell's appointment, "many of which are closely related to technological developments. The engineer, architect, and scientist have been so successful in the latter field, and have come naturally and in such relatively large numbers into positions of great social responsibility, that they realize the need of increasing attention to two aspects of the training of their successors: first, development of a high sense of responsibility and understanding in social matters [and] second, development of those spiritual qualities that bring permanent satisfaction in living." The importance of the post, combined with its intangible qualities compared with science, engineering, and architecture, led Compton to fret about what title to use. Dean of liberal studies, humanities, humanistic studies, liberal arts, general studies, liberal and humanistic studies, social science and humanities, or non-professional studies? Compton suspected that "liberal arts" would "probably give rise to insinuations that we are attempting to encroach on the field of the liberal arts college or university," while any compound name struck him as "too long" and "non-professional" sounded "negative and does not carry . . . publicity value and educational stimulus." Compton settled on dean of humanities as the best but far-from ideal option, considering that "many people . . . feel the name has been so abused as to be undesirable." Burdell left MIT in 1938 to become director of Cooper Union in New York and was succeeded as dean by U.S. diplomat and Latin-American specialist Robert Caldwell, Compton's old college friend from Wooster, the one who in 1909 had helped persuade him that Princeton was a fine place to go for graduate study.

To round out his administrative team, Compton appointed a series of assistants. These joined his staff, one at a time, from among recent MIT graduates, "the cream of the managerial output of the Institute." The first assistant, Carroll Wilson ('32), came aboard right after graduating. Each assistant stayed, typically, two or three years before moving on to careers elsewhere, replaced by some other gifted young graduate. Wilson would later serve as first general manager of the U.S. Atomic Energy Commission and, in 1959, return to MIT on the faculty in industrial management. Allen Horton ('36) was Compton's assistant from 1936 to 1939.

Also drawn into Compton's inner circle was James Killian, in a looser, more informal way than either Wilson or Horton. Killian was neither a scientist nor an engineer—like Wilson and Horton, he came out of Course XV, business and engineering administration—but Compton sensed in him someone whose communication skills and insights on MIT's role in the world could prove valuable. As a student, Killian had slogged through coursework in thermodynamics, machine design, electrical engineering—"I machined," he wrote years later, "an elegant but useless specimen in the now-discontinued machine shop"—and ended up doing poorly enough in mathematics and applied mechanics that he had to repeat these subjects in summer school. Even though he remained identified by choice with the Class of 1926, he did not graduate until January 1929, a lag that often reminded him of Robert Frost's observation that education can mean just "hanging around until you've caught on." Killian was drawn to the social sciences and especially to his business management professor, Erwin Schell, whom he described as "an enthusiasm-amplifier and an optimist, a spur, and above all a personality of exceptional force, warmth, and light," a man with an expansive, imaginative way of looking at the world, human nature, and social values. This spirit suffused Killian's own work on *Technology Review*, whose scope and reach broadened under his editorship. No longer a parochial alumni mouthpiece, it emerged as a bona fide science magazine with the potential to compete with the likes of *Science*, *Scientific Monthly*, and *Popular Mechanics*.

Compton, for one, was impressed. At his first meeting with Killian in April 1930, he had brought up a piece by Joseph Mayer, "Can Americans be scientists?" published in the *Review*'s March issue. He let Killian know how much he appreciated this sort of provocative, accessible approach to scientific topics and soon sought Killian's assistance, over and beyond his work as *Review* editor, in recasting MIT's image. Too many, Compton said, knew nothing about the Institute. Some thought of it as a trade school, others as a refuge for eggheads, social misfits, or the technically super-talented. Compton wanted Killian's help in sculpting a more balanced portrait, as well as in forging dynamic ties between the Institute and its alumni.

Their first plan involved a set of publicity brochures. The project got Killian out of the *Review* office and into the Institute corridors. He poked his head into classrooms and offices, talked with faculty, staff, and students, gained as complete a sense of the place as he could get. He and Compton brainstormed about how to reach out with better clarity and finer nuance, how to improve what Compton referred to as Tech's "advertising medium." Compton wanted especially to draw attention to graduate work, to interdisciplinary ties between science and engineering, and to the Institute's efforts in both these areas. Killian, in turn, created a series of bulletins and brochures that mirrored Compton's priorities. Earlier efforts along these lines had been hasty, dull, dun-covered affairs. But Killian's productions were sharp, with lots of eye-catching graphics—aerial views of the Institute, laboratory interiors, inventions and equipment (the Van de Graaff accelerator, Bush's differential analyzer, Edgerton's high-speed camera), student lounges—accompanied by clear, succinct text likely to grab a reader's attention, the skills that Killian had honed as *Review* editor.

Both men kept a check on the other to keep their message on-point. Notes flew back and forth. "Why hammer the 'national institution' idea? Harvard has worked that to death. . . . Don't say it directly, ever. Let it creep in . . . by indirection." "How about the discovery idea?" "MIT is a trustee for the public, holding in trust a stock of brains and ability. For this reason it is more interested in long-run basic research than in immediate applied research. The second is important, welcome, and readily done. The first will in the long run bring greater good to the community." Compton placed a giant question mark next to one of Killian's ideas. Killian had wanted to include language in one bulletin—"We should also be very happy to have you suggest on the enclosed business reply card the names of any boys who are of the right timber for Technology"—that Compton vetoed as clumsy, crude, and liable to misinterpretation. Killian was as candid about Compton's lapses in judgment or style.

As they worked together to craft a coherent message, they grew to know and trust each other, to build on shared ideas, to polish a sensible public-relations strategy. They schemed up ever more ingenious ways to project news about the Institute. "I agree with you," Compton wrote Killian in 1934, "that it would be highly desirable to have the *Technology Review* placed in club cars, particularly on the eastern roads. I have had occasion to notice the large extent to which it is read in the club cars of the New Haven railroads, and I believe its unique nature in this field gives it an appeal and an importance which should make a strong argument for its inclusion in the list of magazines carried in the railroad club cars." Sometimes these ideas materialized, sometimes not, but they kept percolating. Killian, along with news director Jim Rowlands, assisted Compton with a proposed book, *Put Science to Work! A National Program.* It was never published, but working on it helped refine various ideas about science, technology, and education that grew central to Compton's vision for the Institute, and for national policy. From time to time, Killian played ghost-writer. Along with Rowlands again, he composed a piece—"Science, source of hope"—signed by Compton and published in *The Rotarian*. Compton, embarrassed to receive payment for this, sent the money on to his assistants with the comment, "Obviously this check should not go to me, for I contributed nothing except the waste basket out of which you picked some ideas and facts which made the article appropriate to my name."

Killian forged a good relationship, too, with Compton's gruff, intimidating vice president. Bush's office sat directly across from Compton's. As a student Killian had avoided abrasive, hardcore teachers, gravitating instead toward gentler, patient men like Jayson Balsbaugh, so he may have felt a bit cowed in Bush's presence at first. But Bush took a liking to him. Killian ran errands, made himself generally useful. The dynamic worked well. Killian was an avid listener, Bush loved to talk. Killian became a sounding board for Bush's sometimes wild ideas on science and education. "I can't help inventing," Bush kept saying, not only in reference to his theories but to his work on the differential analyzer and other projects. Invention, that eureka moment, gave him the same thrill that he imagined poets felt in capturing emotions, the human spirit, through perfect harmony of language, phrasing, and rhythms. The two spent time together at Bush's summer home in Dennis, on Cape Cod, where Killian helped out in the garden. One summer they planted fruit trees and, during coffee or lunch breaks, Killian listened as Bush went on about radical new courses of study and experimental colleges. Gradually, Killian moved from listener to listener-participant.

Far from dampening his enthusiasm, or inclining him toward caution, the nationwide depression of the 1930s pushed Compton into proactive mode. In the fall of 1932, just as his restructuring plan took effect, the economic downturn—endowment depleted, student enrollment in decline, teaching staff laid off, and hiring frozen—reached the point where no one knew where his next paycheck would come from. Compton announced a salary reserve plan in which 10 percent of each professor's salary above five hundred dollars would be withheld and deposited, along with taxed receipts from outside consulting, into an emergency pool—a rainy-day fund to be drawn on in rough(er) times, if such materialized. Compton felt, he said, like the man who told his wife about killing his dog. "Was he *mad*?" she asked—"Well," replied the man, "he wasn't exactly pleased." While the joke did not lift anyone's gloom, it helped soften the blow. Other schemes during this difficult period included the Technology Loan Plan, announced by Gerard Swope in June 1930 to provide financial assistance to needy students. Donors included George Eastman along with alumni Charles Hayden ('90), Alfred P. Sloan Jr. ('95), and four du Ponts—Coleman ('84), Pierre ('90), Irenée ('97), and Lammot ('01).

Between 1930 and 1934 operating income dropped by nearly $400,000, from $3,030,000 to $2,647,000. Student enrollment fell about 22 percent, from 3,209 in 1930 to 2,507 in 1934, while the teaching and research staff suffered a 15 percent decline (from 563 in 1930 to 498 in 1933). A slow recovery took place after 1935, so that by 1940 there were 3,138 students, just shy of 1930 levels, and 681 teaching and research staff. Withheld reserve funds were returned to the faculty intact—no one lost a penny, and some ended up appreciating the scheme as a kind of enforced savings plan. The teaching and research staff grew at a faster pace than the student body, reaching in 1940 a level 21 percent higher than 1930 numbers. In 1936 Compton outlined his fledgling reforms, none of which was he willing to scale back, and framed the challenge ahead as primarily financial. "The future vigorous and healthy growth of the Institute depends upon the finding of some means whereby these intellectual babies . . . may be nourished and developed to healthy maturity. The Institute is in somewhat the position of an athlete who has undergone a severe course of training and is now at the peak of condition to perform." With this in mind, he launched a campaign to raise twelve and a half million dollars for expansion.

The physical consolidation of the campus, a cherished goal ever since Maclaurin's time, was finally realized in the late 1930s when the old Rogers Building on Boylston Street in Boston was sold to New England Mutual Life and the School of Architecture moved to the Cambridge campus. In April 1937 Compton contracted with Welles Bosworth, original architect of the Cambridge campus, to design a structure in Bosworth's neoclassical style on Massachusetts Avenue between MIT's Pratt School of Naval Architecture and Guggenheim Aeronautical Laboratory. The new building, capped by a smaller version of the original Great Dome, cost $1,338,000 and was ready for occupancy in 1938. After some wrangling over inscriptions and ornaments, a committee comprised of Compton, Bush, Samuel Prescott, and professor of English Henry Pearson decided that the lettering over the entrance—soon to become MIT's primary point of access, replacing the doors in the Great Court—would consist of William Barton Rogers Founder under Massachusetts Institute of Technology, with School of Architecture engraved off to the left. Suggestions for lettering on the frieze around the interior base of the dome shifted from Dedicated to the advancement of science and its practical applications in arts and industry and Dedicated to the advancement of science and its application to industry and the arts in December 1937, to Built for the advancement and development of science its application to industry art agriculture and commerce, in February 1938, to the final choice—Established for advancement and development of science its application to industry the arts agriculture and commerce. Bosworth's plan for four statues in the lobby reached only as far as construction of concrete pedestals (still present but empty more than seventy years later). Abraham Flexner, influential cofounder of the Institute for Advanced Study at Princeton, promoted the statuary idea, citing "the inspiration it would give to young scientists to see the field of science presented in that way." But Compton vetoed a suggestion by Columbia's president Nicholas Murray Butler that Sophocles would be a better choice than Socrates to represent the humanities. Nor did Compton much like the all-too-many options floated with respect to science. "As a scientist it is hard to know from which epoch to pick the distinguished representative. From our near modern era I suppose that Faraday would be outstanding; back of him probably Isaac Newton and then Galileo." Without consensus, and with money in short supply, the plan was abandoned.

Compton's vital priorities, anyway, were research and the curriculum. Two trends that had begun in the 1920s—the popularity of pure science courses, and the growing number of students working toward advanced degrees—continued in the 1930s. Whereas in 1930 about 80 percent of students opted for engineering courses, 12 percent for science, and 6 percent for architecture, in 1939 about 78 percent opted for engineering, 18 percent for science, and 4 percent for architecture (a small number in both years were unclassified). The trend in graduate education was sharper. The number of graduate students in 1930, 523, comprised 16 percent of the student body, while in 1939 that number had increased to 720, or 23 percent. Growth with respect to higher degrees earned was even more striking. While the number of undergraduate degrees remained constant, the number of master's degrees awarded rose by 26 percent and the number of doctorates almost tripled. A total of 1,609 master's degrees and 165 doctorates were awarded between 1921 and 1930, 2,162 master's degrees and 472 doctorates between 1931 and 1940. The first master's degree in city planning (M.C.P.) went to James McKeever in 1936. The first Sc.D. in meteorology went to Chaim Pekeris in 1934, in ceramics to Earl Wilson in 1935, and in petroleum engineering to Eldon Dunlap in 1937. Of the 258 Ph.D.'s awarded between 1931 and 1940, 146 were in chemistry, 48 in physics, 25 in mathematics, 22 in geology, and 17 in biology. Of the 214 Sc.D.'s, 78 were in chemical engineering, 30 in electrical engineering, 25 in metallurgy, 18 in physics, 13 in mechanical engineering, 12 in civil engineering, 7 in ceramics, 6 in meteorology, 5 in chemistry, 5 in aeronautical engineering, 4 in geology, 4 in mining engineering, 3 in mathematics, 2 in sanitary engineering, 1 in electrochemical engineering, and 1 in petroleum engineering. "This increase in postgraduate education and research," Compton wrote in 1936, "is the Institute's answer, and is industry's demand to meet the problem of ever increasing technical specialization. Not only are students with postgraduate training more readily employable, but in some branches of engineering and nearly all branches of science it is now very difficult even to get a start in a career without such training. The Institute's leadership in the field of technological education therefore depends, in a very important manner, upon its leadership in postgraduate training and research."

This emphasis on graduate education helped shape campus residential policies. The dormitory system, already inadequate for undergraduates, shifted to accommodate graduate students. Part of the dormitory behind the president's house was given over to graduate students in 1933, then converted entirely to a graduate house in 1934 with accommodations for 206 students under the supervision of chemistry professor Avery Ashdown. Most undergraduate and graduate students still lived off campus and the 5:15 Club, founded in 1933, provided a popular social outlet for commuters, especially for those who did not belong to fraternities. In 1937, MIT purchased and converted the old Riverbank Court Hotel into Graduate House with accommodation for around 370 students and with Ashdown, again, as housemaster. After Graduate House opened in September 1938, the old dormitory was reconverted to undergraduate use exclusively as Senior House, with Walter Wood ('17) as housemaster.

Efforts to improve social conditions focused, too, on extracurricular life. Added to the list of usual sports in the 1930s were pistol shooting and skiing. A sailing pavilion erected in 1936 housed 48 dinghies and offices and classrooms for the Nautical Association. Sailing quickly became MIT's most popular recreation. The association began with around 70 members and counted 520 by 1938. A swimming pool opened in August 1940. Other new athletics facilities included the Barbour Field House (1934), with squash courts, locker rooms, and showers; the Briggs Field House (1939), for track and other field sports; and Rockwell Athletic Cage (1948), with enclosed track and facilities for indoor sports. Among the new student professional groups were chapters of the Associated General Contractors of America, 1932; Pi Tau Pi Sigma, honorary Signal Corps fraternity, 1934; and Eta Kappa Nu, honorary electrical engineering society, 1939. A club called Grupo de habla española appeared in the early 1940s, its purpose being to spread Spanish culture. Latin American House was established as a living group in 1943. The Outing Club had "no objective except pure fun." "Contrary to the opinion of many," *Technique* editors wrote in 1941, "Tech is not devoid of social life. Very few weekends go by without at least one function. Barn dances and cowboy frolics join with the more formal parties in providing good entertainment for all who like to dance." After fifteen years in hibernation, the Tech Show was revived in 1948.

The trend toward geographic diversity, in evidence since the early 1900s, gained momentum in the 1930s. Of six national regions—North Atlantic, South Atlantic, South Central, North Central, Western, and Territories and Dependencies—only one, North Atlantic, showed a decline in student numbers in 1940 as compared with 1931. Most students still came from the North Atlantic—79 percent in 1931, 70 percent in 1940—with the proportionate reduction compensated for by increases in enrollment from other parts of the country, especially the North Central and Western regions. The foreign contingent, meanwhile, showed moderate growth, from 182 in 1931 (6 percent of the student body) to 224 (8 percent) in 1940. The number of countries represented rose as well, from 38 in 1931 to 47 in 1940. In 1931, the largest group of foreign students came from Canada (34), followed by the Soviet Union (25), China (17), Cuba (14), and Mexico (13). Compton, a committed internationalist in science and politics, worked hard to open opportunities for students from overseas. In 1941 most foreign students still came from Canada (37), followed by China (26), the Philippines (18), India (14), Cuba (13), Turkey (12), and Brazil (11). The number of Soviet students dropped precipitously after 1932 and by 1940 there were none, reflecting shifts in national and international politics during the 1930s. Margaret Compton recalled an incident in 1931 or 1932 when two dozen Soviet students sang Russian songs after supper at the president's house—"One of our faculty wives, who was Russian, hung her head as she sat in the back of the room and said, 'You wouldn't applaud if you could understand what they're saying.'"

The number of black students dropped steeply during the 1930s, off by a factor of two or three from the decade before. A mere handful, half a dozen or so, earned bachelor's degrees. These included John Caulder ('31) in mechanical engineering, Arthur Jewell ('32) in general engineering, Frederick Drew ('34) in general science, George Hines ('36) in civil engineering, James Ames ('37) in chemistry, and Joseph Dunning ('37) in aeronautical engineering. Jewell and Ames went on to earn master's degrees, in 1934 and 1948 respectively, while Caulder was a graduate student in 1936–37 but left without a further degree. The lone black doctoral student, William Knox, was awarded the Ph.D. in chemistry in 1935. In response to a survey by the American Council on Education in 1940, MIT characterized itself as a "Technological and scientific school, for men and women, privately controlled; non-sectarian. Prevailing race White. Location urban." Among the tiny group of blacks to earn undergraduate degrees in the 1940s were William Bowman ('44) in mechanical engineering, Lucien White

('44) in chemical engineering, Thomas Meeks ('47) in physics, Victor Ransom ('48) in electrical engineering, and Yenwith Whitney ('49) in aeronautical engineering. Henry Hill, awarded the Ph.D. in chemistry in 1942, would become, in 1977, the first black president of the American Chemical Society.

Black students, excluded by racial covenants from membership in the usual fraternities, often joined black-only, national societies such as Kappa Alpha Psi and Omega Psi Phi. George Hines and James Ames, both Kappas, socialized offsite with fraternity brothers from Harvard, Boston University, and other nearby campuses. MIT's nonfraternity groups, on the other hand, generally welcomed blacks into their ranks. Ames was on the varsity track team and Dunning on the freshman hockey team, while Drew joined the Liberal Club and the freshman tug-o'-war and rifle teams. Thomas Meeks was treasurer of the Rocket Research Society. But there were pockets of racial insensitivity, too, as when the Tech Show for 1931 featured white actors in blackface aping spirituals and cannibal dances. Compton himself lapsed on occasion, as in 1933 when he approached Erwin Schell for advice on detecting deceptive or fraudulent claims by companies such as A.I.G. Chemical Corporation. "If I remember correctly, when those who sold these securities were subjected to an investigation, their defense was that the 'nigger in the woodpile' was clearly stated in the advertisement and that any investor walked into it with all information before him. Mr. Buffum [of the Chemical Foundation] is interested to find out whether your . . . men can detect this 'nigger.' I will confess that I myself have been unable to do so, but I am not an expert in such matters." In February 1935, MIT's Catholic Club staged a popular farce entitled *A Nigger in the Woodpile.* Such discourse was tolerated, if not officially sanctioned. The *Tech*, meanwhile, editorialized in 1934 about fascist and Nazi ideology in Europe exacerbating already high levels of racial intolerance in America. "If Fascism makes headway in this country, the Negro, the foreigner and the Jew will be the object of special attacks." But the *Tech*'s liberal Jewish editor, Paul Cohen ('35), did not necessarily speak for the MIT community as a whole.

Jewish students faced growing antagonism in the 1930s, just as Cohen predicted. There were the usual gratuitous comments about Jews and money. William Power ('32), the efficient young manager of MIT's exhibits at the Chicago World's Fair, included tasteless digs of this sort in his correspondence with Compton's assistant Carroll Wilson: "Jewry is still in fashion" (referring to a bit of bargaining that got a price down) and, regarding reimbursement for expenses, "Here are a few bills, my dear Shylock, which you may exhort the keeper of the coin to pay by day before yesterday." Some years later, when Frank Jewett told Compton that Gerard Swope was "a merchant by nature and an exceedingly good merchant," it was clear that Jewett had Swope's ethnicity in mind.

But these were mild examples, next to the emotions that surfaced in April 1935. That month, a number of student societies planned an antiwar strike conference to galvanize opposition to Nazi- and fascist-inspired assaults on civil liberties and on ethnic groups (Jews, gypsies) and political movements (communists, socialists) in Germany, Italy, and elsewhere. Among the societies involved were the Liberal Club, Debating Society, Menorah Society, and several professional clubs. Also active were MIT student members of leftist outside groups such as the National Student League and the League for Industrial Democracy. Students on the other end of the political spectrum, however, threatened a counterdemonstration announced by leaflets decorated with swastikas and signed by a group calling itself Tech Militarist and Anti-Semitic Society. This material was dismissed as a bad joke until April 4, when nine MIT students entered the bedroom of conference organizer Robert Landay ('38), held him down, and cut off his hair. On April 7 a larger group, estimated at around twenty, broke in and assaulted another conference organizer, Robert Newman ('36), who described what happened: "I was bound, tied, and gagged by these 'potential fascists,' and before I knew any more I went 'out.' When I came to they had almost completed their dastardly task of shaving off my hair and mustache. They then cut a swastika in the remaining hair (they left about ¼ to ⅛ an inch) and after warning that there had better not be a strike, they went off." A *Boston Globe* headline for April 10—"Tech fascists shave heads of pacifists"—caused quite a stir, with calls for disciplinary action against the perpetrators in defense of "the good name of the Institute."

While such events were rare, they pointed to a surge in ethnic tensions that sensitized some in the MIT community to risks facing the nation and the world. Paul Cohen's successor as *Tech* editor, Arthur York ('37), called Americans hypocritical for tolerating terrorism in their own country—wholesale lynching of blacks in the South—while wringing their hands over Hitler's violence toward Jews and other minorities. "The bullying of the black in the South is closely parallel to the persecution of the Jews in Germany; Americans who criticize Hitler's policy condone at home the same practices on an even more helpless race." By 1939, meanwhile, Karl Compton joined his brother Arthur and other scientists in a campaign by the American Association of Scientific Workers to boycott aspects of German science "promulgating racial hatred."

Like blacks, Jewish students were excluded from the regular fraternities. While few new frats appeared on campus in the 1930s—Phi Delta Theta, for example, in 1932—these, like the established groups, adhered to restrictive covenants on race and religion. But Jews, at least, had the numerical mass to organize living groups of their own. Richard Feynman recalled being rushed by two of the Institute's Jewish fraternities while he prepared to enroll as a first-year student in 1935. "The summer before I went to MIT I was invited to a meeting in New York of Phi Beta Delta. . . . In those days, if you were Jewish or brought up in a Jewish family, you didn't have a chance in any other fraternity. Nobody else would look at you." Feynman pledged Phi Beta Delta, even though he "wasn't particularly looking to be with other Jews," and in spite of heavy-handed efforts by another Jewish fraternity, Sigma Alpha Mu, to get him to join their ranks.

The segregation of Jews from Christians in residential and social settings reinforced ethnic strains at a time when such were already on the rise. To compensate, local synagogues and shuls organized through college menorah clubs to bring Jewish students into the homes of local congregation members on holy days. Yet even Compton, tolerant though he was in many respects, considered ethnicity—particularly Jewish origin—a reasonable factor in setting hiring limits, or quotas, when a satisfactory balance might be thrown off. For more than a year he sought an academic appointment for MIT doctorate and Albert Einstein's assistant, Nathan Rosen, somewhere other than MIT, Rosen's ethnicity apparently his prime disqualifier. "Rosen . . . is quiet," Compton wrote to Gaylord Harnwell of the University of Pennsylvania's physics department in February 1939, "has not exhibited the undesirable Jewish characteristics, and is well liked and respected here. I would not hesitate in considering him favorably for a position on our own staff except for two factors: first, we already have several Jews in our Mathematics and Physics Departments, and I question the advisability of adding another, and second, we are already well equipped with young men in the general field of Rosen's interests." Similar letters went to John Ross, of Clarkson College of Technology in Potsdam, New York, in April; to John Davis, of Lynchburg College, in May; to Arthur McLean, of Portia Law School, Boston, in September; and to Harvey Davis, of Stevens Institute of Technology, in June the following year. As Compton wrote to Ross: "I would say that he is a Jew of the finest type, cultured, sensitive, modest and well liked. So far as I have been able to see, he completely lacks the traits of aggressiveness or of appearance which are responsible for the Jewish problem." And to John Davis: "I would have no hesitation in adding him to our staff in either mathematics or physics, were it not that we already have as large a proportion of Jews in these two departments as I think it desirable to carry."

Neither department, however, counted more than three or four Jewish faculty members in a group of 17 in mathematics and 26 in physics. Wiener, Philip Franklin, Norman Levinson, and Samuel Zeldin were the only Jews in mathematics. Levinson's appointment as instructor in 1937 had come about via a short but sharp struggle. When Vannevar Bush declined to hire him apparently on ethnic grounds, distinguished British mathematician G. H. Hardy was said to have embarrassed him into changing his mind: "Tell me, Mr. Bush, do you think you're running an engineering school or a theological seminar? If it isn't [the latter], why not hire Levinson?" Compton voiced concern that hiring more Jewish faculty would attract into physics and mathematics "an abnormal number of Jewish students." This ethnic preoccupation grew as he pressed Rosen's case. In his last letter on the subject, to Harvey Davis, Compton states: "Dr. Rosen has none of the objectionable qualities that sometimes go with this race. He is very modest, generously cooperative, sensitive, and a thorough gentleman. His wife also is a talented and cultured woman with considerable musical background. The fact that she is also not of an objectionable Jewish type is illustrated by the fact that Professor [George] Harrison recently told me that he felt that Mrs. Rosen would be a real asset in any college group and that, while he assumed that she is also Jewish, he is not sure." Rosen went on to teach at the University of North Carolina, Chapel Hill, before moving to Israel in 1953 to join the Technion in Haifa and later becoming president of Ben-Gurion University. Compton's feelings in this respect also showed in 1943, when he complained to Cornell University president E. E. Day about several hundred Army Specialized Training Program (ASTP) students assigned to MIT for war-related coursework: "Judging by their names, well over half of the entire group are Jews. From the standpoint of academic qualifications, they are altogether a very sorry lot, with backgrounds such as drygoods salesman, wood worker, etc. There are not over two or three whom we would care to have out of the entire lot. . . . It looks to me as though the entire program is a colossal flop."

Women students, conversely, made notable headway during the 1930s. Numbers hovered between 42 and 56 each year, a moderate increase over the previous decade—on the order of 20 percent—but still a minuscule 1–2 percent of the student body. Women continued to lean toward biology and public health, architecture, and chemistry, but they could be found in increasing numbers in engineering fields. In 1930, aeronautical engineering registered more women (4) than any other course besides biology (18), architecture (13), and chemistry (5), a trend possibly inspired by the exploits of legendary aviator Amelia Earhart. MIT's first woman graduate in aeronautical engineering, Isabel Ebel, earned the bachelor's degree in 1932, while that same year Hilda Lyon earned a master's in the same field. The first woman doctorate in geology was Katharine Carman in 1933, the first in biology Helen Breed in 1937. Mechanical engineering was the only field to register no women in the 1930s.

Compton showed a certain receptiveness toward women scientists in 1935, when he floated the idea of bringing in Wanda Farr, of the Boyce-Thompson Institute, to help build MIT's program in colloid chemistry. Whether this was to have been a faculty appointment or a staff position (probably the latter), the suggestion never caught on and Farr, an expert on cellulose research, remained with Boyce-Thompson before joining the firm American Cyanamid in the 1940s.

Compton backed a series of visiting appointments for women from other (mostly women's) colleges, including physicist Rose Mooney of Sophie Newcomb College in 1939–40 and, in the spring of 1941, chemist Helen Jones, a Ph.D. (1925) of MIT, from Wellesley College. The distinguished Jewish algebraist and theoretical physicist Emmy Noether, hailed by David Hilbert and Albert Einstein as the most important woman mathematician ever, visited MIT shortly after being dismissed from her post at the University of Göttingen following the German government's purge of Jews and political undesirables from the civil service in 1933. "The department should have given her an appointment," wrote Dirk Struik, "I do not know why it did not work out." The twin drawbacks of gender and ethnicity may have factored in, overriding the certainty that Noether's work equaled or surpassed that of the Institute's finest mathematicians, including Norbert Wiener and Dirk Struik. Noether found a position, instead, at all-female Bryn Mawr College, where she died in 1935. MIT would not appoint its first woman faculty member, Elspeth Rostow, assistant professor of history, until 1952.

Of the female academic staff at MIT in the 1930s (9 in 1931–32, 12 in 1939–40), all but one—Frances Stern, lecturer in biology—were research associates, research assistants, technical assistants, or librarians. Architecture librarian Florence Stiles ('22) was tapped by Compton in 1939 for an Institute-wide role in managing women's affairs. He met with women's association officers that April and listened to their concerns: "They feel that the greatest single need is to have some woman of tact, judgment and general competence to whom the girls can go for advice or help on personal matters, as distinguished from curriculum or disciplinary matters. The Women's Association knows that Miss Stiles has informally been very effective in this capacity with the women students in Architecture. They suggest therefore that she be given some such title as 'adviser to women students' so that these students from other departments than Architecture may feel free to call on her." Compton appointed Stiles within two weeks of this meeting.

Stiles's duties, aside from counseling, revolved around use and upkeep of the Margaret Cheney and Emma Rogers rooms, chaperoning, and housing. She raised enough awareness that in 1945 MIT bought a house at 120 Bay State Road to accommodate a dozen or so women under the supervision of a house mother, Margaret Alvord, widow of a former Boston University professor. Earlier that year Stiles had told Compton of lifestyle concerns by department-store magnate J. C. Penney and his wife, concerning their daughter, a freshman: "Mrs. Penney is afraid that her daughter will develop into a queer sort of person interested only in her work and she, therefore, is very much interested in housing accommodations as a good influence and balance wheel." Stiles struggled with such issues, and when she resigned three years later to join an architectural firm in Bar Harbor, Maine, she left with mixed feelings. "The Institute," she told Compton's administrative assistant Robert Kimball ('33), "has really been most generous to me considering MIT is a man's world." The Penneys' daughter, incidentally, Mary Frances ('47), later Mrs. Phillip Wagley, would earn a doctorate in physical chemistry at the University of Oxford in 1950, teach at all-female Smith College from 1950 to 1953, and serve as headmistress of St. Paul's School for Girls in Brooklandville, Maryland, from 1966 to 1978. In 1970 she would become the first woman on the MIT Corporation and, in 1984, the first to serve as president of the MIT alumni association.

The 1930s and early to mid-1940s saw women students drawn increasingly into the mainstream of MIT's extracurricular life, a trend most marked with respect to campus publications. Ida Rovno ('39), the lone female staffer on the *Tech* in 1936, was joined by several other women in 1937, including Ruth Raftery ('38), Jeanne Pearlson ('40), and Ruth Berman ('40), and in 1938 by Pearl Rubenstein ('39) and Anne Schivek ('39). Roberta Kohlberg ('46) was treasurer of the *Tech*, 1943–46. Rovno worked on the staffs of both *Tech* and *Voo Doo* in 1938, while Jeanne Kitenplon ('38) and Dorothy Betjeman ('40) signed up for *Voo Doo* that year. Rovno was elected to *Tech*'s managing board in her senior year.

Women found a welcoming atmosphere in several campus professional groups. M. Elsa Gardner ('33) and Nancy Klock ('39) were both members of the Aero Engineering Society, with Gardner serving on the staff of *Aero Digest* in 1933. Among the women who held elected office were Anne Schivek and Marjorie Quinlan ('41), vice presidents of the Chemical Society; Frances Emery ('39), Janet Norris ('42), and Judith Turner ('48), secretaries of the Architectural Society; Barbara Laven

('40), secretary of Dramashop; Jeanne Pearlson, vice president of the Mathematical Society; Leona Norman ('41), president of the Sedgwick Biological Society; Frances Ross ('42), secretary-treasurer of the Sedgwick Biological Society; Mary Sullivan ('45) and Rosemary Durnan ('48), secretaries of the Catholic Club; Virginia Ferguson ('47), chairman of the Walker Memorial Committee; and Virginia Tower ('47), secretary-treasurer of the Mathematical Society. Marjorie Siff ('44) was on the debating team in 1943–44. Opportunities opened up even in athletics, traditionally an all- or mostly-male domain. Isabel Ebel competed in women's rifle-shooting in 1932 and Domina Spencer in fencing in 1939, while a Miss Li—possibly Lien-Fung Li ('44)—was on the pistol-shooting team in 1939. Sailing drew women in substantial numbers, among them Ruth Pfeiffer, Ruth Berman, Frances Emery ('39), Anne Person ('39), Domina Spencer, Edith Cameron ('40), Eloise Humez ('42), Gloria Kay ('43), Anne Lyons ('44), Jacquelyn MacLean ('45), and Mary Jeffries ('45).

Outside MIT, Compton played the role of scientific statesman at the highest levels. He had served as president of the American Physical Society in the mid-1920s and would be elected president of the American Association for the Advancement of Science in the mid-1930s. In between he became a moving force behind the creation of the American Institute of Physics (AIP), an amalgamation of the nation's physics societies. AIP's goal was to move forward—mini-League-of-Nations style—with a common voice in certain areas while preserving member societies' unique identities. Compton was most interested in AIP's educational value, in its potential, for example, to counter the alarming rise of anti-science strains of thought and culture. In January 1934, he brought together Caltech's Robert Millikan and other distinguished physicists for a public conference in New York to create "a backfire against those who are preaching such doctrines as 'Science has had its day and made a mess of things, and now the economists and sociologists will take hold,' or the pronouncements of the technocrats which have certainly had an effect in making many people, who would otherwise be inclined to support science, hesitant or inclined to throw their support in other directions." He hoped, also, to get President Franklin Roosevelt to attend.

By this time, Compton had Roosevelt's ear. The summer before, on shipboard with his family heading to Maine for a seaside vacation, an urgent radiogram arrived from MIT: "Word received that you have been appointed chairman of committee to reorganize federal government." Not even remotely possible, Compton chuckled to himself. As it turned out, Roosevelt wanted him to chair a new federal unit called the Science Advisory Board (SAB). The board had originated in a request from Roosevelt's agriculture secretary, Henry Wallace, for advice from the National Research Council (NRC) on reorganizing the U.S. Weather Bureau. NRC chairman Isaiah Bowman took this as an opening to make the council and its parent body, the National Academy of Sciences (NAS), more helpful with respect to crafting and guiding national science policy. Roosevelt signed an executive order creating the SAB in July 1933. Compton accepted the chairmanship; filling out the original board were NAS president William Campbell, Carnegie Institution president John Merriam, Robert Millikan, Geological Society of America president Charles Leith, Bell Telephone Labs president Frank Jewett, J. G. White Engineering Corporation president Gano Dunn, and General Motors research director Charles Kettering.

This, the federal government's first systematic venture into science policy, quickly sank into political turf wars and miscommunication about mission, mandate, and scope. There was no budget and no office (meetings were held at the NAS), members served without compensation, and government officials ignored the group. Members offered unsolicited, often unwelcome advice and put together research proposals that required spending on the order of sixteen million dollars. Harold Ickes, director of the Public Works Administration (PWA), brushed the proposals aside—depression-era laws, he said, restricted public money to construction projects. At one point Roosevelt offended the board by unilaterally appointing two new members, neither of whom was a member of the NAS.

The board fizzled out in 1935, with little to show for itself. Compton managed over two years to eke out enough resources to support a few studies and to issue reports on the American patent system, government support for science overseas, land-use and soil-conservation, and federal mapping services. But he left disappointed, with a sense of failure, unaware that the SAB had laid important groundwork for the government's mobilization of science and scientific personnel during the Second World War, just a few years away. The experience gave him a distaste for politics and darkened, at least temporarily, his optimistic outlook: "I am in sympathy with some of the Roosevelt policies and not with others. On the whole, however, I am inclined to agree with those who consider him a national menace, because with the best of intentions, but without having thought things through or secured competent technical advice, he is likely to plunge confidently into huge mistakes for which the country will pay heavily. This tendency has already been shown on numerous occasions." Compton was no avid New-Dealer, in other words, and felt ambivalent then and later as to whether or not the benefits of government involvement in science and other areas outweighed the risks: sacrifice of autonomy, control, and independence, surrender to uninformed political operatives and government bureaucrats.

The range of Compton's commitments, both within MIT and outside, would have strained a physically fitter man. But healthy Compton was not, in spite of his athletic background and muscular frame. He arrived at the Institute in 1930 nursing a stomach ulcer kept in check by his easygoing disposition and a regimen of midmorning milk. A mishap in 1931 or 1932 laid him up for several weeks: while carrying his daughter Jean on his back at the president's house, he fell downstairs and cracked a vertebra. His choice was either to confine himself to an uncomfortable cast or swear off all movement for several weeks. He chose the latter, with Margaret testifying to his stubborn willpower: "That's one thing you won't have to worry about," she told Institute medical director George Morse, "if he decides not to move, he won't move, period." Compton made the best of it, using the interval that he was immobile and free from interruptions to catch up on paperwork. But in 1936, following a time of peak stress—the economic depression, reforms at MIT, struggles over the SAB—he suffered what appeared to be a mild stroke. When the episode kept recurring, his doctors diagnosed arterial hardening aggravated by the cigar-smoking habit that he had picked up in Paris during the First World War. His friends grew accustomed to his swearing off smoking and then, a few days later, reappearing somewhere calmly puffing on one big cigar after another.

While Compton never gave up his cherished cigars, he did make some effort to lighten his work load. "A few weeks ago," he wrote in 1936 to the president of an organization from whose board he had decided to resign, "I was taken ill with certain symptoms which indicated that I must reconcile myself to a considerable reduction of activity for some time to come, and I am under medical orders to drop out of every possible activity which is not immediately connected with my responsibilities at M.I.T." Vannevar Bush stepped in with additional advice in the fall of 1937.

> "You know, K.T., you and I would live longer if we had a place in the country where we went out and sawed wood weekends."
>
> "All right, you go find two places and take your first choice, and I'll take the second."

Bush found a map, drew a circle around the Boston area, and charted locales based on proximity to Boston (within one-and-a-half hours of campus, by car) and price (the depression had wreaked havoc with their savings). He found something that would not work for him, he said, but might for Compton.

The place was near Jaffrey Center, New Hampshire, almost exactly seventy miles from MIT. It was a sturdy house, a bit bleak, with no electricity or running water, stove heat, a rambling half-tumbled-down barn off to one side. Compton fell in love with it and passed papers two weeks later, in early December. The family moved in the following July and spent as much time as they could there—weekends in spring and fall, occasionally midwinter, two or three Christmases over the years. "It was very much a family party," Margaret recalled, "which was one of the intents of it, that the children used to an official house with servants would get used to what we consider a normal family home." They grew vegetables, puttered with machinery, rode around on a Ford-Ferguson tractor, hiked, dug wells—a welcome respite from MIT, Cambridge, and the nation's capital.

The generous, kindly Compton relied on Bush, the tough, some said cold-blooded straight-talker, to handle the distasteful side of administration: denying requests, breaking bad news on budgets, defusing personnel problems, restructuring—sometimes doing away with—departments and programs regardless of sentiment, tradition, or any factor not essential to the Institute's needs. But as Bush's reputation spread, so too did the risk of losing him. He rejected a number of college presidencies, assuring Compton that he would "rather be second frog in the M.I.T. pool than first in a smaller puddle." Then, in 1938, he accepted the presidency of the Carnegie Institution of Washington. Compton did his utmost to keep him in place—"Look, let me go to the Corporation and ask them to make you president instead of me; then I will be chairman . . . and you can run the Institute"—but Bush stood firm. So as not to lose him entirely, the MIT Corporation elected him to life membership.

Compton chose Edward Moreland to replace Bush as dean of engineering, but finding a new vice president would prove more difficult. Bush put in a strong word for James Killian. "I don't think Karl Compton knew Killian at all well," he recalled, so "I urged on him that he study Killian because I thought he had some extraordinary qualifications." Compton, more familiar with Killian than Bush realized, looked in at the offices of *Technology Review* in October 1938 and asked Killian to join his staff as executive assistant. No one thought of Killian as vice-presidential material—too young and inexperienced, "magnificently unprepared" as Killian later remarked—but both Compton and Bush saw him as energetic, patient, and loyal, a capable adjunct to carry out Bush's routine duties.

The offer surprised almost everyone, no one more than Killian himself. His background and preparation—editor of the alumni magazine, informal public-relations aide, part-time teacher at Simmons College—hardly seemed ideal for this level of responsibility at a place like MIT. But he accepted anyway and started on January 1, 1939. Later on he would joke, in imitation of a couplet from *H.M.S. Pinafore* (he and his wife Elizabeth were lifelong Gilbert and Sullivan fans), about writing and public relations as his most strategic attributes:

> I flourished that pen with such manifest joy
> That K.T. appointed me his office boy.

The appointment letter mentioned a probationary term of six months. Killian was "a puppy," as he described himself, "not yet housebroken." The executive committee, meanwhile, urged Compton to find a permanent replacement for Bush without delay.

Killian quickly wrapped up business at the *Review*. The January 1939 issue was his last as editor. He had spent thirteen years there and churned out 115 issues. His successor, Frederick Fassett, associate professor of English, was the first nonalumnus to hold the job. Killian, reluctant to cut ties altogether, stayed on as editorial associate. Even though he would have little time to give, he remained for two more decades, through March 1959.

Waiting for him on his desk, his first day at his new job, was a note in Bush's scrawl. "May your occupation of this desk bring you a full share of satisfaction and enjoyment; the sense of true accomplishment which renders tolerable the many irritations of an executive officer." Killian expected something along the lines of doing one's duty, or of living up to expectations. But this instead, a calming, relaxed voice that finished with "Don't forget to put your feet on the desk once in a while, drive everyone out, and take time to think."

Killian was not the type to throw his feet up on a desk, or to shoo people out of a room, but he cherished the note and its contents turned into a kind of mantra, as new challenges flew his way. Gone were the days when he could sit behind a desk obsessing over word choice, print layout, or article sequencing. He barely knew science and engineering, certainly not as Bush did, but he had one advantage over him: a nonthreatening, diplomatic way with people. Among his first assignments was to help prepare the Institute's budget. Bush had always done this by fiat. Here it is, take it (to leave it was never an option). But Killian crafted a different approach. He put together a budget and review committee consisting of the deans of science, engineering, architecture, and humanities. This would have struck Bush as cumbersome or inefficient, like begging for trouble. But it worked well under Killian's gentle, persuasive eye. It rid the atmosphere of any suspicion that certain schools and departments, even some individuals, were favored over others, and it built confidence that the administration wanted advice, if not necessarily across-the-board consent, from faculty. Killian's way felt more participatory, compared with Bush's authoritarian style. Killian liked to say that as a newcomer, and a relative novice, he did not dare follow in Bush's footsteps, for no one would have listened to him. But beneath the obvious contrasts lay clear differences over how best to run an academic institution.

Killian faced a steep learning curve. The fact that he was neither a scientist nor an engineer caused him much self-doubt at the start. Had he been foolish to take on a role that he was not qualified for? There were times, Killian fretted, when Compton "hardly knew what to make of his anomalous assistant [and] when it seemed to him that the arrangement would not fly." Much of this was beginner's nerves. Compton never appeared to doubt Killian's ability or judgment. In January 1939, he told Bush that Killian "has started in fine shape, and has already proved that he will be a great help." The events of 1939–40 conspired, anyway, to blur any doubts. Compton, drawn into wartime strategy on a national level, came to rely on Killian to hold MIT together and, ultimately, to help coordinate its role in national defense.

Spirits were mixed on the evening of March 5, 1940, when Compton joined several hundred alumni at the Waldorf-Astoria in New York to celebrate his first decade as MIT president. Corporation member Harlow Shapley captured the ambivalent mood: "As a happy augury on this important occasion I have arranged a display of all the major planets across the western sky, from Mars above Bethlehem and Pittsburgh to Venus overhanging Hollywood, and ordered that they continue to follow your brilliant example and continue to radiate beneficently on a planet that much needs light and beneficence"—this last phrase an allusion to the war raging in Europe.

Compton was already deep in discussion about the future. How soon would the U.S. government grasp that the European conflict posed a threat to America? How much effort would it take to tap into the nation's scientific potential, so as not to be left fighting a new war—if it came to that—with old weaponry? Compton met regularly with Harvard president James Conant, Vannevar Bush, Frank Jewett, and others on these matters. Bush, typically, did not wait long to move past talk to action. Through a friend who happened to be a relative of

President Roosevelt's, Bush got the ear of Harry Hopkins, one of the chief architects of the New Deal and a close adviser to the president. Hopkins, swayed by Bush's arguments about wartime preparedness, arranged for him to meet with Roosevelt on June 12, 1940. Roosevelt quickly saw the value of an independent agency overseen by the nation's best scientific minds, whose mission was to brainstorm, develop ideas, invent, design, and test devices for military deployment. The agency, Bush insisted, must have its own budget as well as authority to carry on projects in cooperation with the military establishment, but ultimately outside its control. Roosevelt and Bush, cognizant of the political minefields involved, both considered the risks well worth taking. Roosevelt signed an executive order creating the National Defense Research Committee (NDRC) and appointed Bush as chairman.

Bush telephoned Compton, Conant, and Jewett with the news right after his meeting with Roosevelt. They all agreed to serve under him. Another scientist, Richard Tolman ('03) of Caltech, was added to the group, and Bush selected as government representatives U.S. patent commission director Conway Coe, rear admiral Harold Bowen as the Navy's representative, and brigadier general George Strong to represent the Army. He set up office space for the NDRC at Carnegie Institution headquarters, 1530 P Street, in the heart of the nation's capital. A few wondered why Compton, backed by his experience as head of the abortive SAB and, relative to Bush, an even-tempered, non-abrasive personality, was not selected as chair. The simple answer was that Bush, not Compton, had taken the lead, gone to Roosevelt, and convinced him. From the standpoint of managerial style, too, Bush was perceived as the better choice. "A new organization launched," one observer wrote, "in the whirlpool of emergency Washington needed someone with his shrewdness and self-assurance to guide it, someone who would never be torn by self-doubt once his mind was made up, and whose reaction to opposition would be increased determination."

The oversight committee was quickly dwarfed by its offshoots, by the multiple projects and programs required to carry out the work. To start with, Bush set up four divisions: Division A (armor and ordinance), under Tolman; Division B (bombs, fuels, chemical warfare), under Conant; Division C (communications and transportation), under Jewett; and Division D (detection problems, controls, instruments), under Compton. A fifth division dealing with patents and inventions was added later, under Conway Coe. Each division branched off into several sections. Compton's division had four sections: D-1 (detection), D-2 (controls), D-3 (instruments), D-4 (heat radiation). He chose Alfred Loomis, Wall Street banker, physicist, and member of the MIT Corporation, to take charge of D-1. D-2 went under Warren Weaver, of the Rockefeller Foundation; D-3 under George Harrison, professor of physics at MIT; and D-4 under Alan Bemis, research associate in meteorology at MIT. Among other MIT faculty delegated to sectional work were Edward Bowles and Samuel Caldwell (electrical engineering), Tenney Davis (chemistry), and Warren Lewis and Thomas Sherwood (chemical engineering). In 1941, over two dozen faculty members were granted leaves of absence for defense work, among them Julius Stratton (physics), Francis Bitter (metallurgy), and John Trump (electrical engineering).

Compton's likeable, positive personality proved invaluable in recruiting these men, pre-Pearl Harbor, into government service as so-called dollar-a-year public servants who kept their regular jobs while throwing themselves into the defense effort. He worked closely with them to attract scientists, engineers, and technicians from industry, academia, wherever the best could be found. Tensions between the different groups—scientists and engineers, universities and companies—ran high because of cultural differences and traditional jealousies, as well as competitive factors. One wild card was the military reaction. Leaders at the top—General George Marshall, Admiral Harold Stark, General Henry ("Hap") Arnold—welcomed the NDRC, but cooperation proved spottier in the lower echelons, where attitudes ranged from tolerance, to suspicion, to open hostility. Everyone muddled along while Compton smoothed ruffled feathers. After Pearl Harbor, the NDRC was reorganized and grouped along with the Committee on Medical Research under a new umbrella, the Office of Scientific Research and Development (OSRD). Bush became OSRD director, with Conant taking his place as head of the NDRC. Compton's new oversight responsibilities included radar, radar countermeasures, guided missiles, optics, and physics. He was assisted as division vice chair by physicist Alan Waterman, who in 1951 would become the first director of the National Science Foundation.

Once the United States entered the war, Compton was more off campus than on for the duration. He commuted by rail between Boston and Washington at least twice a week. There was much travel elsewhere, too, to recruit staff, set up research projects and labs, inspect facilities, review progress, break bureaucratic log-jams, and boost morale. "The job is more interesting than I expected," Compton told Killian, "and also more complicated." He committed MIT to an unprecedented level of involvement: "To each adjustment and modification in our program, we have applied the test, 'Will it assist our institution in making its maximum contribution now to the winning of the war?' . . . This attitude is vital, since the supreme and swift mobilization of all national resources required by the war permits no partial measures and no temporizing. It is especially vital since we are a technological institution in a war where technological superiority spells the difference between defeat and victory." The balance could shift, and fast—some saw MIT as a teaching institution that carried on a lot of defense research, others considered it a research institution that carried on a little teaching. It was either, sometimes both, and without question defense-absorbed rather than merely defense-oriented. Compton forged that balance. In his view, MIT could contribute to the national defense in three ways: by taking on government research contracts, by offering special courses, and by training personnel for wartime service. Traditional priorities would have to wait.

Compton's grueling schedule—nonstop meetings, more than a hundred thousand miles traveled each year—was interrupted every other weekend or so by a day-long visit, two days at the most, to his New Hampshire retreat. As he ran the tractor, hoed the garden, hiked, and enjoyed simple family pleasures, he recharged his inner batteries for the next round. In the summer of 1942, when the nation's rubber supplies were cut off by the Japanese invasion of the East Indies, Roosevelt added to his load by appointing him, along with Conant and Bernard Baruch, as a Rubber Survey Committee to devise an emergency plan. Baruch, among Roosevelt's closest economic advisers, chaired the committee, which among other things mandated a national 35 m.p.h. speed limit. "We wanted 40," Compton recalled, "but because we realized that every red-blooded American feels that he has the God-given right to travel at 5 m.p.h. over the speed limit, we set it at 35."

These and other demands could have overwhelmed Compton had James Killian not been in place to oversee what went on at MIT. As the frequency of Compton's absences picked up after July 1940, Killian became the go-to leader on campus just a year after taking on the role of executive assistant. He learned as he went along, enough science and technology, he hoped, to keep up with basics and to acquire a feel for the issues. He harbored no illusions about connecting, somehow, to technical intricacies beyond the level of a well-informed layman. But he knew that to do his job, he would have to be very well informed. Later, he would appreciate the way Winston Churchill described his own predicament—"I knew nothing about science, but I knew something of scientists . . . and I had much practice in handling things I did not understand."

Killian depended a good deal on Edward Moreland, a genial fellow Southerner who had succeeded his former teacher and business partner, Dugald Jackson, as head of electrical engineering at MIT in 1935, and then became dean of engineering when Bush went to the Carnegie Institution in 1938. A personality more different from Jackson and Bush would be hard to find. Moreland, like Killian, was calm, collected, nonflamboyant—and the two forged a close, productive working relationship. Moreland's office sat across from Killian's, and Killian often sought him out for advice on technical substance and fine points of institutional politics. What impressed him most, he said, were Moreland's "skills in resolving diverse points of view, aided by the way his face could light up with an utterly engaging smile when making tough administrative decisions . . . his power to radiate a gentle warmth and his considerate courtesy in dealing with his colleagues." In 1942, when Bush pulled Moreland to Washington, D.C., to work on national defense, Killian felt his absence keenly. Compton almost lost Killian to the military draft that same year, but secured a 3-b deferment by arguing Killian's indispensability to MIT:

> As my Executive Assistant, Mr. Killian is the key executive officer of our institution. He has been in this position for a sufficient number of years to be thoroughly acquainted with our administrative problems, with the

> personnel, with the details of our budget, and with our problems of public relations and contacts with governmental and business concerns. There is no other member of our staff who has the background to perform these duties. . . . Were he to be called to military duty, I and various of my administrative colleagues . . . would have to reduce their contributions to the war effort to help carry the administrative responsibilities now handled by Mr. Killian. . . . Because of these facts, it seems clear that Mr. Killian's best contribution to our national war effort will be secured by permitting him to continue his present activities which are essential and for which he is practically irreplaceable under existing conditions.

Killian helped organize and coordinate MIT's largest wartime project—the Radiation Laboratory, or Rad Lab as it was commonly called—in the face of Compton's early doubts about the Institute's ability to handle it. After the Manhattan Project, the Rad Lab was the government's next-largest, next-most-essential wartime research effort. Its primary goal was to develop microwave radar operating and detection systems for combat use. Compton, as head of the NDRC's Division D, held primary responsibility for the lab's placement. At first he wanted a government facility or research institution, preferring to stay away from academic and commercial groups—academic ones because the project was so huge that it might swallow educational programs, commercial ones for fear that proprietary issues could complicate security needs. First on Compton's list was the Carnegie Institution's terrestrial magnetism department. He dropped the idea, however, as the department was already involved in another NDRC project. The next option, Bolling Field near Washington, D.C., came about by arrangement with the Army Air Corps. This was not an ideal location—no lookout over the Atlantic Ocean, although plenty of airplanes for practice—but it appeared feasible until the Air Corps switched plans and space was no longer available. Next, Bush and Alfred Loomis laid out a case for MIT, arguing that the Institute had a number of advantages: technical expertise in its highly sophisticated communications and microwave theory programs, proximity to the ocean, and access to airport facilities in East Boston.

Compton agreed to explore the idea, and the more he thought the more he saw the sense in it. He was not the only doubter. Robert Millikan worried that concentrating too many of the nation's finest physicists in one place might make the Rad Lab, and MIT with it, too tempting a target for saboteurs, or for enemy attack. While Millikan held little sway, not being a member of the NDRC's inner circle, Frank Jewett exerted enormous influence and shared some of his concerns as well. Mostly, however, Jewett believed that his own outfit, Bell Laboratories, was a more suitable venue: better facilities, research in this area already far along, personnel familiar and comfortable with secrecy issues, communications a central corporate focus. The choice of MIT placed strains on the relationship between Compton and Jewett, and these deepened as the Rad Lab grew into one of the nation's wartime triumphs. Jewett and Compton "did not get along very well," Bush recalled, "and the reason was fundamentally this. Under Compton there was assembled at the Radiation Laboratory probably the hottest crowd of physicists on electronic gadgetry and the like that was ever put together. A lot of them were prima donnas; there were a couple of thousand people in the Laboratory and it was quite a show. Frank Jewett had been the fellow who had really put the Bell Laboratories together, and he'd directed it. It was the pride of his whole career. It was just too much for him to think that a crowd of youngsters could do something better than the Bell Labs could."

At the start Compton telephoned Killian, who sized up space options and returned the call within a few hours. MIT had about 11,000 square feet to offer, much of it by vacating Edward Bowles's lab in electrical engineering—more a conversion than an emptying out, since Bowles's group fed seamlessly into the project. Killian discovered, too, that the project could make use of the National Guard's hangar at the East Boston airport. There was no way to forecast how much space would be needed in the long term, but here was a start (eventually, about sixty times more space would be needed to accommodate equipment, materials, and staff).

The lab opened modestly with twenty staffers on November 11, 1940, and grew rapidly. Between 1940 and 1945 its 3,900 employees, under the directorship of physicist Lee DuBridge and associate directors F. Wheeler Loomis (no relation to Alfred Loomis) and physicist I. I. Rabi, worked on microwave theory, operational radar, systems engineering, long-range navigation, and control equipment, and helped design and deploy the first radar systems small enough for use in airplanes. Meanwhile, MIT was chosen to run a radar school to train Army and Navy officers, enlisted men, and civilians in the theoretical and practical aspects of radar. Through the end of the war, 8,800 personnel passed through that school. Rad Lab devotees bragged that while the atomic bomb *ended* the war, radar *won* it.

Compton's wartime role took him further and further afield. In 1943, Bush appointed him chief of a new unit, the Office of Field Services, with Alan Waterman as his assistant. This work took them into the Pacific theater for meetings with General Douglas MacArthur and other top brass. Killian kept in touch by cablegram, whenever possible, but the running of MIT fell almost entirely to him. In recognition of his expanding, critical role, the executive committee voted in April 1943 to promote him. The exact title was left to Compton, who canvassed the deans on what they thought suitable—vice president, executive vice president, administrative vice president, executive director, or liaison vice president. Compton settled in the end on executive vice president.

Since 1939 Killian had gone from fairly routine decision making on budget and personnel matters to complicated negotiations over contracts, grants, land purchases, and academic policy. Research funds flowing into MIT swelled from $3 million in 1939 to $40 million in 1945, besides 75 OSRD contracts valued at more than $100 million. Killian also coordinated a vast influx of Army and Navy personnel for short-term coursework, offset by a 60 percent drop in the civilian student population (3,138 in 1940–41, 1,198 in 1944–45). MIT's V-12 Unit, organized in June 1943 to train naval officers, produced several hundred graduates within two years. The number of foreign students rose (195 in 1941, 237 in 1945) with China the largest group, followed by India, the Philippines, and Brazil. Membership in the Chinese Students' Club more than doubled in three years, from 52 in 1942 to 93 in 1944 and 136 in 1945. This surge raised some alarms, with John Bunker, Harry Goodwin's successor as dean of the Graduate School, commenting at one point: "Chinese students have been always among our best students and we are glad to have them but it must be evident that there is a numerical proportion beyond which it is unwise to extend admissions and this proportion has already been exceeded by accepted Chinese."

In wartime the last German and Japanese students, one each, attended in 1941. Yasuo Tani withdrew after Killian asked the director of U.S. naval intelligence for an opinion on "admission of foreigners from unfriendly countries to courses having a relation to our defense activities" (Tani had registered for coursework in mathematics and engine design). Even though the war department, in 1942, permitted enrollment of Japanese-Americans in colleges and universities, MIT took a harsh stand in anticipation of risky outcomes. One Hawaiian student with American citizenship, George Yamashiro ('42), was tagged as a problem case because his father was Japanese. "We have too much important work going on here to justify us in taking any unnecessary chances even though we may be imposing a hardship on some innocent individual," wrote admissions director B. Alden Thresher on the administration's behalf in April 1942. "For this reason it is now our settled policy for the duration of the war not to accept any students of Japanese ancestry whether citizens or not." In August 1943, on a more positive note, Killian boldly laid out a "memorandum on new opportunities and new objectives at M.I.T."—library building, faculty club, recreational center, dormitory complex, and a host of academic reforms—that could not be acted on until the end of global hostilities, but that would become a blueprint for action in the postwar period.

Just as others had groomed him for an administrative role, Killian looked to groom one or two others. Julius (Jay) Stratton emerged as a trusted adviser. No one predicted that he would turn into one of MIT's most thoughtful critics of academic policy. He had always come across as immersed in science, happiest in his lab or at Round Hill going over ideas and experiments with his colleagues, refining scientific articles, shaping and reshaping the framework that went into his seminal 1941 book, *Electromagnetic Theory*. He was known, too, as a patient, thoroughly involved educator and, following his marriage to Catherine (Kay) Coffman in 1935, as a devoted family man. But by the mid-1930s, he also found himself drawn to issues of governance through his membership on MIT's student-faculty committee. This, as he later told Samuel Prescott, was "really the beginning of my interest in the broader problems of education." In his next assignment, on the staff-administrative committee, he revealed leadership qualities that Compton and Killian took note of. As the committee's public spokesman, he drafted and refined a much-admired proposal for faculty tenure.

Stratton kept abreast of MIT affairs while he was away on war duty, mostly in Washington, D.C. He and Killian corresponded on contract negotiations and procurement policy. Stratton brought the perspective of a profound thinker, a philosopher, an intellectual whose range extended into realms of cultural experience, historical perspective, and social responsibility that Compton and Killian had not thought about quite so deeply. As ties between universities, industry, and the government tightened, MIT's leadership began to plan as early as 1943 for the war's aftermath. With Compton preoccupied in Washington, Killian drafted documents on long-range policy and circulated them among a tiny group of advisers, mostly MIT insiders. Stratton was part of this group.

Killian's proposal for government contracts drew a pointed response from Stratton in October 1943. While others made minor suggestions for wording or consistency, Stratton approached the problem on a different level—deeper, conceptual, grounded in institutional goals, culture, expediency versus intrinsic value, all informed by a careful reading of MIT's charter and how the Institute had evolved. He critiqued Killian's attempt to capture MIT's mission in a single phrase: advancement of engineering art.

> Let us agree that our primary function is educational. In fact, our entire function is educational when interpreted broadly. To prevent stagnation of our teaching, to ensure a constant flow of new ideas and new material, it is essential that a vigorous program of pure research be maintained. . . . But MIT is an engineering as well as a scientific institution, and a similar argument must be made for the necessity of pushing forwards the frontiers of engineering art. Unfortunately, these frontiers aren't clearly marked and unless we do a little boundary work I am afraid that we may easily stray over into territory occupied by industry. Perhaps we should, on occasion, but I think we ought to know when and why. It seems to me that the excerpt cited from the charter adds to the complexity of the problem, for it states among other things that we should aid, by suitable means, 'the advancement, development and *practical application* of

> science in connection with arts, agriculture, manufactures and commerce.' I strongly suspect that in view of the state of the arts and manufactures in 1861 the Founders meant just what they said. The development of a first-class harvester by the Mechanical Engineering Department of that epoch would doubtless have been looked upon as a practical contribution to society offsetting a lot of nonsense with test tubes and batteries, and justifying some solid financial support. Times have changed, and I submit that anyone interpreting literally that charter phrase about practical application of science in manufactures might anticipate conflicts. As with our [U.S.] Constitution, the interpretation must evolve with the times. . . . I do believe that it will be profitable to pursue this matter of our relation to industry further, in order to arrive at a clearer statement of policy.

Even as Stratton played the role of adviser, thought-provoker, and philosopher, Compton and Killian wondered if he would return to MIT after the war—a question asked about many academics drawn into war service. Government and industry, anxious to build their own teams of skilled personnel, would prove competitive in this regard. So in August 1944, when John Slater proposed establishing at MIT a peacetime, civilian successor to the Radiation Laboratory, all eyes turned to Stratton. "It was obvious," Slater recalled, "that Jay was the right one to lead it. He had had government experience during the war, he had been connected with the War Department . . . and had gotten to know how to wangle things out of Washington and about how to wangle things out of MIT." Stratton's intellectual background was ideal, too. His work, not only at the Rad Lab but for a decade or more before that, had bridged physics and electrical engineering in the very ways that Slater had in mind. Slater's concept, what turned into the Research Laboratory of Electronics (RLE), was a good one, Stratton felt, but did he want to spend the next decade in administration? The idea of getting back to research, writing, teaching, theoretical physics—or, maybe into a whole new career in government or private industry—held much appeal as well.

The first obstacle came in the form of an offer from Bell Laboratories. Compton's old friend and sometime nemesis Frank Jewett was about to retire as chairman of the board with Ralph Bown, research director, considered a likely replacement. Bell Labs wanted Stratton in Bown's old slot, which involved supervising the organization's research program: much of the Murray Hill outfit and all of Holmdel and the Deal Test Site. Here was an opportunity for Stratton to build on areas of special appeal to him, particularly basic research in electronics. Stratton, Bell Labs felt, combined "in a very unusual way excellent scientific knowledge with executive and administrative ability."

It was a tempting offer, which Stratton leaned toward accepting. He only let Slater, Compton, Killian, and Bowles in on the secret. But Slater felt confident that Stratton would prefer to come home, that he was attached enough to MIT that if the right place could be found for him there, on the right terms, he would forgo the Bell Labs position even at considerable financial sacrifice. Compton and Killian, with Slater's help, concocted a series of strategies to change Stratton's mind.

First, Stratton was brought to a meeting at MIT on August 28, 1944, ostensibly to discuss the general concept behind RLE but also to feel him out on whether he would commit to this new venture—a partnership, essentially, between the physics and electrical engineering departments. The other participants were Compton, Slater, dean of science George Harrison, and Harold Hazen, head of electrical engineering. "The general feeling," wrote Slater, "was that the proposed laboratory provided an unusually favorable chance to develop harmonious cooperation . . . in a field which clearly overlapped the interests of both [departments], and that this cooperation might be hoped to attract industrial support from outside the Institute, to attract good students, and to furnish a pattern for similar cooperative projects in other fields." Hazen was then assigned to coax Stratton back to MIT. Stratton, who returned to Washington after the meeting, got a call from Hazen a day or two later. "In a long-distance telephone conversation," Hazen recalled, "when his sense of uncertainty came through to me, I asked him if it would be any help if I were to hop on the Federal [Boston-to-Washington overnight train] to talk it over with him personally. He said he thought it would." Hazen rushed to Washington and this, according to Stratton, tipped the balance (it "had much to do with influencing my final decision to return"). In transition to the RLE directorship, Stratton headed the Rad Lab's basic research division from October to December 1945.

Compton's last wartime project ended before it began. With the Pacific War still raging in the summer of 1945, he went to Manila with Edward Moreland to organize and head up a Pacific branch of OSRD. They arrived on August 5, a day before the first atomic bomb fell on Hiroshima. The second bomb, dropped on Nagasaki on August 9, was followed within days by negotiations for the Japanese surrender, formally signed aboard the battleship *Missouri* on September 1. Compton wired Washington to cancel preparations for 200 civilian scientists to supplement the 60 already in Manila. He looked forward to heading home. But first, he and Moreland were asked to organize a mission to Tokyo to gather intelligence from Japanese scientists. Killian rushed to Compton a list of all Japanese alumni of MIT, Caltech, Harvard, the University of Chicago, and the University of California, along with last-known occupations and addresses.

Compton, Moreland, and their team were the first Americans to enter Tokyo since the war began. What surprised them most was the friendliness of the inhabitants, which Compton attributed to a culture of politeness, self-discipline, and obedience to the emperor who had instructed them to cooperate. Japanese scientists told of an army and navy in disarray, at loggerheads over policy and field strategy, and of atomic energy used not for weaponry but, primitively, as a coal substitute. Compton negotiated an agreement whereby the U.S. government would not destroy Japanese cyclotrons so long as they were used only for medical research. Soon after reaching home, however, he learned that the war department had arranged to dump them all into the ocean. Compton wrote U.S. war secretary Robert Patterson as scathing a rebuke, "condemning the wanton destruction," as his gentle disposition could muster. "It was an act of utter stupidity which has seriously set back public confidence in the military. . . . It has brought censure and ridicule on us from intelligent people of other countries."

As Compton tried to get back to business as usual at MIT, government affairs kept getting in the way. It would take almost a year to wind up the OSRD. A brief, pleasant interlude in February 1946 found members of the Compton clan gathered in St. Louis to celebrate the inauguration of Arthur Compton, himself a key player in the Manhattan Project, as chancellor of Washington University. Karl Compton, now 58, had been president of MIT for sixteen years, and Wilson

Compton, aged 53, had been president of Washington State College since 1944. The media played up this fraternal triumvirate of academic executives—quadrumvirate, actually, with sister Mary's husband, C. Herbert Rice, serving as principal of Christian College in Allahabad, India, then Forman College in Lahore. Margaret went along to celebrate, possibly one or more of the Compton children as well. By this time Mary Evelyn and her husband Bissell Alderman ('35), an instructor in architecture at MIT for some years, had an eight-year-old daughter. Jean, about to graduate from Connecticut College for Women and preparing to become a math teacher, was engaged to marry Carroll Boyce ('44) later that year. Arthur, a Phillips Exeter Academy graduate, would soon enter MIT as a first-year student. His parents thought he should go away to college, but he insisted on MIT—"What the heck," he told them, "I want to go into science, so why wouldn't I go to Tech?" He would graduate in 1951, teach at the Mount Hermon School in western Massachusetts, and earn a master's degree in education at Harvard. Everyone missed the family patriarch and matriarch, Elias and Otelia. Elias had died in 1938, aged 82. Otelia passed away peacefully in December 1944 at 86, in the old family homestead on College Avenue in Wooster, Ohio. The family treasured the memory of her selection as American mother of the year in 1939, of the reception held at New York's Waldorf-Astoria, and of the award presented by Franklin Roosevelt's mother, Sara (Mrs. James) Roosevelt.

Postwar demands on Compton by the government did not let up. A vow that he had made in 1944 to his son-in-law, not to get roped into postmortem commissions by "the Powers That Be" and "once the war is won . . . the big job is to get things going again back here" at MIT, quickly fell by the wayside. In June 1946, President Harry Truman appointed him a member of the evaluation board for the Joint Chiefs of Staff, and as chair of Truman's personal commission to oversee and assess nuclear detonation and fallout tests in the Bikini Atoll. When reporters probed him for classified information, Compton brushed them off with a (for him) raunchy, off-color joke. "Gentlemen, you remind me of the bishop at the banquet, who in the course of helping himself to salt, spilled it when the cover of the shaker dropped off. To hide his confusion he picked up a pinch and tossed it over his shoulder. Some of the flying salt hit a dowager sitting just behind him at another table and went down the back of her low-cut gown. When the lady realized what had happened she turned around and cried, 'Oh no, Bishop, you can't catch me that way!'" Compton's cheeriness in the face of such a grave mission—he found time, in the midst of it all, to go deep-sea fishing off Kwajalein—offended some politicians, scientists, and ordinary citizens concerned about the consequences of atomic warfare. His support, in fact, for the use of the atomic bomb against the Japanese never wavered. A number of his friends—Christian missionary John Hayes, for example—tried to change his mind, but he stood firm: "I agreed with the decision, and I've taken it in all conscience, and I'll let the matter rest there." On the advice of Bush and Conant, he laid out his arguments in an article, "If the atomic bomb had not been used," published in the December 1946 issue of *Atlantic Monthly*. The number of lives lost in the incendiary raids over Tokyo, he pointed out, was less than that caused by either bomb dropped on Hiroshima and Nagasaki, and Japanese leaders conceded that the bombs had preempted a devastating ground and air assault that would have multiplied casualties many times over.

Before leaving for the Bikini tests, Compton arranged for Killian to be appointed acting president. A few months earlier, the executive committee had changed Killian's title from executive vice president to vice president in hopes of reinforcing his stature as second-in-command and heir apparent. Edward Moreland agreed to become executive vice president as a part-time consultant to both Compton and Killian, while Thomas Sherwood took Moreland's place as dean of engineering. Compton, meanwhile, thought long and hard about passing the presidential reins to someone else, confiding to friends what his father Elias had told him years before, about having stayed past his welcome—and usefulness—as dean of the College of Wooster. He resisted the notion that Killian might succeed him, not because he doubted Killian's ability but because he felt that MIT's interests would be best served with a scientist at the helm. Instead, he recommended outgoing Rad Lab director Lee DuBridge. DuBridge, however, found few supporters at MIT. Some resented his having been chosen over their heads to direct the Rad Lab. Several Institute loyalists—Julius Stratton, Edward Bowles, treasurer Horace Ford, and Corporation members Bradley Dewey, Redfield Proctor, and Frank Jewett—discussed ways to derail the appointment.

"I unburdened my soul," said Bowles about a conversation he had with Ford, "indicating that I thought MIT had had enough of a dose of physics." Bowles met with Jewett, who told him: "Ed, I think I have an answer. As you know, I'm a graduate of the California Institute of Technology. They are in need of a president and I believe I can help by suggesting DuBridge for that job." Jewett's ruse worked. DuBridge went to Caltech to replace the retiring Robert Millikan.

Late in 1946, Truman asked Compton to chair the Advisory Commission on Universal Training, commonly called the Compton Commission. The commission heard testimony for six months, then issued a report in May 1947 unanimously recommending universal military training. The report was rejected in part because postwar fatigue made it difficult to build support for such a plan (only after the Korean War did Congress enact a peacetime selective service system). Compton's disappointment, along with his contentious public debates with peace advocates such as Henry Cadbury of the Harvard Divinity School, made him long for a quick return to MIT. What worried him most, he said in November 1947, was the prospect of a "weak and irresolute America" adrift in a world where threats from foreign powers seemed likely to grow. He took some comfort in knowing that one segment of the report, stating that "nothing could be more tragic for the future attitude of our people, and for the unity of our Nation, than a program in which our Federal Government forced our young manhood to live for a period of time in an atmosphere which emphasized or bred class or racial difference," likely influenced Truman to craft and sign his historic January 1948 executive order ending racial segregation in the armed services. Two years earlier Compton's brother Arthur had described how well scientists and technicians from differing backgrounds—"colored and white, Christian and Jew"—had bonded in the Manhattan Project.

Compton went back to MIT and stayed for about a year. Then, in the fall of 1948, came another request, this one more urgent than the last, from Truman's defense secretary James Forrestal. The nation's military affairs—there was as yet no centralized Department of Defense—were being managed by a civilian group, the Research and Development Board (RDB), chaired by Vannevar Bush. But military resentment over civilian control grew so pervasive that little got done,

Bush's nerves wore thin, and his doctors advised him to step down. Compton took his place out of patriotic duty, not because he wanted the post or felt well qualified for it. The job was so difficult—it took on the aspect, said one observer, of "irresistible forces meeting immovable objects"—that he could hardly manage both it and the MIT presidency. Killian, meanwhile, mulled an offer to become president of Duke University. He was a native North Carolinian and had begun his undergraduate career at Duke before transferring to MIT, so sentimental attachment as well as professional opportunity came into play. In July 1948, as if preparing himself for this new role, he drew up a three-page "memorandum on specifications for a chief executive of a university." With MIT facing the loss of both men, Compton quickly decided to step aside and Killian as quickly decided to stay.

The Corporation voted on October 4, 1948, to accept Compton's retirement and to appoint him chairman. At the same meeting, on Compton's recommendation, Killian was appointed president-designate effective October 15, 1948—the fastest, smoothest transition in MIT's complex, checkered history of shifts in leadership. A new position—provost—was created in February 1949, and on March 10 Killian's choice to fill it, Julius Stratton, was approved by the executive committee.

Compton did not stay long as RDB chairman, just ten months. For a time he lived at the Cosmos Club in Washington, D.C., then took a small apartment at the Wardman Park Hotel. The job wore him out. His diplomatic skills proved useful, but conflicts between the various military agencies, and between them and the civilian sector, were more than he was prepared to cope with. When Forrestal resigned pronouncing himself a failure, then died in May 1949, Compton took both events very hard. He did not get along with Forrestal's successor, Louis Johnson, a deficit hawk known to begin meetings by pounding his desk and demanding news on how many people his advisers had fired that week, or how much fat they had slashed from the budget. Compton felt unproductive yet trapped by duty. In late May he was hospitalized with severe chest pains and on August 14, at lunch with friends in Washington, suffered a stroke. He returned to Washington after several weeks of recuperation with Margaret in Cambridge and at the Jaffrey farm. But in November, he resigned on the advice of doctors. "My deep concern for your health," Truman wrote in reply, "leaves me no alternative but to bow to medical advice and with utmost regret to accept . . . the resignation. . . . I tender thanks to you in the name of the Nation and pray for your speedy restoration to health." Truman chose Compton's friend and ally, William Webster, to succeed him.

Compton's struggles with national policy and personal health problems were offset by the thrill of returning once again to MIT, albeit in a different role. One special bright spot—MIT's Mid-Century Convocation—gave him a boost at a time when he felt close to despair over RDB affairs. The convocation, planned to coincide with Killian's inauguration as president, stretched over three days, March 31–April 2, 1949. Its objective, grandly phrased, was "to explore some of the pressing problems which have been raised by science for the twentieth century." Scholars and public figures came from around the world to exchange views on the impact of modern science. Among the featured speakers were Bernard Baruch, Vannevar Bush, Nelson Rockefeller, Percy Bridgman, Merle Tuve, and Lee DuBridge, but the primary drawing cards were former British prime minister Winston Churchill and U.S. president Harry Truman. Truman dropped out at the last minute, replaced by Harold Stassen, president of the University of Pennsylvania and perennial Republican-primary candidate for U.S. president. Churchill came, he said, on account of his "great admiration for Dr. Compton." Compton greeted him at South Station, then escorted him and his party of friends, family, bodyguards, secretaries, and assistants to their quarters at the Ritz-Carlton. Some sessions had to be held at the Boston Garden, the only venue large enough to accommodate the crowd of invited dignitaries and guests. As MIT did not award honorary degrees (and still holds to this tradition), Killian gave Churchill an honorary lectureship instead. "I shall take this," Churchill joked, "to mean that I can come and lecture you on any subject whenever I wish." He never followed up on that promise (or threat). The next day, April 2, Compton passed the reins of office to Killian, each promising to work together in their new roles.

The partnership continued strong, though with Compton watching mostly from the sidelines and playing thoughtful adviser as needed. Killian's initiatives had grown bolder through the years. In August 1946 he had proposed an institutional self-study to assess, perhaps to reshape, MIT's mission in the postwar era. To carry out the study, a Committee on Educational Survey—often called the Lewis Committee, after its chairman Warren Lewis—was appointed in January 1947. The other members were Ronald Robnett, Richard Soderberg, John Loofbourow, and Julius Stratton. Two subcommittees, one on staff environment and the other on general education, were chaired, respectively, by Rupert Maclaurin and Thomas Sherwood. In its report, published in 1949, the committee laid out a number of major reforms, including creation of a humanities and social studies school "on an equal footing with the existing schools at MIT," that "advancement of knowledge be considered an essential part of its program, that it assume responsibility for planning and administering the program of general education as a part of the common curriculum, and that it offer professional courses leading to graduate as well as undergraduate degrees." The school was established a year later. It was followed in 1952 by the School of Industrial Management (later the Sloan School), made possible by a grant of $5,250,000 from the Alfred P. Sloan Foundation.

Many of the postwar changes that occurred on Compton's watch bore Killian's imprint. It fell to Killian to coordinate a return to campus normalcy, to reintegrate some of the 1,500 students and 175 teaching staff in war service and to cope, as well, with an influx of new students and faculty. New faculty included Charles Coryell and John Sheehan in chemistry; Bruno Rossi, Victor Weisskopf, George Valley, Albert Hill, and Jerrold Zacharias in physics; Gyorgy Kepes in architecture; William Hawthorne and Warren Rohsenow in mechanical engineering; Arthur Ippen in civil engineering; H. Guyford Stever in aeronautical engineering; Jerome Wiesner, Yuk-Wing Lee, Robert Fano, and Henry Zimmermann in electrical engineering; Edward Cochrane in naval architecture; E. Cary Brown in economics; Chia-Chao Lin in mathematics; Elting Morison, Robert Woodbury, Duncan Ballantine, William Locke, and Klaus Liepmann in the humanities; and many more. Groundwork for a capital fundraising campaign, with a target of twenty million dollars, was laid in 1947 via a Killian-inspired brochure,

M.I.T.—A New Era, its "opening guns" (Compton's words) fired a year later, just before Killian's inauguration. There were new academic programs to plan for, new buildings (dormitories, laboratories), athletics grounds, and other facilities expected of a modern university. The twenty-million-dollar campaign surpassed expectations, netting twenty-six million, and new construction went up around campus—Baker House, a dormitory designed by Finnish architect Alvar Aalto; the Dorrance Laboratory of Biology and Food Technology; and Kresge Auditorium and the Chapel, both designed by Finnish architect Eero Saarinen.

The Cold War, meanwhile, imposed new expectations and responsibilities. Defense-related, government-sponsored projects—Project Lexington and Project Charles in the late 1940s and early 1950s—led to the creation of Lincoln Laboratory, whose initial goal, in response to the Soviet Union's detonation of its first atomic bomb (five months after Killian's inauguration), was to develop an electronic continental defense system. The Instrumentation Laboratory set to work on inertial guidance systems for the coming age of intercontinental ballistic missiles. Difficult social and political issues—race, gender, equality, loyalty, patriotism—required Killian to think long and hard, to fend off extreme or inflammatory positions, to craft nuanced approaches, and to adjust to a shifting dynamic between government, industry, and the academic world. Killian looked first to his provost and closest adviser, Jay Stratton, to help address these issues, but also took advantage of Compton's wisdom and experience.

Compton wondered about history repeating itself, as he watched Killian drawn away on federal assignments—to Truman's communications policy board and advisory committee on management in 1950, to the Army's scientific advisory panel (as chair) in 1951, then, starting in 1954, into President Dwight Eisenhower's inner circle of defense experts. For his own part, Compton cut back on trips, speeches, and meetings. His 1949 stroke had slowed him down, leaving him with a slight speech impediment and the inability to detect temperature changes on his right side. But he still pushed himself to travel, to serve on several corporate boards, and to offer advice when asked. A bout with pneumonia laid him low on a six-week trip to England and France in the fall of 1952. In February 1954, he spoke on technical education before a joint committee of Britain's House of Commons and House of Lords. He and Margaret then flew on to Tel Aviv, spoke at Technion, part of a campaign to raise funds for a chemical-engineering building, and toured Hebrew University in Jerusalem. They spent a day at the Sea of Galilee. When someone pointed out the Hill of the Beatitudes, traditional location of the Sermon on the Mount, Compton was overcome by emotion and whispered to Margaret: "It doesn't matter where the sermon was given—but it matters mightily that it *was* given, and has become the standard by which to judge human excellence."

Back home in late May, he turned compost one day at the family's Jaffrey retreat and collapsed from fatigue. He moderated a panel at MIT's Alumni Day on June 14, then headed off to New York. Margaret got a call on June 17 to say that he had been admitted to Cornell Medical Center after suffering a heart attack. She rushed to his bedside. His first question was about the children. All fine, she said, nothing to worry about. "What's in the paper about Oppenheimer?" Physicist and former Manhattan Project director J. Robert Oppenheimer, whom Compton greatly admired, was about to lose his contract as chairman of the Atomic Energy Commission's general advisory committee in the wake of a series of highly charged, Cold-War inspired allegations. Compton rested comfortably for five days, then died of a massive blood clot on June 22. Three days later, following a private service, his ashes were scattered in MIT's Great Court, where on another June morning twenty-four years earlier he had stepped forward, a bit nervously, to give his inaugural address.

Notes

Archival (AC) and manuscript (MC) collections are cited below in abbreviated form by collection number and, if available, box and folder numbers, e.g., MC1:2:3 refers to MC1, William Barton Rogers Papers, 1804–1911, box 2, folder 3, Institute Archives and Special Collections, MIT. For complete citations, see Sources. Also see Sources for bibliographic information on authored and unattributed works, cited here in abbreviated form. MIT Museum citations refer to biography files, available in the Museum's reference collection. The MIT *President's Report* is abbreviated throughout as PR, with the date cited referring to year of report preparation or publication (e.g., PR 1872 indicates the report for academic year 1871–72). Numbers at the left indicate page number in this text.

William Barton Rogers

1 "An Act to incorporate . . . manufactures and commerce": MC1:15:40, An Act to incorporate . . .

1 "I have very little of interest to tell": Rogers, E. 1896, II, 75, William Rogers to Henry Rogers, April 2, 1861.

3 "Your son has . . . promise before him": *Abstract of the Proceedings* 1883, 31.

4 "to impart such knowledge . . . mercantile pursuits": *Baltimore Commercial and Daily Advertiser*, May 13, 1828.

4 "the lofty spirit of my father in me": Rogers, E. 1896, I, 84, Henry Rogers to William Rogers, February 2, 1830.

4 "I am confident . . . our country": MC1:1:12, Joseph Henry testimonial, July 6, 1835.

5 "true science and broad philosophical views": Rogers, E. 1896, I, 159, William Rogers to Robert Rogers, April 19, 1838.

5 "He, of all men . . . the human condition": Walker 1895, 4.

5 "applications . . . in the workshop": Rogers, W. n.d. [1837].

6 "Matters here as usual . . . too dull": Rogers, E. 1896, I, 182, William Rogers to James Rogers, March 8, 1841.

7 "negro serving-man . . . as his master thought": *Abstract of the Proceedings* 1883, 36.

7 "stupid dullness . . . promote my happiness": Rogers, E. 1896, I, 181, William Rogers to Robert Rogers, 1841.

7 "fratricidal war": Rogers, E. 1896, I, 102, William Rogers to James Rogers, February 22, 1833.

7 "even the windows . . . eager listeners": *Abstract of the Proceedings* 1883, 37.

7 "A grander geological theme . . . mysterious mountain chain": MC1:8:130a, John Hayes to Matthias Ross, June 1, 1882.
9 "The true and only practicable object . . . in connection with physical laws": quoted in Bevan 1940, 123.
10 "a type of the best New England womanhood": AC13:24:113l, memorial to Emma Rogers by James P. Munroe and Thomas Livermore.
10 "given to rather extreme opinions and violent expression of them": Gunn 1928–36.
11 "for the purpose . . . common purposes of life": quoted in McAllister, 367–71.
12 "for the use . . . public good": *Abstract of the Proceedings* 1883, 15.
12 "We need . . . domestic and political economy, etc.": quoted in Stratton and Mannix 2005, 159.
13 "I *denounced* no doctrine . . . so fair minded a philosopher as Darwin": MC1:3:35, William Rogers to Henry Rogers, February 21, 1860.
13 "Agassiz's . . . capacity for debate was small . . . should control statements": quoted in Livingstone 1987, 116–17.
14 "The more pity I felt . . . same blood as we are": quoted in Lurie 1988, 257.
15 "We the subscribers . . . Massachusetts Institute of Technology": MC1:16:39, Act of association of an institute of technology, January 11, 1861.
15 "Between ourselves I know . . . might come in and weaken it": Rogers, E. 1896, II, 75, John Andrew to William Rogers, March 9, 1861.
16 "patriotic and firm": Rogers, E. 1896, II, 72, William Rogers to Henry Rogers, March 5, 1861.
16 "The times are not favorable . . . education and the arts": Rogers, E. 1896, II, 129, William Rogers to C. C. Jewett, August 7, 1862.
17 "as its leading object . . . Industrial Sciences and Arts": quoted in Stratton and Mannix 2005, 349.
17 "can be carried . . . Buildings and arrangements": quoted in Stratton and Mannix 2005, 350.
17 "the instant reply . . . with this independence": Rogers, E. 1896, II, 163–64, William Rogers to William Walker, May 4, 1863.
18 "more especially to the unbooked knowledge": MC431 s.II:7: Curriculum notes, comment by George Emerson at Corporation meeting, 1863.
18 "from the very elements . . . scientific engineer": Rogers, E. 1896, II, 186, William Rogers to Henry Rogers, January 19, 1864.
19 "In pursuing this object . . . generous scientific culture": MC431 s.II:7: Curriculum materials, *Scope and Plan*, 3
19 "I knew how exceedingly able . . . perfect in all its parts": MC1:4:50, John Runkle to William Rogers, January 15, 1865.
19 "nearer what it is intended . . . not for show": Rogers, E. 1896, II, 217–18.
20 "great and growing demand . . . industrial interests of this community": MC431 s.II:7: Curriculum notes, January 19, 1865.
21 "Mathematics, with practice . . . French Language": *Boston Evening Transcript*, February 20, 1865.
22 "*Organized the School! . . . a memorable day!*": Rogers, E. 1896, II, 224.
23 "One day a student . . . the ungainly thing": Richards, R. 1936, 38.
23 "a triumph of oratorical art . . . . upon their minds": "Sketch of Prof. William B. Rogers" 1876, 611.
23 "He always showed . . . of the child": *Abstract of the Proceedings* 1883, 27.
24 "cold as an icicle": James, H. 1930, I, 68.
24 "sweetness and sympathy": *Abstract of the Proceedings* 1883, 10.
24 "Professor Rogers . . . the good of the School": *Abstract of the Proceedings* 1883, 9–10.
26 "I have imagined . . . only thirty-one years of age": Eliot 1920, 433.
26 "through actual handling . . . shoulder to shoulder with the boys": Richards, R. 1936, 60–61.
26 "blackboards . . . always carefully prepared": *Abstract of the Proceedings* 1883, 10–11.
27 "a wonderful labyrinth": Richards, R. 1936, 36.
27 "ceased to be a plague spot and became a delight": Richards, R. 1936, 37.
28 "It was a long tramp . . . where they terminated": Bowditch n.d., I, 3–4.
29 "in what was intended . . . during daylight hours": Bowditch n.d., I, 5.

29 "It stands upon . . . purposes of the Institute": *Technique 1901*, 7.
31 "The Harvard men . . . between the two institutions": Richards, R. 1936, 48.
31 "in aid of the practical studies of the School": PR 1872, 74–75.
32 "the slaves . . . *forever free*": Rogers, E. 1896, II, 133, William Rogers to Henry Rogers, September 26, 1862.
32 "has brought within the folds . . . our American negroes": Rogers, E. 1896, II, 172, William Rogers to Henry Rogers, August 16, 1863.
33 "persons of either sex": quoted in Mannix n.d., "Women," 15.
33 "We would add . . . enjoyment of all": quoted in Mannix n.d., "Women," 13.
33 "with the present condition . . . organization of the classes": quoted in Mannix n.d., "Women," 23.
33 "seriously embarrassing": MC1:4:56, William Rogers to Nathaniel Thayer, February 4, 1867.
33 "They didn't want . . . a little diplomacy": Stinson [1911].
33 "lady assistant librarian . . . to take charge of the library and study-room and keep order therein": Stinson [1911].
34 "in the nature of an experiment": quoted in Mannix n.d., "Women," 25.
34 "belonging to any sex": Rogers, E. 1896, II, 373.
34 "special accommodations for the use of women": MC1:8:129, William Rogers to Ellen Richards, May 12, 1882.
35 "Institute on the brain": Rogers, E. 1896, II, 287, John Runkle to Emma Rogers, February 9, 1869.
35 "The Institute seems . . . connection with it": MC7:1, Emma Rogers to John Runkle, January 23, 1869.
35 "his nerves . . . about the Inst": MC431 s.II:11: Runkle, Emma Rogers to John Runkle, December 15, 1870.
37 "I am well enough . . . a *little* scientific work": quoted in Munroe 1904, 540–41.
37 "perhaps the crowning honor of his life": Rogers, E. 1896, II, 359.
38 "the Nestor of American geology": Rogers, E. 1896, II, 357–58.
38 "[He] had been informed . . . able to attend": Rogers, E. 1896, II, 359.
38 "the question of restoration . . . Declaration of Independence": MC1:7:102, Carl Schurz to William Rogers, May 6, 1880.
38 "I know how much influence . . . you will have": MC1:7:114, C. V. Riley to William Rogers, May 5, 1881.
39 "I am convinced . . . with other institutions": MC7:1, William Rogers to Charles Eliot, February 7, 1870.
39 "any less independent . . . than it is now": MC7:1, Charles Eliot to William Rogers, February 9, 1870.
39 "I could not see . . . repugnance to all such names": quoted in Munroe 1904, 539.
40 "The Institute of Technology . . . scientific schools of the country": Rogers, E. 1896, II, 316, William Rogers to William Galt, December 29, 1873.
40 "It appears to me . . . business red tape": MC431 s.II:11: Runkle, Samuel Kneeland to William Rogers, October 15, 1877.
41 "satisfied . . . satisfactory condition": MC1:5:78, William Rogers to Corporation, June 12, 1878.
41 "friends of Industrial Education": MC1:5:78, John Cummings et al. appeal, June 17, 1878.
41 "the cordial good will . . . conduct of its affairs": MC1:5:78, Corporation committee to Alumni Association, June 17, 1878.
41 "It was mutually . . . and yet survive": MC1:5:82, William Ware to Edward Atkinson, September 19, 1878.
42 "war before he is ready to meet the Enemy": MC1:5:82, Matthias Ross to Edward Atkinson, September 27, 1878.
42 "I cannot think . . . such bombshells as this": MC1:5:82, Robert Richards to Edward Atkinson, September 20, 1878.
42 "never returned from vacation . . . at the present time": MC1:5:82, Robert Richards to William Rogers, September 28, 1878.

42 "matters of discipline . . . the students": MC1:5:82, Alumni Association executive committee to alumni, September 25, 1878.

42 "sees fit to be present . . . entrusted to the President": MC1:5:78, William Rogers drafts, June 12, 1878.

42 "polytechnic in its character . . . auxiliary to the former": MC1:8:123, James Welling to William Rogers, November 12, 1881.

43 "divert into a scientific direction . . . law & politics": MC1:7:109, Charles Venable to William Rogers, January 13, 1881.

43 "the political bush-whacking . . . debates of spring": MC1:7:108, William Warren to William Rogers, December 18, 1880.

43 "Many of our graduates . . . in the newspapers": MC1:7:110, Charles Flint to William Rogers, February 14, 1881.

43 "Not one among us . . . as their father": MC1:8:126, Albert Hall to William Rogers, January 17, 1882.

43 "Though seeming weak . . . well earned laurels": MC1:7:115, A. H. Russell to Emma Rogers, May 28, 1881.

44 "In a high sense . . . a patent indefeasible": *Abstract of the Proceedings* 1883, 22.

45 "It is true . . . built upon the practical": quoted in *Abstract of the Proceedings* 1883, 22–23.

45 "Stephen Hales published . . . 128 grams of bituminous coal—": quoted in *Abstract of the Proceedings* 1883, 23.

45 "his expositions . . . scarcely seemed to come from earth": quoted in Munroe 1923, 226.

John Daniel Runkle

47 "I need not tell you . . . among his colleagues": MC7:1, Emma Rogers to John Runkle, March 24, 1869.

48 "Tell Dr Rogers . . . with him this summer": MC431 s.II:11: Runkle, John Runkle to Emma Rogers, n.d.

48 "It must be a great pleasure . . . elected President": MC431 s.II:11: Runkle, Edward Pickering to William Rogers, May 2, 1870.

48 "reap advantage . . . at the head": MC431 s.II:11: Runkle, Government records, September 19, 1870.

49 "it was a ship . . . change of navigator": MC431 s.II:11: Runkle.

50 "I date my birth . . . senior assistant in the office": Newcomb 1903, 1.

51 "Care should be taken . . . mathematical students": quoted in Bennett 1967, 6–7.

51 "You have one astronomer . . . and no mathematicians": Doolittle 1910, vi.

52 "from a distant state . . . saw the fraud": Newcomb 1903, 85.

52 "I have supposed . . . not realize a dollar": Tyler 1902, 281.

53 "In my own country . . . as your debtors": MC7:1, William Walton to John Runkle, October 5, 1860.

54 "You have . . . easy to you": MC431 s.II:11: Runkle, George Emerson to John Runkle, January 5, 1861.

54 "An attack was made . . . anything else to use": MC7:1, Cornelius Runkle to John Runkle, July 17, 1863.

55 "A kinder gentleman treads not the earth": *Technique 1896*, 244.

56 "[He] would demonstrate . . . amused the class": MIT Museum: Runkle, Edward Rollins to C. H. Davis, December 8, 1899.

56 "I remember very well . . . trouble occurred": MIT Museum: Runkle, Edward Rollins to C. H. Davis, December 8, 1899.

56 "The other morning . . . *Exeunt omnes*": *Spectrum*, November 22, 1873.

57 "a man with a streak of fun . . . occasional twinkle": *Technique 1887*, 125.

57 "repressed . . . thoroughly practical way": MC7, Biographical notes, n.d., 9.

57 "the concrete method of teaching": Eliot 1920, 435.

58 "not because they are necessary, but just to help you fellows": *Technique 1892*, 200.

58 "I learn that Eliot . . . it was settled": MC431 s.II:11: Runkle, John Runkle to William Rogers, July 27, 1870.
59 "agitating the question": quoted in Prescott 1954, 73.
59 "I can see nothing but injury . . . their respective spheres": Rogers, E. 1896, II, 293, William Rogers to John Runkle, February 1, 1870.
59 "if its friends . . . why Amen": MC7:1, Charles Eliot to John Runkle, February 4, 1870.
60 "cannot without a kind of suicide . . . any other institution": MC7:1, William Rogers to Richard Greenleaf, July 26, 1870.
60 "The rule . . . an independent school": MC431 s.II:11: Runkle, Alexis Caswell to John Runkle, August 4, 1870.
60 "The Committee . . . on the part of the Institute": MC431 s.II:11, Runkle, Charles Eliot to John Runkle, March 22, 1871.
60 "I am not afraid . . . everything *but* men": MC431 s.II:11, Runkle, William Atkinson to William Rogers, August 27, 1871.
62 "It has been a great pleasure . . . his awful puns": MC7:1, John Runkle to Catherine Runkle, June 17, [1871].
62 "I owe too much . . . to be tempted": MC7:1, Biographical notes, n.d., 5.
62 "We have made large collections . . . for some time if ever": MC7:1, John Runkle to Catherine Runkle, July 15 & July 19, 1871.
63 "Mr. Runkle seems . . . considerable use to them": quoted in Shrock 1977, 274–75.
63 "full of indoor philosophy . . . grow green in the sky": Fox, S. 1981, 5.
64 "A little reflection . . . theory and practice": PR 1873, 57–58.
65 "The excitements and fatigue . . . greater maturity and value": PR 1876, x.
66 "I was so much occupied . . . I presume": MC7:1, E. S. Hiraka to John Runkle, November 15, 1870.
66 "I have now the honor . . . silk-fabrics &c": MC1:6:92, Tanaka Fujimaro to MIT president, April 9, 1879.
67 "earnest and faithful . . . English language": MC1:8:128, Robert Richards to William Rogers, March 26, 1882.
67 "strong and sometimes turbulent corps": MC431 s.II:11: Runkle, Francis Storer to William Rogers, October 5, 1870.
69 "provided that this position. . . *all their time*": MC7:1, Benjamin Peirce to John Runkle, June 20, 1869.
70 "I am glad to know . . . in theory": MC431 s.II:11: Runkle, William Rogers to John Runkle, August 11, 1873.
71 "obligation . . . a technical education": MC431 s.II:11: Runkle, Committee on Instruction minutes, February 27, 1871.
71 "secure to every student . . . his chief object": *Boston Globe*, January 22, 1875.
71 "not only because the subject . . . spent on logic": *Spectrum*, December 6, 1873.
71 "His high *a priori* abstract methods . . . difficult for us to utilize": MC1:5:77, William Atkinson to William Rogers, February 7, 1878.
72 "We are not a University . . . active business life": PR 1874, 74.
73 "the art of making patterns . . . Oil-Cloths, etc.": PR 1873, 6.
73 "eminently useful addition . . . its kind in America": *Boston Globe*, May 19, 1874.
73 "Nothing more delicate and tasteful can well be imagined": *Boston Globe*, April 7, 1876.
74 "We need to . . . like mere schoolboys": MC1:5:81, John Ordway to William Rogers, August 21, 1878.
74 "regretfully . . . least of the benefits": MC298:1:8, Runkle et al. to Corporation, October 19, 1887; Edward Atkinson to Francis Walker, December 17, 1887.
75 "electricity in the female": MC431, s.II:11: Runkle, Society of Arts.
75 "mechanical humbugs": MC431, s.II:11: Runkle, Society of Arts, December 24, 1874.
75 "confident that some such machine . . . supersede the pen": MC431, s.II:11: Runkle, Society of Arts, April 15, 1877.

75 "the indisposition . . . the exchange": MC431 s.II:11: Runkle, Society of Arts, February 22, 1877.
75 "through inadvertence . . . to keep up its organization": MC431 s.II:11: Runkle, Corporation, February 14, 1877.
76 "required too much study . . . for boys of average ability": MC431 s.II:11: Runkle, Faculty minutes, December 7, 1872.
77 "a fine lot of young fellows . . . sons of ambitious mechanics": Freeman, J. n.d., 43.
77 "our station or abiding place . . . table and desk": Freeman, J. n.d., 40.
78 "We were favored . . . the dining room": PR 1876, 161–62.
79 "a large black tom-cat . . . engraved to another": *Boston Herald*, May 20, 1871.
79 "One cannot . . . while here": *Spectrum*, February 22, 1873.
80 "in conformity . . . such exercises": MC431 s.II:11: Runkle, Faculty minutes, March 26, 1875.
81 "If there were a gymnasium . . . before next winter": *Spectrum*, May 17, 1873.
81 "perhaps a kitchen . . . at cost price": MC431 s.II:11: Runkle, Corporation, April 8, 1874.
81 "The Institute is supposed . . . at least partially, unnecessary": *Spectrum*, April 17, 1874.
83 "that the admission of women . . . at present expedient": AC1, October 12, 1870.
83 "I thought the President . . . my presence in the laboratories": quoted in Slaughter 1980, 3.
83 "I was shut up . . . not allowed to attend classes": quoted in Slaughter 1980, 4.
83 "if we two got a reputation . . . advancement of coeducation": quoted in Slaughter 1980, 4–5.
84 "I hope I am winning a way . . . stronger allies than anything else": quoted in Slaughter 1980, 3.
84 "have proved . . . nor unfit them for housewifely duties": quoted in Slaughter 1980, 15.
84 "I have heard . . . in chemical analysis?": MC431 s.II:11: Runkle.
84 "It really seems hard . . . good education in a specialty": MC7:1, Edward Cabot to John Runkle, August 26, 1872.
85 "The undersigned . . . equal terms to all citizens": Acts—1873, chap. 174, Massachusetts Archives.
85 "a man . . . disturb the harmony of the corporation": *Boston Globe*, February 12, 1873.
86 "extreme women . . . attain the desired end": *Spectrum*, April 5, 1873.
86 "such departments . . . regular work of the School": quoted in Mannix n.d., "Women," 30.
87 "women who may have been . . . the usual diploma": quoted in Mannix n.d., "Women," 33A.
87 "not expedient . . . provided fitly for them [otherwise]": MC1:6:94, Edward Atkinson to William Rogers, October 17, 1879.
87 "the dangers of intermittent coeducation . . . scientific education for women": quoted in Mannix n.d., "Notes," 11–12.
88 "Some of the difficulties . . . first D.S. in chemistry": Richards, R. 1936, 153.
88 "do something to benefit the world . . . intellectual and material wealth": MC1:7:108, John Newell to William Rogers, December 15, 1880.
89 "a broad field . . . genuine capacity": MC1:5:78, Corporation to Marian Hovey, June 24, 1878.
89 "best suited . . . students of both sexes": MC1:6:87, Marian Hovey to William Rogers, December 13, 1878.
89 "Our donation . . . women students are treated": MC1:6:87, Marian Hovey to William Rogers, December 13, 1878.
89 "Can some one else . . . if I am not in it now": MC431 s.II:11: Runkle, John Runkle to William Rogers, January 17, 1875.
90 "I had much rather . . . 'reign' at Troy": MC7:1, T. Sterry Hunt to John Runkle, June 14, 1871.
90 "I understood . . . else we shall fall behind": MC431 s.II:11: Runkle, Edward Atkinson to William Rogers, March 27, 1875.
90 "the somewhat flagging zeal of our Corporation": MC431 s.II:11: Runkle, William Rogers to Edward Atkinson, October 19, 1875.
91 "a man of quite sufficient breadth and culture": MC431 s.II:11: Runkle, Edward Atkinson to William Rogers, October 12, 1875.
91 "want of . . . a firm will and a clear purpose": MC431 s.II:11: Runkle, Edward Atkinson to Marian Hovey, July 17, 1878.
91 "It is this or bankruptcy . . . our invested funds": MC431 s.II:11: Runkle.
91 "the present financial embarrassment of the Institute": MC431 s.II:11: Runkle, Faculty minutes, November 16, 1877.

91 "full of peril . . . present financial depression": MC431 s.II:11: Runkle, Corporation, April 10, 1878.
92 "on the ground of the present financial depression": MC431 s.II:11: Runkle, Corporation, April 10, 1878.
92 "unfavorable impression . . . this time": MC431 s.II:11: Runkle, Edward Atkinson to Marian Hovey, July 17, 1878.
92 "a very successful & pleasant affair . . . for our encouragement": MC431 s.II:11: Runkle, John Runkle to William Rogers, May 28, 1878.
92 "Institute matters . . . not among the probabilities": MC431 s.II:11: Runkle, John Runkle to William Rogers, May 31, 1878.
93 "I am startled . . . not wholly unexpected": MC431 s.II:11: Runkle, William Rogers to John Runkle, June 1, 1878.
93 "I can not let him relinquish . . . an honored place in its history": MC431 s.II:11: Runkle, Corporation, notes by Harry Tyler.
93 "President Runkle has worked . . . professorship in applied mathematics": MC1:5:78, Corporation to Marian Hovey, June 24, 1878.
94 "to avoid all violent exercise . . . eat but few vegetables": MC7:3, August 19, 1878.
94 "He has so little strength . . . work great changes": MC7:3, February 11, 1879.
94 "expose himself . . . would be very acceptable": MC1:6:92, William Rogers to John Cummings, July 11, 1879.
95 "As a teaching laboratory . . . large numbers of students": Runkle 1881, 84.
95 "It is undoubtedly better . . . cannot bear the change": Runkle 1881, 84–85.
95 "The whole morning . . . Moscow school for its model": MC7:3, November 5, 1879.
96 "There is a small . . . make rapid strides in America": Runkle 1881, 88.
96 "we are expecting . . . some of these days": MC1:7:106, John Runkle to William Rogers, September 12, 1880.
96 "gave the impression . . . taught by him": MC431 s.II:11: Rogers, George Osborne to William Rogers, November 2, 1879.
96 "Now you can see . . . has proved my ruin?": MC431 s.II:11: Rogers, John Runkle to William Rogers, September 20, 1880.
97 "preparing him . . . disappointment in store": MC1:7:106, William Rogers letter, correspondent unknown, September 20, 1880.
97 "Considering . . . expressed nothing": MC431 s.II:11: Runkle, William Rogers to Edward Philbrick, September 20, 1880.
97 "I think the action . . . relief & satisfaction": MC1:7:106, William Rogers to Lewis Tappan, September 25, 1880.
97 "I am sorry . . . tuition many times over?": MC431 s.II:11: Runkle, John Runkle to Lewis Tappan, October 30, 1884.
98 "venerable but robust figure . . . you ever came across?": quoted in Davis, E. 1912, 396.
98 "To deal with senescence . . . should have retired years ago": MIT Museum: Pritchett, Robert Bigelow to Abraham Flexner, December 7, 1942.
98 "Our work in pure mathematics . . . students duly qualified": PR 1890, 28.
99 "grateful appreciation . . . your associates": Tyler 1902, 305.
99 "The coast of Maine . . . healthful surroundings": MIT Museum: Runkle, Bertrand Clergue to John Runkle, June 30, 1902.

Francis Amasa Walker

102 "The Classics . . . always a hobby of mine": quoted in Munroe 1923, 215.
102 "a peculiar . . . object of their emulation": Munroe 1923, 3–4.
103 "my superb counselor and best friend": MC1:7:104, Francis Walker to William Rogers, July 5, 1880.
103 "What I do consider . . . chief end of man": MC298:2:3, Francis Walker to Edward Towne, January 31, 1896.

103 "Straws show which way . . . drowning men catch at them": *Springfield Republican*, January 10, 1897.
103 "I shall be out of the woods . . . ask further indulgence": MC431 s.II:11: Walker, Francis Walker to William Rogers, February 5, 1881.
104 "I am aware . . . offered you at Boston": quoted in Munroe 1923, 207.
104 "waltzed away . . . bean-pastures of Boston": quoted in Munroe 1923, 209.
104 "I trust in him . . . his capacities for usefulness": *Abstract of the Proceedings* 1882, 15–16.
104 "My plans are all knocked . . . at present": quoted in Munroe 1923, 217.
104 "We are much gratified . . . its established character": MC1:7:115, John Ordway et al. to William Rogers, May 23, 1881.
105 "To some of us it seemed strange . . . chief business of each one of us": *Boston Transcript*, June 8, 1897, quoting James Crafts.
105 "a great orchestra . . . the great symphony": Wright, C. 1897, 247.
105 "little or no admixture from other than British stock": Munroe 1923, 5.
106 "the only man . . . without any effort": quoted in Munroe 1923, 25.
106 "on account of rowdyism": Munroe 1923, 29.
106 "his eyes . . . peacefully declined": Galpin 1897.
107 "never winced . . . sharp controversy in their mouths": Wright, C. 1897, 253.
107 "pushing forward . . . into good places": quoted in Munroe 1923, 112.
107 "task of Sisyphus . . . intolerance of rascality": quoted in Munroe 1923, 122.
108 "like children . . . demand their presence": quoted in Munroe 1923, 135.
108 "slough off Washington and breathe a pure atmosphere": quoted in Munroe 1923, 143.
108 "a pioneer . . . the scientific method": Chittenden 1928, I, 185.
108 "select course . . . other higher pursuits, to business, etc.": quoted in Chittenden 1928, I, 142.
109 "I shall never forget . . . in dead earnest": quoted in C.-E. A. W. 1897, 41.
109 "a caged lion": *Outlook*, January 16, 1897.
110 "for the beauty . . . technical education in general": PR 1876, 166.
110 "beaten horse, foot and dragoons": quoted in Munroe 1923, 179.
111 "I was born . . . a fearsome thing": quoted in Munroe 1923, 25.
111 "It is fortunate for you . . . by throwing you downstairs": quoted in Munroe 1923, 143.
111 "I told him one time . . . resemble Napoleon strikingly": MC298:10, Memorial scrapbook.
111 "a somewhat punctilious brother . . . rather austere father": quoted in Munroe 1923, 376–77.
111 "I don't remember . . . language is mine": Richards, R. 1936, 94–95.
111 "Nothing is so potent . . . a hard and long fight": MC298:1:11, May 2, 1891.
112 "skedaddling Rebels": quoted in Munroe 1923, 48.
112 "capable of truly Jovian wrath . . . genuine magnetism": Nichols, J. 1928–36.
113 "I thought . . . a big and busy man": quoted in Broderick 1945, 28.
113 "You should be very thankful . . . a strut from a tie": Smith 1954, 61.
114 "to inquire and report . . . the Institute": MC298:1:2, Committee report, March 24, 1882, 1.
115 "I am, myself . . . in the future": MC298:1:9, Francis Walker to Edward Philbrick, May 31, 1888.
116 "against the possibilities . . . internal mismanagement": PR 1884, 19.
116 "enlarge its means of present usefulness": PR 1885, 24.
117 "place us . . . so much needed": PR 1886, 31–32.
117 "It will soon be necessary . . . lack of pecuniary means": PR 1893, 5.
120 "the increasing necessity . . . public health": PR 1889, 45.
121 "It would be difficult . . . promise of usefulness": PR 1892, 42.
121 "As the primary object . . . pupils or their parents": PR 1883, 20.
121 "of a nature . . . man of business": PR 1883, 21.
121 "practical and liberal tendencies . . . practical business pursuits": PR 1883, 20–21.
122 "It is a familiar feature . . . in this school": PR 1883, 22.
122 "repugnant to foppery . . . indolence": PR 1887, 32.
122 "polish the surface . . . mind and character": PR 1886, 7.
122 "the difficult and delicate duty . . . professionally and socially": PR 1890, 23.
122 "Trained, day by day . . . a source of great power": PR 1890, 25.

123 "the students . . . setting timbers and laying track": PR 1888, 33.
124 "the first teacher . . . of our own breeding": PR 1893, 51.
125 "Mr. Brandeis . . . dealing with concrete cases": PR 1895, 71.
125 "Those talks at Tech . . . epoch in my own career": quoted in Mason 1946, 86–88.
125 "Desire will always . . . able to pay": Litchfield 1954, 55.
126 "The history . . . our most valued instructors": PR 1895, 24.
126 "earnest hope . . . full enjoyment of health": PR 1890, 26.
127 "one of the severest blows": PR 1893, 23.
127 "There is something . . . singularly agreeable and inspiring": PR 1885, 23–24.
127 "black sheep": PR 1885, 23.
127 "to secure liberality . . . culture of other institutions": PR 1895, 27.
128 "I think this is doing pretty well . . . in-and-in-breeding is concerned": MC298:2:1, Francis Walker to Mr. Lyon, March 24, 1894.
129 "The General . . . repeated shocks, without complaint": quoted in Munroe 1923, 66.
129 "fruits of the training . . . not equaled on the Continent": quoted in PR 1895, 7–8.
130 "There is nothing here . . . or perhaps better": Walker 1899, 134.
130 "no better and no worse . . . a source of great danger": MC298:2:3, Francis Walker to Edward Towne, January 31, 1896.
130 "immigration and its evils . . . great apprehension and alarm": *Boston Globe*, February 9, 1892.
130 "We are now draining off . . . forests of old Germany": Walker 1895, 12–13.
131 "This great Polish swamp . . . unless we restrict its ingress": Ripley 1899, 372–73.
131 "impossible for white people . . . familiarity breeds audacity": *Boston Globe*, October 26, 1899.
131 "all that we hold true and manly": *Boston Transcript*, January 5, 1898.
132 "is never likely . . . for graduation": PR 1887, 15.
132 "students of this sex . . . laboratory work": PR 1883, 26.
133 "began their professional work . . . at the seashore": PR 1892, 53.
133 "favorably disposed . . . for the position": quoted in Brayer 1996, 165.
133 "by the nature of the case . . . excluded by reason of sex": Munroe 1893, 8.
135 "I see no reason . . . in German and French": MC298:2:3, Francis Walker to Rachel Moore, December 4, 1896.
136 "Every now and then . . . your own proper interests?": MIT Museum: Walker, Francis Walker to George Aborn, January 17, 1887.
136 "The grumblers . . . keep up with the class": MC298:1:8, quoted in C. A. Barton to Francis Walker, November 6, 1887.
136 "more work and study . . . my health or best interests": MC298:1:9, Charles Appleton to Francis Walker, January 12, [1888].
136 "The Institute of Technology . . . if they are to succeed": *Boston Herald*, December 14, 1890.
136 "I am of opinion . . . thoroughly well grounded": MC 298:1:11, Edward Atkinson to Francis Walker, March 10, 1890.
137 "from considerations of health . . . exceptional difficulty": PR 1891, 19.
137 "Since it is naturally regarded . . . work very well": PR 1896, 35.
137 "responsibility for their own lives . . . the community": PR 1896, 38.
138 "a darling masher": *Boston Globe*, February 8, 1886.
140 "How much difference . . . with his fellows": PR 1893, 56–57.
141 "Opportunities for championship athletics . . . athletic career": MC 298:2:3, Francis Walker to Paul Price, November 13, 1896.
142 "a sweet society of fair ones": *Technique 1900*, 263.
142 "opportunities for social intercourse": PR 1896, 37.
143 "We are sorry . . . give it prominent place": *Tech*, November 17, 1887.
144 "an imposing figure . . . presided over the meetings": quoted in Munroe 1923, 269.
144 "quick, decisive, energetic way . . . where it was asked": *Boston Advertiser*, January 7, 1897.
144 "the manner and ways . . . some infernal society": MC298:10, Memorial scrapbook.
145 "was absolutely without foundation . . . any way": *Boston Herald*, January 5, 1897.
145 "Without a word . . . no suffering": *Boston Herald*, January 5, 1897.

145 "I opened the door . . . President Walker is dead": King n.d., 23.
146 "the traveling public . . . the Institute especially": *Boston Herald*, January 5, 1897.
146 "throwing off sheet after sheet . . . service to the community": Haynes 1897, 9.

James Mason Crafts

149 "I feel rather used up . . . advantageous to the Institute": MC431 s.II:11: Crafts, James Crafts to Charles Cross, January 9, 1897.
150 "great joy expressed . . . on the occasion": Crafts and Crafts 1893, 119.
150 "surrounded by glowing eyes . . . around his bedroom": "James Mason Crafts" 1899, 8.
152 "never had a well-equipped laboratory . . . high temperatures and pressures": Crafts 1900, 998.
154 "have proved so easy . . . with them next year": PR 1872, 31.
154 "It is to be hoped . . . make investigations there": PR 1872, 34.
155 "out of jail for five years . . . American Ph.D. mill": Maddocks 1986, 156.
155 "You have been absent . . . it actually embraced": MC431 s.II:11: Crafts, William Rogers to James Crafts, n.d. [1880].
156 "Aside from any question . . . his cooperation will greatly strengthen us": MC1:8:126, Charles Wing to William Rogers, January 28, 1882.
156 "The accession of a chemist . . . history of the Institute": PR 1893, 47.
157 "The Chem. Dept . . . in our own hands": MC121:1:4, Charles Cross to Harry Goodwin, April 2, 1894.
157 "We have no use . . . no reference whatever to consequences": quoted in Ashdown 1928, 919.
157 "a man to have much hold on the students": quoted in Flexner 1943, 69.
157 "he was so bent . . . drop from his nose": Ashdown 1928, 919.
157 "He is a scholar . . . sweet as summer": *Technique 1900*, 258.
158 "desultory . . . serious and scholarly tendency": *Boston Globe*, December 31, 1897; December 30, 1899.
158 "We have here in New England . . . mutual distrust": *Boston Advertiser*, January 25, 1897.
158 "being convinced . . . not inconsistent with their legal obligations": MC431 s.II:5: Harvard negotiations, 1897–98, Francis Lowell to MIT Corporation, April 12, 1897.
159 "It is quite true . . . by a fine esprit de corps": MC298:2:7, James Crafts to Samuel Cabot, July 29, [1897].
160 "diminish undue rivalry": PR 1898, 14.
160 "the school of applied science . . . connected with said University": MC431 s.II:5: Harvard negotiations, 1897–98.
161 "No, I am not the man . . . state his name": *Boston Globe*, August 6, 1897.
161 "this perilous seat": AC1, November 3, 1897.
161 "The chief thing . . . give us theirs": MC431 s.II:5: Harvard negotiations, 1897–98, James Crafts to Augustus Lowell, August 4, 1897.
162 "Our Corporation . . . than had heretofore existed": MC431 s.II:5: Harvard negotiations, 1897–98, Charles Eliot to James Crafts, December 25, 1897.
162 "The union . . . hope for a fruitful celibacy": MC431 s.II:11: Crafts, Silas Holman to James Crafts, January 26, 1898.
162 "the first step . . . extend its resources": MC431 s.II:5: Harvard-Tech alliance 1897–1905, newspaper clipping, January 14, 1898.
162 "The Technology boys . . . absorb the Tech": *Boston Globe*, January 16, 1898.
163 "No radical changes . . . along the usual lines": PR 1898, 32; 1898, 33; 1898, 46; 1899, 36.
163 "as little burdensome as possible": PR 1897, 7.
163 "unerring memory . . . tact and wisdom": MIT Museum: Pritchett, Robert Bigelow to Abraham Flexner, December 7, 1942.
165 "On account . . . impart information": PR 1897, 30.
165 "slovenly and incorrect . . . inexcusably bad": PR 1897, 51.
165 "It is our business . . . a career of professional usefulness": PR 1898, 8.
165 "A college education . . . to make them workers": *Boston Globe*, June 9, 1899.
166 "shorn of decorations . . . businesslike plainness": *Boston Globe*, June 7, 1899.

166 "law of the survival of the fittest": *Boston Globe*, December 30, 1899.
166 "embarrassing evidences of prosperity": PR 1897, 9.
166 "Mr. Lowell's admirable letter . . . simplicity & good taste": MC431 s.II:11: Crafts, James Crafts to George Wigglesworth, July 10, 1899.
167 "It will be long . . . anxieties of such problems": PR 1897, 10.
167 "to meet the want . . . use of students": PR 1898, 11.
167 "Farther up Boylston st . . . loom up to the right": *Boston Globe*, November 29, 1897.
168 "Lost a tooth . . . the clothes I had on": *Technique 1900*, 61.
169 "Never shall we forget . . . the Freshman V": *Technique 1900*, 83.
169 "sports can never . . . more leisurely college life": PR 1899, 14.
169 "Students dwelling together . . . in a large city": PR 1899, 15.
171 "Aesop's fable . . . enjoy your company in that way": *Boston Globe*, February 4, 1899.
171 "Uneasy lies the head . . . whose kingdom is a school": *Technology Review* 2 (July 1900): 220.
171 "an emergency measure . . . help very largely": AC13:12:367, James Crafts to W. H. Lincoln, May 18, 1900.
172 "My reasons for taking this step . . . my most active interest": quoted in Ashdown 1928, 917.
172 "I hope that in years . . . the best way": Crafts 1900, 38.
173 "still much aggrieved . . . in large measure, ignored": MC431 s.II:11: Crafts, James Munroe to Samuel Stratton, July 2, 1924.
173 "As to the matter . . . the Institute would be the gainer": MC431 s.II:11: Crafts, James Crafts to George Wigglesworth, March 24, 1900.
173 "It has been a source . . . private laboratory in the Walker Building": PR 1900, 5.

Henry Smith Pritchett

175 "the duties . . . in other respects": *Boston Globe*, March 8, 1900.
175 "It hardly seems . . . in that direction": *Boston Globe*, January 6 & February 17, 1900.
176 "It is a good omen . . . the great valley of the Mississippi": Massachusetts Institute of Technology 1901, 7.
178 "An American . . . how far will human invention go?": quoted in Flexner 1943, 25.
178 "I went as a young beginner . . . grown famous": quoted in *Technology Review* 7 (January 1905): 27.
179 "I have just come through . . . after one semester's work": quoted in Flexner 1943, 44.
179 "in many ways good for the young fellows": quoted in Flexner 1943, 41.
179 "what the Methodists call a means of grace": quoted in Flexner 1943, 47.
180 "A bureau officer . . . support of all scientific men": AC13:29:460a, Henry Pritchett to W. E. Byerly, February 1, 1898.
180 "After three years of struggle . . . go into the work": MC431 s.II:3: Presidents—Pritchett, Henry Pritchett to George Wigglesworth, April 2, 1900.
181 "I hope that some way . . . its responsibilities": quoted in *Tech*, October 10, 1901.
181 "Is education to have for its object . . . a common civilization": Massachusetts Institute of Technology 1901, 38, 44, 46.
182 "precious opportunities for companionship and friendship": Massachusetts Institute of Technology 1901, 21.
182 "He will not forget . . . committed to his care": Massachusetts Institute of Technology 1901, 26.
183 "I wish to be your friend . . . give you mine": *Boston Globe*, October 27, 1900.
183 "a slight flower among a patch of lusty weeds": quoted in Flexner 1943, 70.
183 "possessed that rare quality . . . great devotion for him": quoted in Flexner 1943, 36–37.
183 "Mr. Hunter . . . enjoy being out there myself": quoted in *Technique 1903*, 246.
184 "a rational system of physical culture": PR 1900, 7.
184 "Only such competitive sports . . . subordinate to it": PR 1900, 8.
185 "by reason of lack of ability . . . technical knowledge": PR 1901, 12–13.
185 "Down with fraternities at Tech! Abolish clannishness!": *Boston Globe*, October 18, 1901.
186 "to prevent them . . . in their homes": *Concerning the Massachusetts Institute of Technology* 1909, 54.

186 "In both of these . . . responsibility to be a gentleman": PR 1905, 16.
187 "every one present . . . a feature of Technology life": *Boston Globe*, December 23, 1901.
187 "I regret . . . temperate and clean living": quoted in Flexner 1943, 71.
187 "the German drinking habit . . . a glass of beer": *Boston Globe*, February 19, 1902.
188 "For it's always fair weather . . . a good song ringing clear": Bullard 1907, 2–3.
188 "troubles smok'd away, with merry comrades near": Bullard 1907, 7.
188 "one of the few things . . . me and insanity": *Technique 1908*, 329.
188 "A change . . . the theory of least squares": *Technique 1904*, 177–78.
188 "rules of life . . . unknown regions of the world": *Boston Globe*, December 19, 1901.
189 "read good books and be manly men": *Boston Globe*, February 14, 1902.
189 "The attendance of students . . . our own fitness": PR 1901, 11.
189 "bright, intelligent looking boys . . . fitting for mining engineers": *Boston Globe*, October 16, 1904.
190 "what some Technology women have done": MC65:3:9: Women survey 1906.
190 "What knowledge of Chemistry . . . that of housekeeping": MC65:3:9: Women survey 1906, Martha Mackay to Ellen Richards, November 4, 1906.
190 "My time . . . children and keeping house": MC65:3:9: Women survey 1906, Mary Lockwood to Ellen Richards, November 16, 1906.
190 "illustrating our work . . . distinguished if we have not": MC65:3:9: Women survey 1906, Alice Tyler to Ellen Richards, n.d.
190 "having taken up the technical courses . . . theory of periodic dusting": *Technique 1903*, 252.
191 "You know my sentiments . . . it's time for them to vote": *Technique 1905*, 277.
191 "Do not be made unhappy . . . whatever work you do": *Boston Globe*, June 13, 1907.
191 "Wants to get a position . . . telephones and wiring houses": AC58, Student record cards, June 3, 1903.
192 "obstinate . . . a chip on his shoulder": AC1, May 12, 1905.
192 "As a race . . . in and out of their families": Laws 1901, 52–53.
192 "the full spirit of Lincoln": quoted in Flexner 1943, 83.
193 "The deeper down . . . highest courage": *Boston Globe*, December 23, 1905.
193 "the great incentive . . . meeting Booker T. Washington": *Technique 1900*, 308.
194 "an audacious proposition": quoted in Sherrill 1940, 151.
194 "seemed almost too good to be true": quoted in Kargon 1982, 95.
195 "The time has come . . . only just beginning to realize": PR 1902, 15; 1903, 18.
195 "The Neponset River . . . can't sit near it": AC13:9:251, Sarah Hughes to Henry Pritchett, June 9, 1902.
196 "had to fly to the dictionary to discover the meaning of": AC13:9:251, Sarah Hughes to Henry Pritchett, August 16, 1902.
196 "work upward . . . unskilled worker and the engineer": PR 1903, 20–21.
197 "spirit most generally prevalent . . . monetary value": AC13:32:559a: Study Committee, 1901–06, William Niles, March 1902.
197 "the Institute should remain . . . professional, graduate and research work": PR 1906, 24–25.
198 "a somewhat greater elasticity": PR 1906, 25.
198 "provide a full course of scientific studies . . . departments of active life": MC431 s.II:7: Course IX, General studies.
199 "My! but those Profs are grumpy": *Technique 1910*, 77.
199 "Home? . . . he won't recognize me": *Tech*, October 10, 1901.
200 "that casino for the young across the Charles": *Tech*, October 10, 1901.
200 "many [students] . . . adequate acquaintance with them": PR 1900, 16.
200 "a live center of intellectual and moral influence": PR 1903, 23.
202 "There's a whisper . . . we *must* be in the van": *Technique 1904*, 300–1.
203 "No one knows better than I . . . our eminent friend needs watching": MC1:8:137, William Sedgwick to Emma Rogers, October 27, 1903.
203 "one of the pleasantest . . . in hearty cooperation": quoted in Flexner 1943, 67–68.
204 "A fellow used to walk . . . place of a call": *Tech*, October 9, 1902.

205 "give to students . . . price of one": *Boston Globe*, October 29, 1904.
205 "the Tech-Harvard merger . . . germ hunting as a business": AC13:16:462, "Report of work done by the publicity bureau in behalf of the Massachusetts Institute of Technology from 1 March to 1 October, 1905."
205 "Here . . . hope of posterity?": MC431 s.II:5: Harvard 1869-, Nathaniel Shaler, June 23, 1905.
206 "what the Institute Faculty really thinks of him": MC431 s.II:8: Tyler, Harry Tyler letter, November 27, 1904.
206 "Dr. P. shows signs . . . increasing intensity": MC431 s.II:8: Tyler, Harry Tyler letters, 1904.
206 "consent to a merger that would sink Tech's individuality": MC431 s.II:5: Harvard-Tech alliance 1897–1905, newspaper clipping, January 29, 1904.
207 "entertain no proposition . . . any other educational body": *Boston Globe*, May 19, 1904.
207 "to ascertain whether any arrangement . . . Massachusetts Institute of Technology": MC431 s.II:5: Harvard-Tech alliance 1897–1905, newspaper clipping, "Tech proposes Harvard merger."
207 "You can't make crimson . . . marching on": *Boston Globe*, June 8, 1904.
208 "We must . . . spirit of easy indolence": *Technique 1905*, 235–36.
208 "The combining of two antagonistic forces . . . with other compounds": *Technique 1906*, 73.
208 "The main and sufficient restraint . . . serve our cause best": MC431 s.II:8: Tyler, Harry Tyler letters, 1904.
208 "We are more and more convinced . . . fair chance": MC431 s.II:8: Tyler, Harry Tyler letters, 1904.
209 "wear and tear considerably": MC431 s.II:8: Tyler, Harry Tyler letter, December 13, 1904.
209 "a good deal . . . as well as many others": MC431 s.II:8: Tyler, Harry Tyler letters, January 1905 and n.d.
209 "I regret . . . plant itself on permanently": AC13:9:251, Sarah Hughes to Henry Pritchett, n.d.
209 "I hate to think . . . with the Tech sort": AC13:9:251, Sarah Hughes to Henry Pritchett, n.d. [1904].
210 "act of defiance . . . high standing of our Corporation": MC1:8:139, James Munroe to Emma Rogers, June 13, 1905.
210 "heavy blow . . . wrong through and through": MC1:8:139, William Sedgwick to Emma Rogers, June 24, 1905.
210 "I felt all along . . . would advance himself": MC1:8:138, Edmund Zalinski to Emma Rogers, December 22, 1904.
211 "this step will not lead . . . Mr. Rogers would have chosen": MC1:8:139, Henry Pritchett to Emma Rogers, June 9, 1905.
211 "a long road to travel . . . many difficult places": MC1:8:139, John Sumner to Emma Rogers, September 9, 1905.
211 "merged or confused . . . with any other institution": MC 1:8:139, William Sedgwick to Emma Rogers, June 24, 1905.
211 "has freed our lips . . . the attacking party": MC1:8:139, Ellen Richards to Emma Rogers, June 11, 1905.
211 "humiliated and dragged into bondage": MC1:8:139, James Munroe to Emma Rogers, June 13, 1905.
211 "dignified belligerency": MC1:8:139, James Munroe to Emma Rogers, July 5, 1905.
211 "the Harvard element . . . in the direction of an alliance": AC13:14:408, Ernest Bowditch to Everett Morss, April 7, 1905.
211 "You'd never expect . . . yell of M.I.T.": *Technique 1915*, 290.
212 "impossible to proceed . . . brought to naught": PR 1906, 22.
212 "It seems clear . . . in its old home": MC1:8:139, Henry Pritchett to Emma Rogers, October 10, 1905.
212 "develop with sound judgment . . . research work": PR 1906, 24–25.
213 "to speak of parting . . . choking sensations": *Boston Globe*, December 30, 1905.
213 "analogous to that of a prime minister . . . foreign technical schools": *Boston Globe*, December 14, 1905.
213 "into touch with the world at large": *Technology Review* 8 (January 1906): 3.
213 "Carnegie knows mighty well he never could get him": *Boston Globe*, February 21, 1903.

215 "recent warm contest . . . bigger, better, and busier Tech": *Boston Globe*, March 30, 1906.
215 "Having awakened . . . honor for all time": *Technique 1915*, 188.
215 "I write to say . . . turned over to Professor Noyes": MC431 s.II:3: Presidents—Pritchett, Henry Pritchett to George Wigglesworth, June 29, 1907.
216 "absolutely contented": Wright, H. 1966, 241.
216 "I am still regretting . . . matters of this kind": quoted in Wright, H. 1966, 241, Arthur Noyes to George Hale, October 4, 1906.
216 "the largest possible proportion . . . work of executing them": Hale 1907, 469.
217 "You're nice boys, but it's too bad you never went to college": MC431 s.II:3: Presidential candidates, Kenneth Bainbridge to Julius Stratton, January 14, 1975.
218 "It is a long way . . . must go forward hand in hand": Cattell 1906, 537.
218 "a man of broader academic training . . . purely technical man": MC431 s.II:3: Presidential candidates, George Hale and George Baldwin report, April 26, 1907.
219 "It certainly is . . . can take it": MC431 s.II:3: Presidential candidates, Benjamin Wheeler to Henry Pritchett, May 16, 1907.
219 "all that has followed . . . full vigor of institutional life": *Technology Review* 18 (July 1916):552.
219 "Bigelow . . . need to spend": MIT Museum: Pritchett, Robert Bigelow to Abraham Flexner, December 7, 1942.
220 "We have been reveling . . . Truly California is a wonderful country": quoted in Flexner 1943, 179.
220 "It is not easy . . . personal liberty of the citizen": quoted in Flexner 1943, 180.

Richard Cockburn Maclaurin

221 "I have no hesitation . . . more highly developed": quoted in Prescott 1954, 226.
221 "The qualifications . . . extremely rare": PR 1908, 23.
222 "progressive policy . . . liberal studies with the professional": *Boston Globe*, October 11, 1908.
222 "a liberal education . . . practice of their profession": PR 1908, 26.
222 "a scientific institution of university scope": *Boston Globe*, January 5, 1908.
223 "a large social or living room . . . entertainments may be held": PR 1909, 25.
224 "all delightful fellows . . . good fellowship": quoted in Pearson 1937, 70–71.
224 "is to produce men . . . a high moral purpose": Noyes 1908, 1, 15–16.
225 "pleasing personality . . . sound commonsense": AC13:12:377, Arthur Noyes to Thomas Livermore, April 29, 1908.
225 "He believes . . . unusual charm to his conversation": quoted in *Technique 1910*, 44.
225 "businesslike power . . . quite of the first rank": AC13:13:385.
226 "I think we've found our man": Woodbury n.d., 5.
226 "quiet, steady gaze . . . seemed to fix and hold a visitor": Sedgwick 1920, 27.
226 "unusual knowledge . . . in all branches of human effort": AC13:13:385, Corporation resolution, January 30, 1920.
226 "a man thoroughly sure of himself": quoted in Pearson 1937, 162.
227 "about to enter . . . in all directions": PR 1909, 9.
228 "talk ranged over . . . a natural wisdom": quoted in Pearson 1937, 31.
229 "There is so much danger . . . spent my youth": quoted in Pearson 1937, 39–40.
229 "the simplest and shortest way of avoiding the narrow mind": quoted in Pearson 1937, 45.
229 "a few more odd letters tacked to my name": quoted in Pearson 1937, 49.
230 "This is . . . hard wall of scientific difficulty": quoted in Pearson 1937, 68–69.
230 "He was patient . . . thoughtless questions": quoted in Pearson 1937, 84.
230 "an elaborate work of art . . . results that they obtain": quoted in Pearson 1937, 80.
231 "I had often heard . . . so solidly & well": MC1:8:150, Richard Maclaurin to Emma Rogers, December 10, 1908.
231 "*We are happy* . . . national weapon": Pearson 1937, 91.
232 "Why isn't *that* a good site for Technology?": Pearson 1937, 90.
232 "They are really most delightful . . . I had expected": quoted in Pearson 1937, 93.
232 "I for one . . . the old Technology": quoted in Pearson 1937, 95.

232 "in such a way . . . the really cultured man": quoted in Pearson 1937, 91.
233 "we should be . . . must be strong enough for independence": quoted in Pearson 1937, 95.
233 "absurdly inadequate . . . what we need": AC13:12:377, Richard Maclaurin to Thomas Livermore, June 18, 1910.
233 "I am very much interested . . . Institute over which you preside": AC13:13:391, Massachusetts legislature—acts and resolves, 1896–1916.
234 "a dignified group . . . the Institute's importance": AC13:3:85, Cambridge campus site committee report, October 27, 1910.
235 "total lack of the 'advertising' opportunities . . . its great institutions": AC13:3:85, Walter Kilham et al. to Richard Maclaurin, December 15, 1910.
235 "I hope that none of the old prejudice . . . correct degree of separation": AC13:8:223, John Freeman to Richard Maclaurin, April 18, 1911.
235 "excellent outlook . . . very conspicuous location": AC13:3:85, Cambridge campus site committee report, October 27, 1910.
235 "a very serious peril . . . so great as to be prohibitive": quoted in Pearson 1937, 106.
235 "Ye're no blate . . . *which should be done*": quoted in Pearson 1937, 108.
236 "careful and competent": AC13:9:246, Pierre du Pont to Richard Maclaurin, June 19, 1916.
236 "under either name or a new one": *Boston Globe*, January 29, 1911.
236 "We could support . . . find the money": quoted in Pearson 1937, 112–13.
236 "who are without influence . . . attracting attention by their noise": AC13:3:72, W. G. Davis to Richard Maclaurin, February 18, 1911.
236 "a blessing and not a burden upon the community": AC13:3:72, William Brooks to Richard Maclaurin, February 24, 1911.
237 "men of the baser sort . . . might be expected to cope": AC13:13:403, Richard Maclaurin to Hiram Mills, April 16, 1912.
237 "Boston is sorry . . . still be in the family": *Boston Globe*, October 12, 1911.
237 "Will bring a bunch of men . . . and come with us": AC13:23:103l, Congress of Technology brochures, 1911.
238 "a jolly evening . . . music, mirth and mystery": AC13:23:103l, Congress of Technology brochures, 1911.
238 "the nonsense of caste . . . idle wealth": *Boston Globe*, April 12, 1911.
239 "Everything seems . . . it's a deserving institution": *Boston Globe*, May 24, 1911.
239 "more nearly . . . demands of modern American life": *Boston Globe*, January 14, 1912.
239 "I would employ . . . no place vacant": *Boston Globe*, March 12, 1912.
240 "recent developments . . . new Technology possible": AC13:13:380, Frank Lovejoy to Richard Maclaurin, February 22, 1912.
240 "the inappropriateness . . . elaborate details": AC13:25:185l, George Eastman to Richard Maclaurin, March 6, 1912.
241 "I can use my money . . . the country": *Boston Globe*, June 15, 1912.
241 "Mr. Smith . . . a great thing quietly and unostentatiously": PR 1913, 19.
242 "strong leaning . . . rather than complication and ornateness": AC13:2:56, John Rockefeller to Richard Maclaurin, February 13, 1913.
242 "trained taste . . . frills of any kind": AC13:2:56, Richard Maclaurin to John Rockefeller, February 6, 1913.
242 "The dirt is not flying . . . absolutely necessary": *Boston Globe*, December 22, 1912.
244 "should take an active part . . . a most valuable aid": PR 1916, 16.
244 "typically Technology—utilitarian . . . richness and coziness well combined": *Boston Globe*, May 8, 1921.
244 "your new palace (or do you call it a shanty)?": AC13:2:44, Albert Bemis to Richard Maclaurin, September 16, 1917.
245 "Let no one ignorant of Geometry presume to enter here": AC13:12:369, Clarence Ward to Richard Maclaurin, February 3, 1915.
246 "most of the students . . . humanity in its limitless phases": AC13:12:369, Richard Maclaurin to Frederick Fish, June 1, 1915.

246 "Alia initia et fine": Pearson 1937, 146.
247 "uprooting of traditions . . . half a century": Woodbury n.d., 10.
247 "to prevent him . . . apt to do": AC13:19:549, Richard Maclaurin to Charles Stone, December 10, 1915.
248 "an institution without nonsense . . . business on hand": Maclaurin 1916, 89.
249 "Hello, New York . . . 130 members and guests": quoted in Prescott 1954, 318.
249 "Come, raise a song of M.I.T. . . . Now all together:—M.I.T.": Institute Committee 1929, 10–11.
250 "a problem . . . without care and patience": PR 1910, 24.
250 "entirely proper . . . lasting benefit": AC13:13:406, Richard Maclaurin to F. Jewett Moore, October 3, 1910.
250 "The cooperative spirit . . . problems that confront us": PR 1910, 26.
250 "anything much less . . . amalgamation": AC13:12:377, Richard Maclaurin to Thomas Livermore, February 3, 1910.
250 "with its greater financial resources . . . out of business": quoted in Pearson 1937, 194.
251 "in danger of being swamped": quoted in Pearson 1937, 195.
251 "You're holding a pistol to our head . . . placing of the ingredients": quoted in Pearson 1937, 196–97.
252 "put an end . . . merger and anti-merger men": *Boston Globe*, January 11, 1914.
252 "an alliance between independents . . . nothing more": quoted in Pearson 1937, 209.
252 "its equal . . . its former policy": *Technique 1915*, 454.
252 "She gets plenty out of it . . . that is not important": quoted in Pearson 1937, 214.
252 "put over a deal . . . anything adequate in return": AC13:19:549, Charles Stone to Richard Maclaurin, January 31, 1916.
253 "The school appears . . . quick and natural growth": PR 1915, 98.
253 "a real power in the land": AC13:22:625, Richard Maclaurin to George Whipple, June 23, 1913.
253 "a school of technical education . . . practical arts": AC13:26:212l, Richard Maclaurin to Frederick Fish, January 24, 1916.
253 "unique and transcendent prestige": AC13:26:212l, Richard Maclaurin to Frederick Fish, October 14, 1915.
253 "feel like an oil can for purposes of lubrication": AC13:27:285l, A. Lawrence Lowell to Richard Maclaurin, April 24, 1916.
254 "provided our men do not play the fool": AC13:27:286l, Richard Maclaurin to A. Lawrence Lowell, June 16, 1916.
254 "Harvard University . . . and the Massachusetts Institute of Technology": AC13:27:286l, George Swain to Richard Maclaurin, November 8, 1916.
254 "real problems practically untouched": AC13:26:212l, Richard Maclaurin to Frederick Fish, October 8, 1918.
254 "as a temporary convenience . . . paid the coal bills": quoted in Pearson 1937, 236–37.
255 "crown of culture . . that reality is": *Boston Globe*, December 19, 1909.
255 "the very breath of life to a scientific school": PR 1910, 13.
255 "flying machines": PR 1914, 16.
256 "the actual difficulties of our industrial life": PR 1912, 23.
256 "the whole field of profitable enterprises": PR 1916, 19.
256 "especial aim . . . actual conditions of life": *Boston Globe*, December 18, 1912.
256 "In this profession . . . learn to swim": AC13:5:125, Visiting committee report, December 6, 1915, 2.
257 "the most successful laboratory . . . in the world": quoted in Kargon 1982, 97.
257 "During the last few years . . . a technological school of the first rank": AC13:15:437, Arthur Noyes to Richard Maclaurin, January 31, 1916.
258 "It is not . . . may have chosen": AC13:24:131l, Visiting committee report, March 13, 1912.
258 "strengthen American industry . . . expand in any direction": quoted in Prescott 1954, 311–12.
258 "May we not hope . . . assume equal importance?": Hale 1916, 803.
259 "I want to talk . . . home of few research men": Whitney 1917, 114.

259 "broad gauge physicist": AC13:11:330, Frank Jewett to Richard Maclaurin, April 14, 1917.
259 "this Institute exists . . . service to industry": AC13:11:330, Richard Maclaurin to Frank Jewett, April 12, 1918.
260 "so bound together . . . suffer with the rest": *Boston Globe*, February 21, 1915.
260 "We are playing . . . buried patriots": AC13:22:628, Willis Whitney to Richard Maclaurin, October 31, 1917.
260 "Personally . . . as much ours as theirs": quoted in Pearson 1937, 260.
261 "When war comes . . . shameless unpreparedness": *Boston Globe*, March 26, 1917.
261 "If you don't raise your boy . . . have to fight for him": *Boston Globe*, March 29, 1917.
261 "At last we are in the war": AC13:17:500, Richard Maclaurin to Robert Falconer, June 7, 1917.
262 "what we are doing . . . warlike ends": AC13:11:333, Richard Maclaurin to Otto Kahn, October 30, 1918.
264 "Inwardly . . . 'militarizing' of the colleges": quoted in Pearson 1937, 268.
264 "It should never be permitted again . . . the next conflict": quoted in Pearson 1937, 267.
266 "glow warmed the hearts . . . lagging upper-class-man": *Technique 1915*, 9.
266 "I don't approve . . . talks of nothing but athletics": AC13:13:385, MIT Alumni Association bulletin, November 27, 1908, 10.
267 "men whose interests . . . electrical engineering": *Technique 1914*, 263.
268 "for the purpose . . . economic and scientific problems": *Technique 1918*, 367.
268 "men of socialistic tendencies": *Boston Globe*, March 17, 1913.
268 "danger of . . . snobbishness": PR 1914, 26.
269 "only Southerners . . . members of the club": *Technique 1911*, 265.
269 "The broader the basis . . . likely to be interested": AC13:20:573, Richard Maclaurin to A. A. Cushman, March 31, 1915.
269 "Tech bible": AC13:20:574, W. M. R. to Herbert Pratt, November 6, 1919.
269 "We do not want mollycoddles . . . we want strong men": *Boston Globe*, October 5, 1913.
270 "the damsels be good to look upon": *Technique 1920*, 192.
270 "hesitate to enter . . . designed to lead": PR 1912, 12.
270 "has a graduate school . . . a few young women": AC13:23:104l, Public relations drafts, 1916.
271 "woman's place . . . that profession": *Boston Globe*, September 16, 1923.
271 "digestible if cooked properly . . . their brick ovens": *Boston Globe*, February 20 & March 20, 1910.
271 "Something will have to be done . . . before settling it": AC13:28:350l, Richard Maclaurin to William Lawrence, September 20, 1916.
272 "as a war measure": AC13:27:286l, Richard Maclaurin et al. to A. Lawrence Lowell, May 1916.
272 "the adequate education of women for advanced scientific work": AC13:13:385, Susan Minns et al. resolution, n.d. [March 10, 1920].
272 "but the nucleus . . . during subsequent years": AC13:22:627, Jasper Whiting to Richard Maclaurin, August 7, 1911.
272 "If the present tendency . . . from foreign parts": PR 1914, 13.
274 "Success would be splendid . . . failure would be disastrous": AC13:8:214, Richard Maclaurin to Desmond FitzGerald, July 2, 1919.
274 "We want to meet the industries . . . at its disposal": *Boston Globe*, January 5, 1920.
275 "If any Tech man . . . no fear of competition": quoted in Pearson 1937, 277.
275 "a good deal of hell . . . an idea reared by the devil himself": Hodgins 1973, 145.
275 "going to the dogs . . . best graduates are losing interest in it": AC13:22:633, E. B. Wilson to Richard Maclaurin, April 12, 1919.
276 "horse sense": AC13:21:616, William Walker to Charles Warren, January 19, 1919.
276 "I can see no reason . . . they be adopted": AC13:21:616, William Walker to Richard Maclaurin, March 21, 1919.
276 "sinister influences . . . figments of the imagination": AC13:15:437, Richard Maclaurin to Arthur Noyes, April 12, 1919.
276 "My instincts and my training . . . its effects": AC13:15:437, Richard Maclaurin to Arthur Noyes, April 12, 1919.

277 "The opportunity . . . an extraordinary one": AC13:9:246, Richard Maclaurin to Irenée du Pont, August 11, 1919.
277 "I am often asked . . . my secretary, Miss Miller, and my wife": quoted in Pearson 1937, 281.
278 "We have gone further . . . do it well and quickly": MIT Museum: Maclaurin, newspaper clipping, January 11, 1920.
278 "What we offer . . . all the more quickly solved": MIT Museum: Maclaurin, newspaper clipping, January 11, 1920.
278 "Bolshevism . . . hell for the nearest port": MIT Museum: Maclaurin, newspaper clipping, January 11, 1920.
279 "would have been vain and foolish": *Boston Post*, January 13, 1920.
279 "It has been an arduous life . . . well supported from the first": quoted in Pearson 1937, 283.

Ernest Fox Nichols

281 "Not the man for us": MC431 s.I:2, Julius Stratton to family, February 4, 1923.
281 "not to rock the boat but just keep her going": MC431 s.II:3: Wilson, E. B., E. B. Wilson to Julius Stratton, September 24, 1956.
282 "Talbot wanted to be president . . . damn well he was": MC55:62:8, Oral history project, 1982, Julius Stratton comment, 8.
282 "carried on . . . during a long term of years": MC8:3:40, Frank Laws to Samuel Stratton, October 30, 1922.
285 "Our old friend Nichols . . . business on the legal side": MC431 s.II:3: Wilson, E. B., E. B. Wilson to Julius Stratton, September 24, 1956.
285 "The strong teacher . . . healing power of truth": *Technology Review* 23 (July 1921): 376.
286 "There seems . . . already done": AC13:14:410, E. B. Wilson to Everett Morss, November 10, 1921.
286 "It made me sad . . . back them up": MC431 s.II:3: Wilson, E. B., E. B. Wilson to Julius Stratton, September 24, 1956.
286 "may figure . . . official public appearances": MC431 s.II:3: Wilson, E. B., E. B. Wilson to Julius Stratton, September 24, 1956.
287 "smilingly confessed . . . relativity idea": quoted in Illy 2006, 132.
287 "Einstein no sooner . . . gravitational phenomena": quoted in Hunsaker and MacLane 1973, 290–91.
288 "Theodore is a good boy . . . won't work with me": quoted in Illy 2006, 225.
288 "At one stroke . . . anxiety of the Alumni and undergraduates": AC13:3:67, Alfred Burton to Everett Morss, March 31, 1921.
288 "interest in broadening . . . engineers and executives": AC13:21:620, E. B. Wilson to Charles Warren, June 23, 1921.
289 "I should like to be called . . . give a little advice": *Technology Review* 23 (July 1921): 380.
289 "position in the world of applied science": *Boston Globe*, June 2, 1921.
290 "a technical school . . . closest touch with industry": *Technology Review* 23 (July 1921): 323.
291 "It is going to be my very pleasing privilege . . . it shall not call in vain": *Technology Review* 23 (July 1921): 328.
291 "now known unwarrantedly as the Harvard Bridge": *Technology Review* 23 (July 1921): 327.
292 "A sufficient time has now elapsed . . . leave to return": AC13:15:431, Ernest Nichols to MIT Corporation, November 3, 1921.
293 "to shoulder . . . his physical breakdown": MC8:3:38, Chester Dows to Samuel Stratton, October 18, 1922.

Samuel Wesley Stratton

295 "Shaping things in orderly fashion": MC8:2:22, Robert Richards to Samuel Stratton, December 23, 1922.
295 "a physicist . . . of entire respectability": AC13:26:277l, E. B. Wilson to Francis Hart, November 22, 1921.

295 "I imagine . . . able to get some results": AC13:22:641, E. B. Wilson to James Angell, March 7, 1922.
296 "exceptionally well poised bearing": AC13:26:2771, E. B. Wilson to Francis Hart, March 6, 1922.
297 "I would rather be president . . . than of the United States": MC8:3:40, Samuel Woodbridge to Samuel Stratton, October 15, 1922.
298 "Dr. Stratton . . . public service at government salaries": *New York Times*, July 12, 1931.
299 "thought of being . . . molding their future": Parris, M. n.d., 14.
299 "not . . . one word in opposition": MC8:3:40, Francis Hart to Samuel Stratton, October 9, 1922.
299 "I know of no better man . . . be done here": MC431 s.I:2, Julius Stratton to family, October 14, 1922.
299 "the sad . . . in search of a leader": MC8:4:41, Francis Benedict to Samuel Stratton, October 12, 1922.
299 "Old Tech's been looking . . . Doctor Samuel Stratton forever": MC8:1:3, Poems, sayings, epigrams, etc.
300 "as an expression of . . . a greater Technology": *Technique 1924* (vol. 38), front matter.
300 "Get behind President Stratton": MC8:3:38, "Preliminary announcement of joint meeting . . ."
300 "Technology has been hungry . . . such confidence as you have": MC8:2:22, Robert Richards to Samuel Stratton, December 23, 1922.
300 "I confess . . . a background of good breeding": Fields 1960, 20.
302 "a beautifully organized . . . human machinery": Prescott n.d., chap. 2, 9.
302 "he was always diffident . . . never a good speaker": Hobbs n.d., 2.
302 "a youth . . . above and below him": quoted in Prescott n.d., chap. 2, 27.
302 "never have been countenanced by . . . wives": Hobbs n.d., 2.
302 "chief of clan . . . fine comradeship": Prescott n.d., chap. 3, 3.
303 "what modern psychologists . . . a father complex": MC8:3:34, Katherine Girling to Glenn Hobbs, [June 1932].
303 "the first display of that wanderlust": Hobbs n.d., 4.
303 "Stratton was a wonderful traveler . . . best use of our time": Hobbs n.d., 6.
304 "with the enthusiasm of youth": Hobbs n.d., 8.
305 "While Michelson . . . a very good team": Hobbs n.d., 11.
305 "My naval service . . . machinery on the ship": Stratton, S. n.d., 3–4.
306 "If we are to advance . . . create original things": *Hearings before Subcommittee of House Committee on Appropriations* (Washington, D.C.: G.P.O., 1902), 70.
307 "I can only say . . . I back up": quoted in Cochrane 1966, 155.
307 "How is that work . . . such-and-such turned out?": quoted in Lobdell 1922, 9.
308 "three gentlemen of Washington . . . measuring watermelons": Green 1905.
308 "proved to be the realization . . . Dinner parties": Parris, W. n.d.
309 "the most awful lies": Parris, M. n.d., 12.
309 "I will miss you greatly . . . their kin and friends": MC8:4:41, Albion Parris to Samuel Stratton, October 16, 1922.
310 "Unfortunately . . . my own interests alone": MC8:2:21, Samuel Stratton to Robert Woodward, February 10, 1914.
310 "I always feel . . . your whole life into it": MC8:3:39, Maude McKelvy to Samuel Stratton, December 7, 1922.
310 "It makes me very sorry . . . a living position": MC8:3:39, August Flegel to Samuel Stratton, October 12, 1922.
310 "*But* . . . become of the Bureau?": MC8:3:40, Herbert Putnam to Samuel Stratton, October 12, 1922.
310 "But what will the poor old Bureau do?": MC8:3:40, Edwin Sweet to Samuel Stratton, October 18, 1922.
310 "I must confess . . . passing of his Round Table": MC8:3:40, W. J. Canada to Samuel Stratton, October 15, 1922.
310 "It serves . . . a living wage": MC8:4:41, Maximilian Toch to Samuel Stratton, October 16, 1922.

311 "to register for the term along with the rest of us": MC431 s.I:2, Julius Stratton to family, February 4, 1923.
311 "I am not going to turn . . . the time is appropriate": *Boston Globe*, January 3, 1923.
311 "We like very much . . . in orderly fashion": MC8:2:22, Robert Richards to Samuel Stratton, December 23, 1922.
311 "He is learning . . . reasons for everything": MC431 s.I:2, Julius Stratton to family, February 4, 1923.
311 "may have been a fine physicist . . . than anything else": MC431 s.I:2, Julius Stratton to family, February 4, 1923.
311 "deliberately and of set purpose . . . not love but hate": Sinclair, U. 1923, ix.
312 "a climber and a toady . . . stay where he was or descend": Sinclair, U. 1923, 68.
312 "This is one of the most marvelous collections . . . interested in electric railways": Sinclair, U. 1923, 64.
313 "no room for Harvard in Cambridge. . . Boston at all": AC13:19:552, Inauguration banquet, June 11, 1923, 3.
313 "one trained in administration . . . render to the public": MC8:3:40, Frank Laws to Samuel Stratton, October 30, 1922.
313 "opportunities to help industry": MC8:3:40, Arthur Nevius to Samuel Stratton, October 24, 1922.
313 "It is a great relief to me . . . a purely business executive": MC8:3:40, Harry Goodwin to Samuel Stratton, October 15, 1922.
314 "For years I have hoped . . . lifted from my shoulders": MC8:3:39, George Hale to Samuel Stratton, November 3, 1922.
314 "There is much to be done . . . nearly perfect": MC8:3:39, George Swain to Samuel Stratton, October 14, 1922.
314 "look at science . . . powerful scientific institution": MC8:3:38, Harlow Shapley to Samuel Stratton, October 13, 1922.
314 "Under your guidance . . . more truly national asset": MC8:3:38, Keivin Burns to Samuel Stratton, November 11, 1922.
314 "the molding of character in the student corps": MC8:3:40, W. B. Tardy to Samuel Stratton, October 12, 1922.
314 "a magic wand": MC8:3:40, Katherine Moynihan to Samuel Stratton, October 12, 1922.
315 "make . . . the best of America's young manhood": MC8:3:40, Virgil Marani to Samuel Stratton, October 16, 1922.
315 "in the old days . . . this sort of thing": MC8:3:39, B. R. T. Collins to Samuel Stratton, October 13, 1922.
315 "to turn out engineers rather than human beings": MC8:3:38, Bassett Jones to Samuel Stratton, November 9, 1922.
316 "Those that had heard . . . a college education": *Technique 1928*, 112.
317 "for the purpose . . . work in this field": PR 1927, 45.
318 "The Institute . . . leaders for these industries": PR 1926, 14.
318 "While our outside relations . . . greatly increased": AC13:9:294, Robert Haslam to Samuel Stratton, May 9, 1924.
319 "our staff . . . the importance of applications": AC13:11:327, Dugald Jackson, "Memorandum on electrical engineering staff at Technology for Mr. Gerard Swope," January 7, 1925.
319 "turn out a fair grade . . . carry on research": AC13:10:309, George Hooper to Samuel Stratton, December 18, 1926.
319 "I know of no other institution . . . southern California": AC13:10:309, Samuel Stratton to George Hooper, January 17, 1927.
319 "The weakest point . . . do the same thing": AC13:19:564, George Hale to Gerard Swope, October 19, 1925.
319 "spirit of research pervade the instruction in science": AC13:22:628: Willis Whitney to C. A. Coffin, January 27, 1926.

319 "working out problems . . . as a whole": AC13:115, Samuel Stratton to Frederick Fish, November 9, 1925.
320 "cutting the throats of schools": AC13:21:619, Edward Warner to Samuel Stratton, July 23, 1929.
320 "It is more and more difficult . . . followers only": PR 1926, 43–44.
321 "fields of pure science . . . perhaps in some cases more so": AC13:17:487, Samuel Stratton to Max Mason, November 8, 1929.
321 "world's greatest men . . . *world vision*": AC13:23:32l, Calvin Rice to Samuel Stratton, November 7, 1924.
322 "a technical institution . . . industry is possible": *Tech*, October 15, 1924.
322 "It will be a great pleasure . . . scientific life in America": AC13:10:302, Werner Heisenberg to Samuel Stratton, July 14, 1928.
323 "Frankly . . . excited over it": AC13:15:434, Charles Norton to Samuel Stratton, June 1, 1927.
323 "the University Press . . . would be valuable here": Crawford and Killian 1928, 1.
324 "I think . . . routine work of our press": AC13:8:228, Samuel Stratton to John Freeman, June 22, 1931.
324 "I take it . . . doing more fundamental work": AC13:2:57, William Bragg to Samuel Stratton, April 23, 1928.
324 "One can easily observe . . . pursuit of pure science": AC13:28:340l, Frederick Keyes to Samuel Stratton, May 12, 1925.
324 "every organic chemist . . . of any pretensions": AC13:28:341l, Frederick Keyes to Samuel Stratton, January 15, 1926.
325 "keep clear of American politics": AC13:22:629, Norbert Wiener to Samuel Stratton, August 12, 1926.
325 "stale and behind the times . . . inducement than salary": AC13:14:410, Samuel Stratton to Everett Morss, March 21, 1923.
325 "the most brilliant man": AC13:29:461l, Samuel Stratton to Harry Goodwin, April 8, 1929.
326 "I am learning . . . choosing wisely": AC13:32:550a, Julius Stratton to Samuel Stratton, July 25, 1926.
326 "My own wish . . . none in the country": AC13:32:550a, Samuel Stratton to Julius Stratton, February 20, 1928.
327 "I would stress . . . many to disappointment": AC13:5:123, Visiting committee report, June 6, 1924.
327 "level of semi-illiteracy": AC13:15:444, Henry Pearson to Samuel Stratton, October 7, 1929.
327 "Be a snob . . . get away with": AC13:17:490, *New York Herald*, June 4, 1929.
327 "a knowledge of how the mind of man works": AC13:6:150, M. F. Brandt, "The Institute course in engineering administration," June 1928.
328 "give . . . common sense in engineering": PR 1923, 10.
328 "humanics . . . practice of an engineering profession": PR 1929, 11.
328 "it would be a good thing . . . literature and philosophy": AC13:11:337, Samuel Stratton to Henry Kendall, June 12, 1926.
328 "treat sex . . . heart of God": AC13:20:574, Technology Christian Association circular, March 24, 1924.
330 "the Institute now regards itself . . . mere routine men": AC13:6:139, Samuel Stratton to Raymond Blakney, March 4, 1927.
330 "some university . . . everyday problems": AC13:7:182, Henry Doherty to Samuel Stratton, November 27, 1930.
330 "stand up under the strain . . . nervous wrecks": AC13:11:335, William Locke to William Kales, March 18, 1924.
330 "she always attributes . . . unfeeling jailers": AC13:11:335, William Kales to Mr. Hall, January 19, 1930.
330 "social and out of hour life": AC13:5:127, Visiting committee report, March 10, 1926.
331 "plan to over-work the boys": AC13:11:335, Samuel Stratton to William Kales, April 2, 1924.

331 "We have consistently looked . . . avoided the latter course": AC13:10:299, Samuel Stratton to Charles Hayden, March 24, 1930.
331 "It seems . . . interested in sport": *Boston Globe*, January 25, 1924.
332 "service to anyone . . . creed or color": *Technique 1922*, 322–23.
332 "forced to cultivate . . . boarding houses": *Technology Review* 29 (December 1926): 94.
332 "We have too long . . . dispel this idea": *Technique 1923*, 318–19.
333 "really high hat musical treat": AC13:22:626, Allan Rowe to Samuel Stratton, April 11, 1927.
333 "We have to run . . . MIT's program": Burchard 1975, 544.
333 "beskirted . . . young men": *Boston Globe*, April 21, 1925.
333 "low and unworthy": AC13:10:306, Franklin Hobbs to Samuel Stratton, April 23, 1928.
334 "We bought sweet spirits . . . our gin was ready": Burchard 1975, 372.
334 "he had never seen . . . D.K.E. house": AC13:8:221, Raymond Hughes to Henry Talbot, May 31, 1924.
334 "flagrant offenses against the Institute's good name": AC13:13:400, Leonard Metcalf to Samuel Stratton, April 14, 1924.
334 "to rush in . . . a sophisticated passage": Burchard 1975, 375–76.
335 "I have attended . . . character of the alumni": AC13:11:330, Frank Jewett to Samuel Stratton, January 9, 1928.
335 "hard-working boys . . . spirit of deviltry": AC13:19:565, Henry Talbot to Samuel Stratton, November 10, 1923.
336 "the sordid aspects of life": *Technique 1930*, 233.
336 "broadening, extending, and stimulating . . . at the Institute": *Technique 1928*, 238.
336 "one who has been through the mill": AC13:6:139, "Mr. Conant's book," newspaper clipping, n.d.
337 "I fear . . . growth of that industry": AC13:8:216, Samuel Stratton to James Angell, May 11, 1928.
338 "I felt uncertain about him . . . mother was English-American": AC13:8:220, William Franklin to Samuel Stratton, October 5, 1929.
338 "The Institute of Technology . . . not very large": AC13:19:558, Samuel Stratton to Charles Richards, December 7, 1928.
339 "Any man . . . pick your own": *Technique 1923*, 144.
339 "Shulits is a Hebrew . . . free from prejudice on racial matters": AC13:8:226–27, John Freeman to Samuel Stratton, August 7 & November 15, 1930; February 20, 1931.
339 "varieties of heathen religions": *Boston Globe*, November 15, 1923.
340 "deserving and ambitious young men of color . . . their capabilities": AC13:18:512, Lincoln Bryant to Samuel Stratton, April 29, 1925.
340 "a very intelligent and energetic Negro": AC13:18:512, George Shattuck to J. L. Tryon, August 3, 1927.
340 "girls for the most part . . . simply fall away": *Boston Globe*, March 5, 1922.
341 "in view of the fact . . . no women members on its faculty": AC13:2:27, Samuel Stratton to American Association of University Women, October 17, 1924.
341 "the girl . . . her pleasures": *Technique 1924*, 144.
341 "I am a sort of step-father . . . group at Technology": AC13:17:496, Allan Rowe to Morris Parris, January 1, 1926.
343 "I don't know . . . won't hurt them": Parris, M. n.d., 16.
343 "His principal claim . . . did not get a fair trial": Burchard 1975, 537.
343 "A more incompetent . . . the less said the better": Hodgins 1973, 209.
344 "I honor you . . . slobber over criminals, and crimes": MC8:2:26, Harry Jordan to Samuel Stratton, August 4, 1927.
344 "the world . . . not for mob rule": MC8:2:26, Franklin Hobbs to Samuel Stratton, August 4, 1927.
344 "fine and unselfish service": MC8:2:26, Miles Sherrill to Samuel Stratton, August 24, 1927.
344 "It is because . . . frenzied, irrational plutocratic group": MC8:2:26, Atherton Hastings to Samuel Stratton, September 29, 1927.

345 "these destructful brutes . . . died to set men free": MC8:2:26, John Barclay to Alvan Fuller, August 4, 1927; Ellen Winsor to Samuel Stratton et al., August 5, 1927.
345 "I was particularly gratified . . . shrewder and less scrupulous men": MC8:2:26, Harold Ickes to Samuel Stratton, October 17, 1927.
346 "This was unfortunate enough . . . alumni or students": Hodgins 1973, 210.
346 "a fairly effeminate male companion-secretary": Burchard 1975, 536–37.
346 "Stratton . . . probably shouldn't be": Bush 1964, 47A.
347 "Okay, if nobody objects": Harrison 1961, 202.
347 "in the declining years . . . his adaptability": Harrison 1961, 202.
347 "The communications system . . . lack of it": Harrison 1961, 201–2.
347 "frittering away the resources . . . important activities": MC431 s.II:3, E. B. Wilson to Karl Compton, March 14, 1930.
347 "not accustomed . . . regretted that sail": MC60:10, George Harrison interview with Horace Ford, 6.
348 "I shall never forget . . . a friendship steadfast, sincere, and sure": MC8:1:6, Samuel Prescott tribute to Samuel Stratton, October 21, 1931, 2–3.
348 "moved deeply . . . many of us youngsters": MC8:4:43, Edward Bowles to Samuel Stratton, March 18, 1930.
348 "efforts to make of this . . . years of service": MC8:4:43, Henry Phillips to Samuel Stratton, March 13, 1930.
348 "devotion to all that is best for the Institute": MC8:4:43, Julius Stratton to Samuel Stratton, March 14, 1930.
348 "splendid services . . . at this Institute": MC8:4:43, Manuel Vallarta to Samuel Stratton, March 14, 1930.
348 "stoutly upheld the view . . . strong foundation of pure science": MC8:4:43, Waldemar Lindgren to Samuel Stratton, June 21, 1931.
349 "The danger that I always foresaw . . . occurred in the twenties": Swope ca. 1955, 82.
349 "The engineering student . . . (whatever that may mean)": *Technology Review* 30 (March 1928): 283.
350 "the right type of man . . . so-called pure scientists": quoted in Kargon and Hodes 1985, 306.
350 "I am sure . . . someone else to do the running!": AC13:16:454, Joseph Powell to Samuel Stratton, May 22, 1930.
350 "As long as it's Karl Compton . . . anybody else": MC60:10, George Harrison interview with Samuel Prescott, 3.
350 "I've got something . . . make a big change": Rosenblith 1982–84, December 22, 1982, 67–68.
351 "The new arrangement . . . undertake research": AC13:9:272, Samuel Stratton to George Hale, February 13, 1931.
351 "Under Admiral Stratton . . . no rivals": MC8:4:43, George Burgess to Samuel Stratton, March 15, 1930.
352 "He would have been far happier . . . long years ahead of him": *Boston Post*, July 26, 1931.
352 "It seldom has fallen . . . His interest—": MC51:43, MIT Corporation correspondence, August-October 1931, clippings on Stratton death, October 19, 1931.
353 "Morris . . . supremely happy": MC8:3:34, Glenn Hobbs to Samuel Prescott, June 29, 1932.
353 "I joked him . . . be happy": MC8:3:34, Samuel Prescott to Glenn Hobbs, July 9, 1932.

Karl Taylor Compton

355 "All knowledge his sphere": AC4:43:15, Godfrey Cabot to James Killian, May 9, 1955.
355 "Dr. A. Compton . . . better than Karl": quoted in Harrison 1961, 204.
355 "on a very high level . . . a good neighbor": quoted in Harrison 1961, 203–4.
356 "As I try . . . objective guide to action": quoted in Harrison 1961, 287–88.
357 "One sunny autumn morning . . . for some weeks": Harrison 1961, 48–49.
358 "One time . . . made the study": AC4:75:11, Karl Compton to B. Alfred Dumm, August 4, 1947.
359 "I had such faith . . . means could be found": quoted in Harrison 1961, 68.
359 "quiet and pleasant . . . remember well": Harrison 1961, 77–78.

360 "while the ship . . . the cook": Harrison 1961, 112.
361 "Research work under Gov. auspices": MIT Museum: Compton, Princeton University War Records, Compton report, June 17, 1920.
361 "Send forty physicists . . . forty at once": quoted in Harrison 1961, 130.
362 "These jobs . . . some speeches": Harrison 1961, 156a.
363 "A feeling of loyalty . . . remain at Princeton": quoted in Harrison 1961, 168.
363 "quite excellent Mr. Oppenheimer": AC13:2:48, Max Born to Samuel Stratton, February 13, 1927.
364 "What I really want . . . as president": Harrison 1961, 179.
364 "I know something . . . a stock and a bond": Killian 1985, 422.
364 "who lets his student . . . in his investigations": Killian 1985, 422.
364 "bid them Godspeed": Killian 1985, 424.
365 "If you will be president . . . big as you wish": Harrison 1961, 185.
365 "Wheresoever thou goest . . . I go also": Harrison 1961, 180–81.
365 "took on a strong slant of immediate practicality": Killian 1985, 423.
366 "the anticipation of technological change . . . for such development": Harrison 1961, 182.
366 "an opportunity . . . research fields generally": Killian 1985, 423.
367 "Jewett, who really knows . . . what the Institute needs": Rosenblith 1982–84, December 22, 1982, 70.
367 "demureness with maturity most effectively": Harrison 1961, 184.
367 "Let me introduce to you the new president of MIT": Harrison 1961, 184.
367 "I admit freely . . . genuine possibility of making progress": Slater ca. 1975, 459.
368 "a very wise and wily old serpent": Harrison 1961, 178.
368 "If Karl had known . . . carrying on his researches": Harrison 1961, 185.
368 "the half-baked ideas . . . great educator was ruined": Killian 1985, 425.
368 "I am looking forward . . . applications of knowledge to life": *Technology Review* 32 (May 1930): 365.
369 "Did we make any mistake . . . a honey the time before": Bush 1964, 50–50A.
369 "The manner of his greeting . . . disposes of his questions": *Technology Review* 32 (May 1930): 355.
369 "Dr. Compton . . . the Institute stands": *Technology Review* 32 (May 1930): 355.
370 "Frankly, I'm terrified . . . this kind before": Harrison 1961, 192.
370 "There is every indication . . . applying this knowledge": Harrison 1961, 193.
370 "one or two of their faithful helpers": MC60:10, George Harrison interview with Margaret Compton, 1955.
371 "rather unheard-of for those days": Slater ca. 1975, 442.
371 "crazy . . . another good physicist gone wrong": Slater ca. 1975, 448.
371 "I believe . . . offer considerable inducements": AC4:129:3, Frederick Keyes to Karl Compton, July 23, 1930.
372 "Sorry . . . next three months": Harrison 1961, 208.
372 "The new work . . . electrical discharges in gases": quoted in Harrison 1961, 224–25.
374 "Look, our new president . . . Ah Hah [A.H.] Compton!": Harrison 1961, 211.
374 "from a truly scientific standpoint": AC4:99:12, Warren Weaver to Waldemar Lindgren, July 24, 1930.
375 "crowning stage . . . program in biology": AC4:98:3, Karl Compton to Herbert Gasser, November 7, 1944.
375 "a lovable little Hungarian": Wiener 1956, 174.
375 "nervous, emotional . . . to my and others' regret": Struik n.d., 15.
376 "Well, look . . . worth anything in five years": Rosenblith 1982–84, Brown 1982, 18.
377 "Samuelson . . . understand science and mathematics": AC4:239:10, E. B. Wilson to Karl Compton, November 5, 1940.
377 "a general feeling in the air . . . social or human point of view": AC4:77:10, Karl Compton to Ralph Freeman, February 17, 1934.
377 "When I was a student . . . closest thing to science I could find": Feynman 1985, 45.

378 "adding prestige . . . body of staff and students": PR 1933, 22.
379 "a practically virgin field": PR 1937, 95.
379 "the finest in the world for their purposes": PR 1932, 21.
380 "Happy is he . . . roar of greedy Acheron": Harrison 1961, 231.
380 "The search for Truth . . . a certain grandeur": Harrison 1961, 442.
380 "Well, I will look . . . you do the same": Harrison 1961, 206.
380 "In carrying forward this program . . . in effect a wise one": *Technology Review* 32 (May 1930): 365.
381 "Dr. Stratton has given orders . . . chemistry and physics sections": Harrison 1961, 233.
381 "two-headed operation": Harrison 1961, 204.
381 "But I thought you wanted . . . fully as much": Bush 1964, 617–18.
381 "ukaz": Bush 1964, 616.
381 "I'm leaving MIT . . . a welter of exceptions": Bush 1964, 616–17.
382 "I'm essentially a mild chap . . . being belligerent": Bush 1964, 2.
382 "The reason was . . . changed my tune completely": Bush 1964, 51.
382 "Karl knew . . . to him there were no villains": Harrison 1961, 254–55.
383 "a fine university . . . but not the whole": Harrison 1961, 252.
385 "make sure . . . great field of the humanities": PR 1937, 135–36.
385 "pretty much in a rut . . . traditional type": AC4:15:13, Karl Compton to Anthony Anable, February 4, 1939.
385 "was doing a better job . . . in this field": Harrison 1961, 307.
385 "Just think . . . any such thing as 'higher Mathematics'": MC423:22, William Hall to James Killian, September 8, 1945.
386 "The country is becoming . . . permanent satisfaction in living": AC4:40:4, Karl Compton, "Dean of humanities," March 22, 1937.
386 "liberal arts . . . so abused as to be undesirable": AC4:40:4, Karl Compton to Philip Stockton, April 2, 1937.
387 "the cream of the managerial output of the Institute": Harrison 1961, 255.
387 "I machined . . . until you've caught on": Killian 1985, 12.
387 "an enthusiasm-amplifier . . . force, warmth, and light": Killian 1965, unpaginated.
388 "advertising medium": AC4:129:7, Karl Compton to James Killian, January 25, 1933.
389 "Why hammer . . . the right timber for Technology": MC423:22, *Technology Review* file, 1930s.
389 "I agree with you . . . railroad club cars": AC4:129:7, Karl Compton to James Killian, January 16, 1934.
389 "Obviously this check . . . appropriate to my name": AC4:188:4, Karl Compton to John Rowlands, February 12, 1941.
390 "I can't help inventing": Killian 1985, 16.
390 "Was he *mad*? . . . wasn't exactly pleased": Harrison 1961, 256.
391 "The future vigorous and healthy growth . . . condition to perform": PR 1936, 20.
392 "Dedicated to the advancement . . . industry art agriculture and commerce": AC4:16:5, Harry Carlson to Vannevar Bush, December 11, 1937; AC4:16:6, Karl Compton to Harry Carlson, February 8, 1938.
392 "the inspiration . . . presented in that way": AC4:16:6, Welles Bosworth to Harry Carlson, June 4, 1938.
392 "As a scientist . . . Newton and then Galileo": AC4:33:9, Karl Compton to Welles Bosworth, September 30, 1938.
393 "This increase . . . postgraduate training and research": PR 1936, 22.
394 "no objective except pure fun": *Technique 1944*, 57.
394 "Contrary to the opinion of many . . . all who like to dance": *Technique 1941*.
395 "One of our faculty wives . . . what they're saying": MC60:10, George Harrison interview with Margaret Compton, 1955.
395 "Technological and scientific school . . . Location urban": AC4:9:10, American Council on Education, 1940.

396 "If I remember correctly . . . expert in such matters": AC4:50:9, Karl Compton to Erwin Schell, March 6, 1933.
396 "If Fascism makes headway . . . object of special attacks": *Tech*, February 2, 1934.
397 "Jewry is still in fashion . . . day before yesterday": AC4:51:12, William Power to Carroll Wilson, June 1 & June 2, 1934.
397 "a merchant by nature . . . good merchant": AC4:215:10, Frank Jewett to Karl Compton, December 7, 1942.
397 "I was bound, tied, and gagged . . . good name of the Institute": AC4:16:1, Antiwar conference, 1935.
398 "The bullying of the black . . . even more helpless race": *Tech*, February 25, 1936.
398 "promulgating racial hatred": AC4:8:11, Kenneth Thimann to Karl Compton, March 31, 1939, enclosure.
398 "The summer before . . . with other Jews": Feynman 1985, 30–31.
398 "Rosen . . . general field of Rosen's interests": AC4:186:15, Karl Compton to Gaylord Harnwell, February 14, 1939.
399 "I would say . . . responsible for the Jewish problem": AC4:186:15, Karl Compton to John Ross, April 10, 1939.
399 "I would have no hesitation . . . desirable to carry": AC4:186:15, Karl Compton to John Davis, May 17, 1939.
399 "Tell me, Mr. Bush . . . why not hire Levinson?": Nasar 1998, 137.
399 "an abnormal number of Jewish students": AC4:186:15, Karl Compton to John Ross, April 10, 1939.
399 "Dr. Rosen . . . also Jewish, he is not sure": AC4:186:15, Karl Compton to Harvey Davis, June 20, 1940.
400 "Judging by their names . . . a colossal flop": AC4:18:2, Karl Compton to E. E. Day, March 18, 1943.
401 "The department . . . it did not work out": Struik n.d., 16.
401 "They feel . . . free to call on her": AC4:210:18, Karl Compton, Memorandum of conversation with Mrs. Frederick T. Lord, April 5, 1939.
402 "Mrs. Penney is afraid . . . influence and balance wheel": AC4:210:18, Florence Stiles to Karl Compton, February 3, 1945.
402 "The Institute . . . MIT is a man's world": AC4:210:18, Florence Stiles to Robert Kimball, May 5, 1948.
403 "a backfire . . . support in other directions": AC4:13:3, Karl Compton to Robert Millikan, January 17, 1934.
404 "Word received . . . reorganize federal government": Harrison 1961, 263.
405 "I am in sympathy . . . numerous occasions": quoted in Harrison 1961, 270–71.
405 "That's one thing . . . he won't move, period": Harrison 1961, 250.
406 "A few weeks ago . . . my responsibilities at M.I.T.": quoted in Harrison 1961, 272.
406 "You know, K.T., . . . I'll take the second": Harrison 1961, 273.
406 "It was very much . . . a normal family home": M60:10, George Harrison interview with Margaret Compton, 1955.
407 "rather be second . . . you can run the Institute": Harrison 1961, 271.
407 "I don't think . . . some extraordinary qualifications": Bush 1964, 75A-76.
407 "magnificently unprepared": Killian 1985, 19.
407 "I flourished that pen . . . appointed me his office boy": Killian 1985, 18.
408 "a puppy . . . not yet housebroken": Killian 1985, 19.
408 "May your occupation . . . take time to think": quoted in Killian 1985, 19.
409 "hardly knew . . . arrangement would not fly": Killian 1985, 20.
409 "has started in fine shape . . . will be a great help": AC4:42:9, Karl Compton to Vannevar Bush, January 9, 1939.
409 "As a happy augury . . . light and beneficence": quoted in Harrison 1961, 311–12.
410 "A new organization launched . . . increased determination": Harrison 1961, 313.

412 "The job is more interesting . . . also more complicated": AC4:129:9, Karl Compton to James Killian, August 14, 1942.
412 "To each adjustment . . . defeat and victory": PR 1942, 5.
412 "We wanted 40 . . . set it at 35": Harrison 1961, 331.
413 "I knew nothing . . . things I did not understand": Killian 1985, 21.
413 "skills in resolving . . . dealing with his colleagues": Killian 1985, 21.
413 "As my Executive Assistant . . . irreplaceable under existing conditions": AC4:129:9, Karl Compton to Local Board #106, June 8, 1942.
415 "did not get along very well . . . better than the Bell Labs could": Bush 1964, 374.
416 "Chinese students . . . exceeded by accepted Chinese": AC4:39:17, John Bunker to Chih Meng, December 10, 1945.
417 "admission of foreigners . . . our defense activities": AC4:15:13, James Killian to W. S. Anderson, January 29, 1941.
417 "We have too much important work . . . whether citizens or not": AC4:126:1, B. Alden Thresher to Charles Fitts, April 6, 1942.
417 "memorandum . . . new objectives at M.I.T.": AC4:129:9, James Killian to Karl Compton, August 20, 1943.
417 "really the beginning . . . problems of education": MC431 s.III:8, Julius Stratton to Samuel Prescott, December 2, 1958.
418 "Let us agree . . . a clearer statement of policy": AC4:210:34, Julius Stratton to James Killian, October 23, 1944.
419 "It was obvious . . . wangle things out of MIT": AC186:4:33, RLE 25th anniversary dinner 1966, Slater comments, 3.
420 "in a very unusual way . . . administrative ability": MC431 s.III:25, MIT—Research Laboratory of Electronics, John Slater to Karl Compton, August 23, 1944.
420 "The general feeling . . . cooperative projects in other fields": MC431 s.III:25: John Slater, "Memorandum regarding the establishment of an electronics laboratory in the departments of physics and electrical engineering, Massachusetts Institute of Technology," August 28, 1944.
420 "In a long-distance telephone conversation . . . thought it would": Hazen 1976, chap. 3, 40.
420 "had much to do . . . decision to return": MC431 s.III:8, Julius Stratton to Harold Hazen, December 23, 1958.
421 "condemning the wanton destruction . . . other countries": quoted in Harrison 1961, 369.
422 "What the heck . . . why wouldn't I go to Tech?": Harrison 1961, 429.
422 "the Powers That Be . . . get things going again back here": AC4:5:6, Karl Compton to Bissell Alderman, November 28, 1944.
422 "Gentlemen, you remind me . . . you can't catch me that way!": Harrison 1961, 377.
423 "I agreed with the decision . . . matter rest there": Harrison 1961, 378.
424 "I unburdened my soul . . . DuBridge for that job": Bowles 1985, 2–3.
424 "weak and irresolute America": Compton 1947, 11.
424 "nothing could be more tragic . . . racial difference": *A Program for National Security: Report of the President's Advisory Commission on Universal Training*, 80th Congress, 1st session, May 29, 1947; excerpt posted at www.trumanlibrary.org.
424 "colored and white, Christian and Jew": Branson 1952, 132.
425 "irresistible forces meeting immovable objects": Harrison 1961, 389.
425 "memorandum . . . chief executive of a university": AC4:129:11, July 30, 1948.
426 "My deep . . . restoration to health": quoted in Harrison 1961, 411–12.
426 "to explore . . . science for the twentieth century": quoted in Harrison 1961, 397.
426 "great admiration for Dr. Compton": quoted in Harrison 1961, 398.
426 "I shall take this . . . whenever I wish": quoted in Harrison 1961, 403.
427 "on an equal footing . . . undergraduate degrees": quoted in Killian 1985, 81.
428 "opening guns": AC4:129:11, Karl Compton to James Killian, November 22, 1948.
429 "It doesn't matter . . . judge human excellence": Harrison 1961, 439.

Sources

This list identifies key source materials, both print and non-print. Unless otherwise specified, manuscript (MC) and archival (AC) collections cited are in the Institute Archives and Special Collections, MIT, as are cited MIT theses and dissertations.

Archives and manuscripts: MIT

AC1, Faculty records, 1865– .
AC4, Office of the President (Compton and Killian) records, 1930–1959.
AC13, Office of the President records, 1883–1930.
AC20, School of Humanities and Social Science Dean's Office records, 1933–1990s.
AC58, Student record cards, 1903–1951.
AC65, Office of the Chairman (Compton) records, 1948–1954.
AC134, Office of the President (Julius Stratton) records, 1957–1966.
AC186, Research Laboratory of Electronics records, 1944–2000.
AC272, Executive Committee of the MIT Corporation minutes, 1883–1995.
AC278, MIT Corporation records, 1862– .
AC333, Office of the Vice President (Bush) records, 1932–1938.
AC344, Francis A. Walker Memorial Service Committee records, 1897.
MC1, William Barton Rogers papers, 1804–1911.
MC2, Rogers Family papers, 1811–1904.
MC3, William Barton Rogers II papers, 1817–1919.
MC7, John Daniel Runkle papers, 1853–1880.
MC8, Samuel Wesley Stratton papers, 1881–1934.
MC23, Carl Richard Soderberg papers, 1914–1979.
MC24, Gordon Stanley Brown papers, 1920–1984.
MC25, Harold Eugene Edgerton papers, 1889–1990.
MC29, Carroll Louis Wilson papers, 1926–1983.
MC45, Robert Jemison Van de Graaff papers, 1928–1948.
MC51, John Ripley Freeman papers, 1827–1952.
MC55, Walter Alter Rosenblith papers, 1930–1991.
MC60, George Russell Harrison papers, 1916–1973.
MC65, Association of MIT Alumnae papers, 1900–1980.
MC75, Philip McCord Morse papers, 1927–1980.
MC76, John Ely Burchard papers, 1923–1975.

MC121, Harry Manley Goodwin papers, 1888–1936.
MC143, Vannevar Bush oral history, 1964.
MC189, Typescript of "A physicist of the lucky generation," by John C. Slater, ca. 1975.
MC238, Karl Taylor Compton memorial tribute, 1954.
MC298, Francis Amasa Walker papers, 1862–1897.
MC351, Margaret Hutchinson Compton interview, 1971.
MC416, Karl Taylor Compton papers, 1906–1961.
MC423, James Rhyne Killian papers, 1923–1988.
MC431, Julius Adams Stratton papers, 1800s–1990s.
MC446, Alfred Perkins Rockwell papers, 1867–1876.
MC460, Margaret Stinson interview, [1911].
MC558, Henry Smith Pritchett papers, 1891–1924.
MC588, Richard Cockburn Maclaurin papers, 1892–1908.
MC650, John Machlin Buchanan papers, 1938–2007.

Archives and manuscripts: non-MIT

Acts—1873, chap. 174, Act in addition to an Act incorporating the Massachusetts Institute of Technology, approved April 8, 1873; petitions, etc. Massachusetts Archives, Boston, Mass.

Edward Atkinson papers, 1847–1905. Massachusetts Historical Society, Boston, Mass.

Henry Smith Pritchett papers, 1876–1939. Manuscripts Division, Library of Congress, Washington, D.C.

William Barton Rogers correspondence, 1843–1844. Special Collections Department, University of Virginia Library, Charlottesville, Va.

Amasa Walker papers, 1823–1902. Massachusetts Historical Society, Boston, Mass.

Francis Amasa Walker papers, 1878–1896. Manuscripts Division, Library of Congress, Washington, D.C.

Other materials, published and unpublished

Abstract of the Proceedings of the Society of Arts for the Twentieth Year, 1881–1882, Meetings 271 to 287 Inclusive. Boston: W. J. Schofield, 1882.

Abstract of the Proceedings of the Society of Arts for the Twenty-first Year, 1882–1883, Meetings 288 to 303 Inclusive. Boston: W. J. Schofield, 1883.

An Act to incorporate the Massachusetts Institute of Technology, and to grant aid to said institution and to the Boston Society of Natural History, approved April 10, 1861. MC1, box 15, folder 40, facsimile.

Adams, Sean Patrick. "Partners in geology, brothers in frustration: The antebellum geological surveys of Virginia and Pennsylvania." *Virginia Magazine of History and Biography* 106 (Winter 1998): 5–34.

"The administration of President Pritchett." *Technology Review* 8 (January 1906): 1–6.

Angulo, A. J. *William Barton Rogers and the Idea of MIT.* Baltimore: Johns Hopkins University Press, 2009.

Angulo, A. J. "William Barton Rogers and the Southern sieve: Revisiting science, slavery, and higher learning in the old South." *History of Education Quarterly* 45 (Spring 2005): 18–37.

Ashdown, Avery. "James Mason Crafts." *Journal of Chemical Education* 5 (August 1928): 911–21.

Babson, Roger W. *Actions and Reactions: An Autobiography of Roger W. Babson.* New York: Harper & Brothers, 1935.

Baker, William A. *Massachusetts Institute of Technology Department of Naval Architecture and Marine Engineering: A History of the First 75 Years*. Cambridge, Mass.: MIT, 1969.

Bartlett, Eleanor L. "The writings of Karl Taylor Compton." *Technology Review* 57 (December 1954): 89–92, 98, 100, 102, 104, 106, 108, 110, 112.

Baxter, James Phinney III. *Scientists Against Time*. Boston: Little, Brown, 1946.

Beckley, Lawrence. "Professional research staff at the Massachusetts Institute of Technology: A report tracing the evolution of the role of professional research staff at M.I.T." Unpublished manuscript, 1977, MC431, series II, box 1, Division of Industrial Cooperation.

Benjamin, Marcus. "The early presidents of the American Association [for the Advancement of Science]." *Science* 10 (November 24, 1899): 759–66.

Bennett, Albert A. "Brief history of the Mathematical Association of America before World War II." *American Mathematical Monthly* 74 (January 1967): 1–11.

Bevan, Arthur. "William Barton Rogers, pioneer American scientist." *Scientific Monthly* 50 (February 1940): 110–24.

Bever. Michael. *Metallurgy and Materials Science and Engineering at MIT, 1865–1988*. Cambridge, Mass.: MIT, 1988.

Billings, John S. "Biographical memoir of Francis Amasa Walker, 1840-1897." Read before the National Academy of Sciences, April 17, 1902. *Biographical Memoirs of the National Academy of Sciences* 5 (1902): 214–18.

Blackwood, James R. *The House on College Avenue: The Comptons at Wooster, 1891–1913*. Cambridge, Mass.: MIT Press, 1968.

Boston Advertiser.

Boston Evening Transcript.

Boston Globe.

Boston Herald.

Boston Post.

Boston Transcript.

Bowditch, Ernest W. Unpublished memoirs, n.d., 2 vols., author's possession (partial copy).

Bowles, Edward L. Interview, July 9, 1985. Unpublished transcript, MIT Museum: Bowles.

Bowles, Edward L. Interview by Roslyn Romanowski, March 17, 1982. Unpublished transcript, MIT Museum: Bowles.

Branson, Herman. "The Negro and scientific research." *Negro History Bulletin* 15 (April 1952): 131–36, 151.

Brayer, Elizabeth. *George Eastman: A Biography*. Baltimore: Johns Hopkins University Press, 1996.

Broderick, John T. *Willis Rodney Whitney: Pioneer of Industrial Research*. Albany: Fort Orange Press, 1945.

Brown, F. C. "Samuel Wesley Stratton: 1861-1931." *Science* 74 (October 30, 1931): 428–31.

Brown, Gordon S. Interview by Alex Pang, July 24, 1985. Unpublished transcript, MC24, box 32.

Brown, Sanborn, and Leonard M. Rieser. *Natural Philosophy at Dartmouth: From Surveyors' Chains to the Pressure of Light*, chapter 10, "The end of an era," 113–21. Hanover, N.H.: University Press of New England, 1974.

Bruce, Robert V. *The Launching of Modern American Science, 1846–1876.* New York: Alfred A. Knopf, 1987.

Buchanan, John M. "Recollections of the history of the Department of Biology at MIT, 1941-1988." Unpublished manuscript, n.d., MC650, box 3, folder 8.

Buderi, Robert. *The Invention that Changed the World: How a Small Group of Radar Pioneers Won the Second World War and Launched a Technological Revolution.* New York: Simon & Schuster, 1996.

Bullard, Frederic Field, ed. *Tech Songs: The M.I.T. Kommers Book.* Boston: Oliver Ditson, 1907.

Burchard, John E. *Q.E.D.: MIT in World War II.* New York: Wiley, 1948.

Burchard, John E., ed. *Mid-Century: The Social Implications of Scientific Progress. Verbatim Account of the Discussions Held at the Massachusetts Institute of Technology on the Occasion of its Mid-Century Convocation, March 31, April 1 and April 2, 1949.* Cambridge, Mass.: Technology Press, 1950.

Burchard, John E. "Trips to Corinth." Unpublished manuscript, 1975, MC76, box 7, folders 9–16.

Bush, Vannevar. Interview by Eric Hodgins, 1964. Unpublished transcript, MC143, boxes 1–2.

Bush, Vannevar. *Pieces of the Action.* New York: William Morrow, 1970.

B. W. B. G. [Bertram W. B. Greene]. "Historical sketch, the Massachusetts Institute of Technology." *Technique 1900,* 37–55. Boston: Frank Wood, 1899.

Centennial Commemoration of William Barton Rogers, December 7, 1904. Boston: Geo. H. Ellis, 1905; repr. from *Technology Review* 7 (January 1905): 26–48.

C.-E. A. W. [Charles-Edward A. Winslow]. "Francis Amasa Walker." *Technique 1898,* 33–53. Boston: Frank Wood, 1897.

Cajori, Florian. *The Teaching and History of Mathematics in the United States.* Washington, D.C.: G.P.O., U.S. Bureau of Education Circular no. 3, 1890.

Carlson, W. Bernard. "Academic entrepreneurship and engineering education: Dugald C. Jackson and the MIT-GE cooperative engineering course, 1907-1932." *Technology and Culture* 29 (July 1988): 536–67.

Cattell, J. McKeen. "The presidency of the Massachusetts Institute of Technology." *Science* 24 (October 26, 1906): 537–38.

Chandler, Alfred D., Jr., and Stephen Salsbury. *Pierre S. du Pont and the Making of the Modern Corporation.* Washington, D.C.: Beard Books, 1971.

Chittenden, Russell H. *History of the Sheffield Scientific School of Yale University, 1846–1922.* New Haven: Yale University Press, 1928.

Clarke, Robert. *Ellen Swallow: The Woman Who Founded Ecology.* Chicago: Follett Publishing, 1973.

Coates, Charles P. "The contribution of John Daniel Runkle to American education." *Journal of Educational Research* 15 (April 1927): 280–83.

Cochrane, Rexmond C. *Measures for Progress: A History of the National Bureau of Standards.* Washington, D.C.: U.S. Department of Commerce, National Bureau of Standards, 1966.

Compton, Karl T. "If I were starting out today." *American Magazine,* August 1950.

Compton, Karl T. "Inaugural address." *Science* 71 (June 13, 1930): 593–96.

Compton, Karl T. "It happened this way." *Tech,* 7 April 1950, 1, 2; 11 April 1950, 1, 6; 14 April 1950, 1, 4; 18 April 1950, 1, 4; 25 April 1950, 1, 3; 28 April 1950, 1, 4.

Compton, Karl T. *Massachusetts Institute of Technology: "Tomb of the dead languages."* New York: Newcomen Society of England, American Branch, 1948.

Compton, Karl T. "Why is universal military training necessary?" Manuscript, 1947, AC4, box 13, folder 5.

Concerning the Massachusetts Institute of Technology. Cambridge, Mass.: MIT Undergraduates, 1909.

Condon, Edward U. "Dr. Karl Taylor Compton, president of the American Association [for the Advancement of Science]." *Scientific Monthly* 40 (February 1935): 188–91.

Cooke, Josiah Parsons. "William Barton Rogers." *Proceedings of the American Academy of Arts and Sciences* 18 (May 1882-May 1883): 426–38.

Crafts, James M. "Friedel memorial lecture." *Journal of the Chemical Society Transactions* 77 (1900): 993–1019.

Crafts, James M. "President Crafts to the alumni, annual dinner, December 29, 1899." *Technology Review* 2 (January 1900): 30–39.

Crafts, James M. *A Short Course in Qualitative Analysis, with the New Notation.* New York: John Wiley, 1869.

Crafts, James M., and William F. Crafts, comps. *A Genealogical and Biographical History of the Descendants of Griffin and Alice Crafts, of Roxbury, Massachusetts, 1630–1890.* Northampton, Mass.: Gazette Printing, 1893.

Crawford, John D., and James R. Killian, Jr. "The advisability of establishing a press at the Massachusetts Institute of Technology." S.B. thesis, MIT, 1928.

Cross, Charles R. "James Mason Crafts, 1839-1917." *Biographical Memoirs of the National Academy of Sciences* 9 (1919): 159–77.

Dana, Gorham. "John D. Runkle." *Proceedings of the Brookline Historical Society for 1953,* 11–14. Brookline, Mass.: Brookline Historical Society, 1953.

Davis, Edward H. "The Runkle presidency—1868-1878." *Technology Review* 14 (July 1912): 391–96.

Davis, Tenney L. "The Department of Chemistry." In *A History of the Departments of Chemistry and Physics at the M.I.T., 1865–1933,* 1–16. Cambridge, Mass.: Technology Press, 1933.

"The death of chairman Stratton: His notable career and the tributes paid to him." *Technology Review* 34 (November 1931): 84–88, 102.

Dewey, Davis R. "Francis A. Walker as a public man." *Review of Reviews* 15 (February 1897): 166–71.

"Dies mourning Edison's death: Dr. Samuel W. Stratton, chairman of board and former president of Technology, stricken while chatting with Post reporter." *Boston Post,* October 19, 1931.

Doolittle, C. L. "Simon Newcomb, F.R.S., LL.D., D.C.L." *Proceedings of the American Philosophical Society* 49 (October-December 1910): iii–xviii.

"Dr. Ernest Fox Nichols: Technology's next president." *Technology Review* 23 (April 1921): 139–42.

Dunbar, Charles F. "The career of Francis Amasa Walker." *Quarterly Journal of Economics* 11 (July 1897): 436–48.

Dunbar, Charles F. "Francis Amasa Walker." *Proceedings of the American Academy of Arts and Sciences* 32 (July 1897): 344–54.

Edgerton, Harold Eugene. Autobiographical fragments. MC 25, box 1, folders 2–9.

Eliot, Charles W. "President Eliot and M.I.T." *Technology Review* 22 (July 1920): 430–38.

Etzkowitz, Henry. *MIT and the Rise of Entrepreneurial Science*. New York: Routledge, 2002.

Feynman. Richard P. *"Surely You're Joking, Mr. Feynman!" Adventures of a Curious Character*. New York: W. W. Norton, 1985.

Fields, Alonzo. *My 21 Years in the White House*. New York: Coward-McCann, 1960.

Five Years at the Radiation Laboratory. Presented to Members of the Radiation Laboratory by the Massachusetts Institute of Technology, Cambridge, 1946. Cambridge, Mass.: MIT, 1947.

Flexner, Abraham. *Henry S. Pritchett: A Biography*. New York: Columbia University Press, 1943.

Forbes, A. and J. W. Greene. *Rich Men of Massachusetts: Containing a statement of the reputed wealth of about fifteen hundred persons, with brief sketches of more than one thousand characters*. Boston: W. V. Spencer, 1851.

Fox, Philip. "Ernest Fox Nichols." *Astrophysical Journal* 61 (January 1925): 1–16.

Fox, Stephen. *John Muir and His Legacy: The American Conservation Movement*. Boston: Little, Brown, 1981.

"Francis Amasa Walker." *Journal of Political Economy* 5 (March 1897): 228–32.

"Francis Amasa Walker: Tributes from President Dwight and Professor Farnam." *Yale Alumni Weekly*, January 7, 1897, 1.

Freeman, Eva C., ed. *MIT Lincoln Laboratory: Technology in the National Interest*. Lexington, Mass.: Lincoln Laboratory, MIT, 1995.

Freeman, John Ripley. "Autobiography." Unpublished manuscript, n.d., MC51, box 1, folders 1–38.

Galpin, S. A. "Recollections of General Walker." *Christian Register*, January 14, 1897, 20–21.

"General Francis A. Walker." *Publications of the American Statistical Association* 5 (March 1897): 221–22.

Gerstner, Patsy. *Henry Darwin Rogers, 1808–1866: American Geologist*. Tuscaloosa: University of Alabama Press, 1994.

Goldblith, Samuel A. *Of Microbes and Molecules: Food Technology, Nutrition and Applied Biology at M.I.T., 1873–1988*. Trumbull, Conn.: Food & Nutrition Press, 1995.

Goldblith, Samuel A. *Samuel Cate Prescott: M.I.T. Dean and Pioneer Food Technologist*. Trumbull, Conn.: Food & Nutrition Press, 1993.

Goodstein, Judith R. *Millikan's School: A History of the California Institute of Technology*. New York: W. W. Norton, 1991.

Goodwin, H. M. "The Department of Physics." In *A History of the Departments of Chemistry and Physics at the M.I.T., 1865–1933*, 17–34. Cambridge, Mass.: Technology Press, 1933.

Goodwin, H. M. "Richard Cockburn Maclaurin (1870-1920)." *Proceedings of the American Academy of Arts and Sciences* 69 (February 1935): 518–21.

Gordon, George A. "A personal appreciation of Dr. Maclaurin." *Technology Review* 22 (July 1920): 367–69.

Green, George Washington Randolph Lee. "European trip." Unpublished manuscript, March 24, 1905, MC8, box 9, folder 97.

Gunn, Sidney. "James Savage, 1784-1873." *Dictionary of American Biography*. New York: American Council of Learned Societies, 1928–36.

Guttag, John V., ed. *The Electron and the Bit: Electrical Engineering and Computer Science at the Massachusetts Institute of Technology, 1902–2002*. Cambridge, Mass.: Electrical Engineering and Computer Science Department, MIT, 2005.

Hadley, Arthur Twining. "Francis A. Walker's contributions to economic theory." *Political Science Quarterly* 12 (June 1897): 295–308.

Hale, George Ellery. "A plea for the imaginative element in technical education." *Technology Review* 9 (October 1907): 467–81.

Hale, George Ellery. "The national value of scientific research." *Technology Review* 18 (November 1916): 801–17.

Hapgood, Fred. *Up the Infinite Corridor: MIT and the Technical Imagination.* Reading, Mass.: Addison-Wesley, 1993.

Harris, Leon. *Only to God: The Extraordinary Life of Godfrey Lowell Cabot.* New York: Atheneum, 1967.

Harrison, George R. "Karl Compton and American physics." *Physics Today* 10 (November 1957): 19–22.

Harrison, George R. "Karl Taylor Compton: A biography." Unpublished manuscript, 1961, MC60, box 10.

Haynes, Henry W. *Memoir of Francis Amasa Walker.* Worcester, Mass.: Charles Hamilton, 1897; repr. from *Proceedings of the American Antiquarian Society,* April 1897.

Hayward, Carle R. "Reminiscences of forty-three years of metallurgy instruction at M.I.T." Unpublished manuscript, ca. 1949, MC431, series II, box 2, Mining and metallurgy.

Hayward, Silvanus. "Gen. Francis A. Walker." *New-England Historical and Genealogical Register* 52 (January 1898): 69–72.

Hazen, Harold Locke. "Memoirs: An informal story of my life and work." Unpublished manuscript, 1976, reference collection, MIT Archives.

"Henry Smith Pritchett." *Century Memorials.* New York: Century Association, 1940, 59–61.

"Henry Smith Pritchett." *Technique 1902,* 43–58. Boston: Carl H. Heintzemann, 1901.

"Henry Smith Pritchett." *Technology Review* 2 (April 1900): 105–09.

Herreshoff, L. Francis. *Capt. Nat Herreshoff, the Wizard of Bristol: The Life and Achievements of Nathanael Greene Herreshoff, together with an account of some of the yachts he designed.* New York: Sheridan House, 1953.

Hobbs, Glenn. "Material for Dr. Stratton's biography for Professor Prescott." Unpublished manuscript, n.d., MC8, box 1, folder 7.

Hodgins, Eric. "Tough old boys at M.I.T." *Science* 152 (June 10, 1966): 1458.

Hodgins, Eric. *Trolley to the Moon: An Autobiography.* New York: Simon & Schuster, 1973.

Home, R. W., and Morris F. Low. "Postwar scientific intelligence missions to Japan." *Isis* 84 (September 1993): 527–37.

Hunsaker, Jerome, and Saunders MacLane. "Edwin Bidwell Wilson, April 25, 1879–December 28, 1964." *Biographical Memoirs of the National Academy of Sciences* 43 (1973): 285–320.

Illy, Joseph, ed. *Albert Meets America: How Journalists Treated Genius during Einstein's 1921 Travels.* Baltimore: Johns Hopkins University Press, 2006.

The Improbable Achievement: Chemical Engineering at MIT. Cambridge, Mass.: Department of Chemical Engineering, MIT, 1979.

"The inauguration of President Nichols." *Technology Review* 23 (July 1921): 306–26.

Institute Committee, MIT Undergraduate Association, comp. *Technology Songs.* Boston: Oliver Ditson, 1929.

"The Institute under President Crafts." *Technique 1901,* 278–81. Boston: Frank Wood, 1900.

"An interview with Dr. Nichols." *Technology Review* 23 (July 1921): 374–78.

James, Henry. *Charles W. Eliot, President of Harvard University, 1869–1909.* Boston: Houghton Mifflin, 1930.

James, Marquis. *Alfred I. du Pont: The Family Rebel.* Indianapolis: Bobbs-Merrill, 1941.

"James Mason Crafts." *Technique 1899*, 38–41. Boston: Frank Wood, 1898.

"James Mason Crafts." *Technology Review* 1 (January 1899): 5–12.

Jarzombek, Mark. *Designing MIT: Bosworth's New Tech.* Boston: Northeastern University Press, 2004.

"John Daniel Runkle, LL.D." *Proceedings of the New-England Historic Genealogical Society at the Annual Meeting, 14 January, 1903, with Memoirs of Deceased Members, 1902*, lviii–lix. Boston: New-England Historic Genealogical Society, 1903.

"John Daniel Runkle and his share in the development of Technology." *Technique 1901*, 37–59. Boston: Frank Wood, 1900.

Kaiser, David, ed. *Becoming MIT: Moments of Decision.* Cambridge, Mass.: MIT Press, 2010.

Kane, Henry B. *MIT in the Twenties.* Cambridge, Mass.: MIT Alumni Association, 1968.

Kargon, Robert, and Elizabeth Hodes. "Karl Compton, Isaiah Bowman, and the politics of science in the Great Depression." *Isis* 76 (September 1985): 300–18.

Kargon, Robert H. *The Rise of Robert Millikan.* Ithaca: Cornell University Press, 1982.

Kennelly, Arthur E. "Samuel Wesley Stratton, 1861-1931." *Biographical Memoirs of the National Academy of Sciences* 17 (1936): 253–60.

Killian, James R., Jr. *The Education of a College President: A Memoir.* Cambridge, Mass.: MIT Press, 1985.

Killian, James R., Jr. "Karl Taylor Compton: the man." *Technology Review* 57 (December 1954): 87–88, 116, 118.

Killian, James R., Jr. "Tribute to a teacher: Erwin Haskell Schell, 1889-1965." Pamphlet, 1965, MIT Museum: Killian.

Killian, James R., Jr. "William Barton Rogers." *Technology Review* 60 (December 1957): 105–08, 124, 126, 128, 130.

King, Ellen. "Technology from the basement of Rogers in the nineties." Unpublished manuscript, n.d., MC431, series II, box 11.

Kinnahan, Thomas P. "Charting progress: Francis Amasa Walker's 'Statistical Atlas of the United States' and narratives of western expansion." *American Quarterly* 60 (June 2008): 399–423.

Kohlstedt, Sally Gregory, Michael M. Sokal, and Bruce M. Lewenstein. *The Establishment of Science in America: 150 Years of the American Association for the Advancement of Science.* New Brunswick, N.J.: Rutgers University Press, 1999.

"Late Dr. Stratton eulogized as man, scientist, executive." *Federal News*, November 7, 1931.

Laughlin, J. Laurence. "Francis Amasa Walker." *Journal of Political Economy* 5 (March 1897): 228–236.

Laws, J. Bradford. "A statistical social study of the negroes of Sinclare Central Factory and Calumet Plantation, Louisiana." S.B. thesis, MIT, 1901.

Lécuyer, Christophe. "Academic science and technology in the service of industry: MIT creates a 'permeable' engineering school." *American Economic Review* 88 (May 1998): 28–33.

Lécuyer, Christophe. "The making of a science based technological university: Karl Compton, James Killian, and the reform of MIT." *Historical Studies in the Physical and Biological Sciences* 23 (1992), part 1: 153–180.

Lécuyer, Christophe. "MIT, Progressive reform, and 'industrial service,' 1890-1920." *Historical Studies in the Physical and Biological Sciences* 26 (1995), part 1: 35–88.

Leslie, Stuart W. *The Cold War and American Science: The Military-Industrial-Academic Complex at MIT and Stanford.* New York: Columbia University Press, 1993.

Leslie, Stuart W. "Profit and loss: The military and MIT in the postwar era." *Historical Studies in the Physical and Biological Sciences* 21 (1990), part 1: 59–85.

Letters of James Savage to His Family. Boston, 1906, privately printed.

Lewis, Warren K., John R. Loofbourow, Ronald H. Robnett, C. Richard Soderberg, and Julius A. Stratton. *Report of the Committee on Educational Survey to the Faculty of the Massachusetts Institute of Technology.* Cambridge, Mass.: Technology Press, 1949.

Liepmann, Klaus. "Music at M.I.T.: A short history of music at the Massachusetts Institute of Technology." Unpublished manuscript, n.d., reference collection, MIT Archives.

Lindsay, Debra. "Intimate inmates: Wives, households, and science in nineteenth-century America." *Isis* 89 (December 1998): 631–52.

Litchfield, P. W. *Industrial Voyage: My Life as an Industrial Lieutenant.* Garden City, N.Y.: Doubleday, 1954.

Livingstone, David N. *Nathaniel Southgate Shaler and the Culture of American Science.* Tuscaloosa: University of Alabama Press, 1987.

Lobdell, H. E. "Samuel Wesley Stratton: An interview." *Technology Review* 25 (November 1922): 7–10.

Lowell, Francis C. *Memoir of Francis A. Walker, LL.D.* Cambridge, Mass.: John Wilson and Son, 1900; repr. from *Proceedings of the Massachusetts Historical Society*, November 1899.

Lurie, Edward. *Louis Agassiz: A Life in Science.* Baltimore: Johns Hopkins University Press, 1988.

Maclaurin, Richard C. "A national opportunity and a national duty." *Stone & Webster Journal* 19 (August 1916): 89–93.

Maddocks, Melvin. "Harvard was once, unimaginably, small and humble." *Smithsonian* 17 (September 1986): 140–42, 144, 146, 148, 150, 152, 154, 156, 158–60.

"Made M.I.T. national institution." *Boston Evening Transcript*, October 19, 1931 [re Samuel W. Stratton].

Mannix, Loretta H. "Notes for the story of Ellen Swallow." Unpublished manuscript, n.d., MC431, series II, box 16, Richards, Ellen Swallow.

Mannix, Loretta H. "Some notes on the health of William B. Rogers." Unpublished manuscript, n.d, author's possession.

Mannix, Loretta H. "Women in the founding period." Unpublished manuscript, n.d., MC431, series II, box 16, Women.

Marcus, Benjamin. "The early presidents of the American Association [for the Advancement of Science]." *Science* 10 (November 24, 1899): 759–66.

Maslanka, John S. *A Century of Technology . . .* Buffalo, N.Y.: William J. Keller, 1961; repr. from *Technique 1961*, 8–41.

Maslanka, John S. "William Barton Rogers's conception of an institute of technology." S.B. thesis, MIT, 1961.

Mason, Alpheus Thomas. *Brandeis: A Free Man's Life.* New York: Viking Press, 1946.

Massachusetts Institute of Technology. *Inauguration of Henry Smith Pritchett as President*. Boston: George H. Ellis, 1901.

Massachusetts Institute of Technology. *Meetings Held in Commemoration of the Life and Services of Francis Amasa Walker*. Boston: MIT, 1897.

Massachusetts Institute of Technology. *President's Report*, 1872– .

Massachusetts Institute of Technology Alumni Association War Records Committee. *Technology's War Record: An Interpretation of the Contribution Made by the Massachusetts Institute of Technology, Its Staff, Its Former Students and Its Undergraduates to the Cause of the United States and the Allied Powers in the Great War, 1914–1919*. Cambridge: Murray Printing, 1920.

"Massachusetts Tech." *Fortune* 14 (November 1936): 107–14, 132, 134, 136, 138, 140.

Mattill, John. *The Flagship: The M.I.T. School of Chemical Engineering Practice, 1916–1991*. Cambridge: David H. Koch School of Chemical Engineering Practice, MIT, 1991.

McAllister, Ethel M. *Amos Eaton, Scientist and Educator*. Philadelphia: University of Pennsylvania Press, 1941.

McLennan, Roy. "The Career of Richard C. Maclaurin. I—The Early Years." Unpublished manuscript, 1975, MC431, series II, box 3.

McLennan, Roy. "The Career of Richard C. Maclaurin. II—The Later Years." Unpublished manuscript, 1975, MC431, series II, box 3.

"Meeting 288. In memory of William Barton Rogers, LL.D., late president of the Society." *Abstract of the Proceedings of the Society of Arts for the Twenty-First Year, 1882–1883, Meetings 288 to 303 Inclusive*, 5–41. Boston: W. J. Schofield, 1883.

"Mercantile Library Association." In *Sketches and Business Directory of Boston*. Boston, 1860.

Miller, John Anderson. *Yankee Scientist: William David Coolidge*. Schenectady, N.Y.: Mohawk Development Service, 1963.

Morse, Philip M. *In at the Beginnings: A Physicist's Life*. Cambridge, Mass.: MIT Press, 1977.

Morse, Philip M. "Karl Taylor Compton 1887-1954: Scientist, educator, statesman." *Cosmos Club Bulletin* (February 1980), suppl.

Munroe, James P. *A Life of Francis Amasa Walker*. New York: Henry Holt, 1923.

[Munroe, James P.] *Massachusetts Institute of Technology, Boston. A Brief Account of Its Foundation, Character, and Equipment Prepared in Connection with the World's Columbian Exposition*. Boston: MIT, 1893.

Munroe, James P. "William Barton Rogers, founder of the Massachusetts Institute of Technology." *Technology Review* 6 (October 1904): 501–550.

Nasar, Sylvia. *A Beautiful Mind: The Life and Mathematical Genius of Nobel Laureate John Nash*. New York: Simon & Schuster, 1998.

Newcomb, Simon. *Reminiscences of an Astronomer*. Boston: Houghton Mifflin, 1903.

Newton, Bernard. *The Economics of Francis Amasa Walker: American Economics in Transition*. New York: Augustus M. Kelley, 1967.

Nichols, E. L. "Ernest Fox Nichols." *Biographical Memoirs of the National Academy of Sciences* 12 (1929): 99–131.

Nichols, Ernest F. "Richard Cockburn Maclaurin as a colleague." *Technology Review* 22 (July 1920): 364–67.

Nichols, Jeannette P. "Francis Amasa Walker, 1840-1897." *Dictionary of American Biography*. New York: American Council of Learned Societies, 1928–36.

"Noted engineer began his life as farm boy." *New York Times*, July 12, 1931 [re Samuel W. Stratton].

Noyes, Arthur Amos. "A talk on teaching." Given at a conference of members of the instructing staff of the Massachusetts Institute of Technology on March 20, 1908. MIT Museum: Noyes.

Objects and Plan of an Institute of Technology. Boston: John Wilson and Son, 1860.

Owens, Larry. "Engineering the perfect cup of coffee: Samuel Prescott and the sanitary vision at MIT." *Technology and Culture* 45 (October 2004): 795–807.

Owens, Larry. "MIT and the federal 'angel': Academic R&D and federal-private cooperation before World War II." *Isis* 81 (June 1990): 188–213.

Owens, Larry. "Vannevar Bush and the differential analyzer: The text and context of an early computer." *Technology and Culture* 27 (January 1986): 63–95.

Page, Leigh. "Ernest Fox Nichols." *Dictionary of American Biography*. New York: American Council of Learned Societies, 1928–36.

Pang, Alex Soojung-Kim. "Edward Bowles and radio engineering at MIT, 1920-1940." *Historical Studies in the Physical and Biological Sciences* 20 (1990), part 2: 313–37.

Park, Charles F. *A History of the Lowell Institute School, 1903–1928*. Cambridge, Mass.: Harvard University Press, 1931.

Parris, Morris A. "Origin of the Stratton family." Unpublished manuscript, n.d., MC8, box 1, folder 4.

Parris, Worden. "Camping trip of 1903." Unpublished manuscript, n.d., MC8, box 1, folder 7.

Pauling, Linus. "Arthur Amos Noyes, September 13, 1866–June 3, 1936." *Biographical Memoirs of the National Academy of Sciences* 31 (1958): 322–46.

Pearson, Henry Greenleaf. *Richard Cockburn Maclaurin: President of the Massachusetts Institute of Technology, 1909–1920*. New York: Macmillan, 1937.

Peterson, T. F. *Nightwork: A History of Hacks and Pranks at MIT*. Cambridge, Mass.: MIT Press, 2003.

Pifer, Alan. *The Carnegie Foundation for the Advancement of Teaching: A Notable Year*. New York: Carnegie Foundation, [1966]; repr. from *Annual Report 1965–66*.

Pollard, Ernest C. *Radiation: One Story of the M.I.T. Radiation Laboratory*. Durham, N.C.: Woodburn Press, 1982.

Prescott, Samuel C. Biography of Samuel Wesley Stratton. Unpublished, partial manuscript, n.d., MC8, box 1, folder 8.

Prescott, Samuel C. "Samuel Wesley Stratton (1861-1931)." *Proceedings of the American Academy of Arts and Sciences* 69 (February 1935): 544–47.

[Prescott, Samuel C.] "Sketch of the Department of Biology and Public Health." Unpublished manuscript, 1933, AC4, box 31, folder 7.

Prescott, Samuel C. *When M.I.T. Was 'Boston Tech,' 1861–1916*. Cambridge, Mass.: Technology Press, 1954.

"President Maclaurin." *Boston Transcript*, January 16, 1920.

"President Maclaurin, the man." *Boston Herald*, January 17, 1920.

Prévost, Jean-Guy. "Controversy and demarcation in early-twentieth-century demography: The rise and decline of Walker's theory of immigration and the birth rate." *Social Science History* 22 (Summer 1998): 131–58.

Pritchett, Henry S. "The astronomer as engineer." *Journal of the Worcester Polytechnic Institute* 2 (November 1898): 1–14.

Pritchett, Henry S. "Beginnings of the Carnegie Foundation." *Early Papers of the Foundation*; printed by the Foundation in 1935 for private distribution in commemoration of the centennial of Andrew Carnegie's birth.

Pursell, Carroll W. Jr. "The anatomy of a failure: The Science Advisory Board, 1933-1935." *Proceedings of the American Philosophical Society* 109 (December 10, 1965): 342-351.

R. E. R. [Robert E. Rogers]. "His life and works." *Technology Review* 22 (January 1920): 20–22. [re Richard Maclaurin]

R.L.E.: 1946 + 20. Cambridge, Mass.: Research Laboratory of Electronics, MIT, 1966.

R. P. F. "In memoriam. Francis Amasa Walker." *Annals of the American Academy of Political and Social Science* 9 (March 1897): 173–77.

Research Laboratory of Electronics, MIT. Transcript, twentieth anniversary dinner, May 1966, AC186, box 4, folder 33.

"Richard Cockburn Maclaurin." *Technique 1910*, 42–44. Boston, 1909.

Richards, Ellen Henrietta Swallow, comp. "Newspaper clippings on the Harvard-Tech alliance." MC431, series II, box 5.

Richards, Robert H. *Robert Hallowell Richards: His Mark.* Boston: Little, Brown, 1936.

Richards, Theodore W. "James Mason Crafts (1839-1917)." *Proceedings of the American Academy of Arts and Sciences* 53 (September 1918): 801–04.

Ripley, William Z. *The Races of Europe: A Sociological Study*. New York: D. Appleton, 1899.

Rogers, Emma Savage, ed. *Life and Letters of William Barton Rogers, Edited by His Wife with the Assistance of William T. Sedgwick.* Boston and New York: Houghton, Mifflin, 1896.

Rogers, William Barton. "For the establishment of a school of arts. Memorial of the Franklin Institute of the state of Pennsylvania, for the promotion of Mechanic Arts, to the legislature of Pennsylvania." Unpublished manuscript, n.d. [1837], MC1, box 1, folder 14b.

Rosenblith, Walter A. Oral history project, 1982–1984. Unpublished transcripts, MC55, box 62. Interviewees include James R. Killian Jr., H. Guyford Stever, Jerome Wiesner, Gordon Brown, Julius Stratton, Victor Weisskopf, Francis O. Schmitt, Harold Edgerton, and Jerrold Zacharias.

Ross, Earle D. "The great triumvirate of land-grant educators: Gilman, White, and Walker." *Journal of Higher Education* 32 (December 1961): 480–88.

Ruffner, W. H. "The brothers Rogers." *University of Virginia Alumni Bulletin* 5 (May 1898): 1–13.

Runkle, John D. *New Tables for Determining the Values of the Coefficients, in the Perturbative Function of the Planetary Motion, Which Depend Upon the Ratio of the Mean Distances.* Cambridge, Mass.: Metcalf, 1856.

Runkle, John D. "Technical and industrial education abroad." *Abstract of the Proceedings of the Society of Arts for the Nineteenth Year, 1880–1881, Meetings 256 to 270 Inclusive*, 83–95. Boston: Bailey Combination Type and Printing, 1881.

Ruschenberger, W. S. W. "A notice of William Barton Rogers." *Proceedings of the American Philosophical Society* 31 (July-December 1893): 254–57.

Ruschenberger, W. S. W. "A sketch of the life of Robert E. Rogers, M.D., LL.D., with biographical notices of his father and brothers." *Proceedings of the American Philosophical Society* 23 (January 1886): 104–46.

Savage, Howard J. *Fruit of an Impulse: Forty-Five Years of the Carnegie Foundation, 1905–1950.* New York: Harcourt, Brace, 1953.

[Schmitt, Francis O.] "History of biology at MIT." Unpublished manuscript, 1952, AC4, box 31, folder 7.

A Scientist Speaks: Excerpts from addresses by Karl Taylor Compton during the years 1930–1949 when he was president of the Massachusetts Institute of Technology. Cambridge, Mass.: MIT Undergraduate Association, 1955.

Scope and Plan of the School of Industrial Science of the Massachusetts Institute of Technology. Boston, 1864.

Sedgwick, William T. "Gentleman and scholar." *Technology Review* 22 (January 1920): 29–31. [re Richard Maclaurin]

Sedgwick, William T. "Richard C. Maclaurin: Some notes on the course and the contribution of a great life." *Boston Transcript,* January 17, 1920.

Servos, John W. "The industrial relations of science: Chemical engineering at MIT, 1900-1939." *Isis* 71 (December 1980): 531–49.

Sherrill, Miles S. "Arthur Amos Noyes (1866-1936)." *Proceedings of the American Academy of Arts and Sciences* 74 (November 1940): 150–55.

Shillaber, Caroline. *Massachusetts Institute of Technology School of Architecture and Planning 1861–1961: A Hundred Year Chronicle.* Cambridge, Mass.: MIT, 1963.

Shrock, Robert R. *The Geologists Crosby of Boston: William Otis Crosby (1850–1925) and Irving Ballard Crosby (1891–1959).* Cambridge, Mass.: MIT, 1972.

Shrock, Robert R. *Geology at M.I.T. 1861–1965. A History of the First Hundred Years of Geology at Massachusetts Institute of Technology. I. The Faculty and Supporting Staff.* Cambridge, Mass.: MIT Press, 1977.

Shrock, Robert R. *Geology at M.I.T. 1861–1965. A History of the First Hundred Years of Geology at Massachusetts Institute of Technology. II. Departmental Operations and Products.* Cambridge, Mass.: MIT Press, 1982.

Sinclair, Bruce. "Harvard, MIT, and the ideal technical education." In *Science at Harvard University: Historical Perspectives,* ed. Clark A. Elliott and Margaret Rossiter, 76–95. Bethlehem, Pa.: Lehigh University Press, 1992.

Sinclair, Bruce. "Inventing a genteel tradition: MIT crosses the river." In *New Perspectives on Technology and American Culture,* ed. Bruce Sinclair, 1–18. Philadelphia: American Philosophical Society, 1986.

Sinclair, Upton. *The Goose-Step: A Study of American Education.* Pasadena, Calif.: Upton Sinclair, 1923.

"Sketch of Prof. William B. Rogers." *Popular Science Monthly* 9 (September 1876): 606–11.

Slater, John C. "History of the M.I.T. physics department, 1930-48." Unpublished manuscript, n.d., MC75, box 1.

Slater, John C. Interview by Charles Weiner, February 23, 1970. Unpublished transcript, Center for History of Physics, American Institute of Physics.

Slater, John C. "A physicist of the lucky generation." Unpublished manuscript, ca. 1975, MC189, box 1.

Slaughter, Sarah. "The first few decades of women at MIT." Unpublished manuscript, 1979, MC431, series II, box 16, Women.

Slaughter, Sarah. "Women at MIT: 1871-1900." Unpublished manuscript, 1980, reference collection, MIT Archives.

Smallwood, W. M. "The Agassiz-Rogers debate on evolution." *Quarterly Review of Biology* 16 (March 1941): 1–12.

Smith, Earl L. *Yankee Genius: A Biography of Roger Babson*. New York: Harper & Brothers, 1954.

Soderberg, C. Richard, and George W. Swett. "The mechanical engineering department." Unpublished manuscript, n.d., MC23, box 28.

Spectrum, 1873–74. [MIT student newspaper]

Speer, William. "Some aspects of military training and conscientious objection at M.I.T. 1865-1958." Unpublished manuscript, 1975, MC431, series II, box 7, Military instruction.

Spencer, Joseph Jansen. "General Francis A. Walker: A character sketch." *Review of Reviews* 15 (February 1897): 159–66.

Springfield Republican.

Stinson, Margaret D. Interview by Ellen Swallow Richards. Unpublished transcript, [1911], MC 460.

Stout, Robert. *"A life to inspire." A gifted New Zealander: The splendid record of Richard Cockburn Maclaurin*. Wellington, N.Z.: New Zealand Times, 1920.

Stratton, Harriet Russell. *A Book of Strattons: Being a collection of Stratton records from England and Scotland, and a genealogical history of the early colonial Strattons in America, with five generations of their descendants*. New York: Grafton Press, 1908.

Stratton, Julius A. "Karl Taylor Compton, 1887-1954." *Biographical Memoirs of the National Academy of Sciences* 61 (1992): 39–57.

Stratton, Julius A. "Karl Taylor Compton: educator and administrator." *Technology Review* 57 (December 1954): 85–86.

Stratton, Julius A., and Loretta H. Mannix. "Chronology of the Institute." Unpublished manuscript, 1986, author's possession.

Stratton, Julius A., and Loretta H. Mannix. "Chronology of the Massachusetts Institute of Technology." Unpublished manuscript, 1993, author's possession.

Stratton, Julius A., and Loretta H. Mannix. "Chronology of William Barton Rogers." Unpublished manuscript, n.d., author's possession.

Stratton, Julius A., and Loretta H. Mannix. *Mind and Hand: The Birth of MIT*. Cambridge, Mass.: MIT Press, 2005.

Stratton, Julius A., and Loretta H. Mannix. "William Barton Rogers, the first leader of M.I.T." Unpublished manuscript, n.d., author's possession.

Stratton, Julius A., and Loretta H. Mannix. "Rogers and the College of William and Mary." Unpublished manuscript, n.d., author's possession.

Stratton, Samuel W. "Naval and military service." Unpublished manuscript, n.d., MC8, box 9, folder 97.

Struik, Dirk J. "The MIT Department of Mathematics during its first seventy-five years: Some recollections." Manuscript, n.d., MC431, series II, box 2, Mathematics.

Swain, George F. "Technical education at the Massachusetts Institute of Technology." *Popular Science Monthly* 57 (July 1900): 257–84.

Swope, Gerard. Interview by Harlan B. Phillips, ca. 1955. Unpublished transcript, Oral History Research Office, Columbia University.

Talbot, Henry P. "James Mason Crafts, LL.D." *Technology Review* 19 (November 1917): 653–55.

The Tech, 1881– . [MIT student newspaper]

Technique, 1885– . [MIT student yearbook]

Technology and Industrial Efficiency. A series of papers presented at the Congress of Technology, opened in Boston, Mass., April 10, 1911, in celebration of the fiftieth anniversary of the granting of a charter to the Massachusetts Institute of Technology. New York: McGraw-Hill, 1911.

Technology Review, 1899– . [MIT Alumni Association journal]

Thomson, Elihu. "Dr. Maclaurin's service as president of the Massachusetts Institute of Technology." *Technology Review* 22 (July 1920): 361–64.

"Those Comptons." *Boston Herald*, May 14, 1930.

Trowbridge, Augustus. "Ernest Fox Nichols." *Science* 59 (May 9, 1924): 415–16.

Turner, Louis A. "Karl T. Compton: An appreciation." *Bulletin of the Atomic Scientists* 10 (September 1954): 296, 304.

Tyler, Harry W. "The educational work of Francis A. Walker." *Educational Review* 14 (June 1897): 55–70.

Tyler, Harry W. "John Daniel Runkle." *American Mathematical Monthly* 10 (August-September 1903): 183–85.

Tyler, Harry W. "John Daniel Runkle." *Proceedings of the American Academy of Arts and Sciences* 38 (July 1903): 727–30.

Tyler, Harry W. "John Daniel Runkle, 1822-1902." *Technology Review* 4 (July 1902): 277–306.

Walker, Francis A. "The colored race in the United States." *Discussions in Economics and Statistics by Francis A. Walker, Ph.D., LL.D.*, ed. Davis R. Dewey. New York: Henry Holt, 1899, vol. 2, 127–37; repr. from *Forum* 11 (1891): 501–9.

Walker, Francis A. *The Indian Question*. Boston: James R. Osgood, 1874.

Walker, Francis A. *Memoir of Wm. Barton Rogers, 1804–1882*. Read before the National Academy of Sciences, April 1887; repr. Washington, D.C.: Judd & Detweiler, 1895.

Walker, Francis A. "The place of scientific and technical schools in American education." *Technology Quarterly* 4 (December 1891): 293–303.

Walker, Francis A. "The restriction of immigration." An address to the Manufacturers' Club of Philadelphia at the regular December meeting in 1895. MC298, box 8.

Walker, Francis A. "The technical school and the university." *Atlantic Monthly* 72 (September 1893): 390–95.

"Walker's contribution to economics." *Gunton's Magazine*, February 1897, 89–96.

Wall, Joseph Frazier. *Alfred I. du Pont: The Man and His Family*. New York: Oxford University Press, 1990.

Wayland, Francis. *Report to the Corporation of Brown University, on Changes in the System of Collegiate Education, Read March 28, 1850*. Providence: George H. Whitney, 1850.

Weeks, Edward. *The Lowells and Their Institute*. Boston: Little, Brown, 1966.

Wendell, George V. "Our new president: A man eminently fitted by experience and personal qualities to direct the affairs of the Institute." *Technology Review* 11 (January 1909): 3–7. [re Richard Maclaurin]

Westervelt, Virginia Veeder. *The World Was His Laboratory: The Story of Dr. Willis R. Whitney*. New York: Julian Messner, 1964.

Whitaker, John K. "Enemies or allies? Henry George and Francis Amasa Walker one century later." *Journal of Economic Literature* 35 (December 1997): 1891–1915.

Whitney, Willis R. "Research." *General Electric Review* 20 (1917): 114–20.

Wiener, Norbert. *I Am a Mathematician: The Later Life of a Prodigy.* Cambridge, Mass.: MIT Press, 1956.

Wiesner, J. B., G. G. Harvey, H. J. Zimmerman, and R. A. Sayers. "Twelve years of basic research: A brief history of the Research Laboratory of Electronics." Manuscript, March 1958, reference collection, MIT Archives.

Wildes, Karl L., and Nilo A. Lindgren. *A Century of Electrical Engineering and Computer Science at MIT, 1882–1982.* Cambridge, Mass.: MIT Press, 1985.

Williams, Clarence G. *Technology and the Dream: Reflections on the Black Experience at MIT, 1941–1999.* Cambridge, Mass.: MIT Press, 2001.

Williamson, Harold Francis. *Edward Atkinson: The Biography of an American Liberal, 1827–1905.* Boston: Old Corner Book Store, 1934.

Wilson, E. Bright, and John Ross. "Physical chemistry in Cambridge, Massachusetts." *Annual Review of Physical Chemistry* 24 (October 1973): 1–27.

Wilson, Robert E. "Karl Compton and the Massachusetts Institute of Technology." Unpublished manuscript, n.d., MIT Museum: Compton.

Wise, George. *Willis R. Whitney, General Electric, and the Origins of U.S. Industrial Research.* New York: Columbia University Press, 1985.

Woodbury, David O. *Beloved Scientist: Elihu Thomson, A Guiding Spirit of the Electrical Age.* New York: McGraw-Hill, 1944.

Woodbury, David O. "M.I.T. historical research summary." Unpublished manuscript, n.d., author's possession.

Wright, Carroll D. "Bibliography of the writings and reported addresses of Francis A. Walker." *Quarterly Publications of the American Statistical Association* 5 (June 1897): 276–90.

Wright, Carroll D. "Francis Amasa Walker." *Quarterly Publications of the American Statistical Association* 5 (June 1897): 245–75.

Wright, Helen. *Explorer of the Universe: A Biography of George Ellery Hale.* New York: E. P. Dutton, 1966.

Wylie, Francis E. *M.I.T. in Perspective: A Pictorial History of the Massachusetts Institute of Technology.* Boston: Little, Brown, 1975.

Zachary, G. Pascal. *Endless Frontier: Vannevar Bush, Engineer of the American Century.* Cambridge, Mass.: MIT Press, 1999.

Index